T0336647

Geophysical Monograph Series

Including
IUGG Volumes
Maurice Ewing Volumes
Mineral Physics Volumes

Geophysical Monograph 176

Exploring Venus as a Terrestrial Planet

Larry W. Esposito
Ellen R. Stofan
Thomas E. Cravens
Editors

American Geophysical Union
Washington, DC

Library of Congress Cataloging-in-Publication Data

Exploring Venus as a terrestrial planet / Larry W. Esposito, Ellen R. Stofan, Thomas E. Cravens editors.
  p. cm. -- (Geophysical monograph ; 176)
 ISBN 978-0-87590-441-2
 1. Venus (Planet)--Exploration. 2. Venus (Planet)--Surface. I. Esposito, Larry. II. Stofan, Ellen Renee, 1961- III. Cravens, Thomas E., 1948-

 QB621.E97 2007
 559.9'22--dc22

                    2007045668

 ISBN 978-0-87590-441-2
 ISSN 0065-8448

Front cover image: Three impact craters are displayed in this three-dimensional perspective view of the surface of Venus. The center of the image is located at approximately 27 degrees south latitude, 339 degrees east longitude in the northwestern portion of Lavinia Planitia. The viewpoint is located southwest of Howe Crater, which appears centered in the lower portion of the image. Howe is a crater with a diameter of 37.3 kilometers (23.1 miles) located at 28.6 degrees south latitude, 337.1 degrees east longitude. Danilova, a crater with a diameter of 47.6 kilometers (29.5 miles), located at 26.35 degrees south latitude, 337.25 degrees east longitude, appears above and to the left of Howe in the image. Aglaonice, a crater with a diameter of 62.7 kilometers (38.9 miles), located at 26.5 degrees south latitude, 340 degrees east longitude, is shown to the right of Danilova. Magellan synthetic aperture radar data is combined with radar altimetry to develop a three-dimensional map of the surface. Rays cast in a computer intersect the surface to create a three-dimensional perspective view. Simulated color and a digital elevation map developed by the U.S. Geological Survey are used to enhance small-scale structure. The simulated hues are based on color images recorded by the Soviet Venera 13 and 14 spacecraft. The image was produced at the JPL Multimission Image Processing Laboratory and is a single frame from a video released at the May 29, 1991, JPL news conference.

PREFACE

Venus has long been recognized as a sister planet to the Earth, as one of a triad of terrestrial planets with an atmosphere (Earth, Venus, Mars), and as a possible analogue for planets circling other stars. Its dense CO_2 atmosphere has provided both a lesson and a warning in our understanding of the effects of greenhouse gases.

The planet was the first to be visited by spacecraft and was the target of numerous American and Soviet space missions launched in the 1960s, 1970s, and 1980s. A long, dry spell in the exploration of this planet was recently broken when ESA's Venus Express went into orbit in 2006 (see the chapter by Taylor). Japan's Venus Climate Orbiter will launch in 2009. The NASA Messenger spacecraft, en route to Mercury, observed Venus closely when it flew by in June 2007. Future missions are under study by NASA and other space agencies. If Venus was the "forgotten planet," it is now receiving well-deserved close attention!

With this revived interest, it was timely to hold the AGU-sponsored Chapman Conference "Exploring Venus as a Terrestrial Planet" on 13–17 February 2006 in Key Largo, Florida. The objectives of that conference were to review the current knowledge of Venus and compare this planet with Earth and Mars; to preview observation plans for Venus Express; and to identify key objectives for future research and space missions. Conference conveners were Larry W. Esposito and Dmitry Titov, and the program committee included Kevin Baines, Alexander Basilevsky, Gordon Chin, David Crisp, Takeshi Imamura, Masato Nakamura, Christopher Russell,

Steven Saunders, Gerald Schubert, Sue Smrekar, Ellen Stofan, Hakan Svedhem, and Fredric Taylor. This monograph consists of invited chapters written by the authors of the major review talks at the Chapman Conference. Financial support for the conference was provided by NASA, ESA, and the University of Colorado's Laboratory for Atmospheric and Space Physics (LASP). More than 120 scientists attended the conference, which emphasizes both the continuing interest in Venus and the importance of Venus to understanding the origin, evolution, and possibly fate, of the terrestrial planets.

We thank the following reviewers: M. Joan Alexander, Sushil K. Atreya, Kevin Baines, Bruno Bezard, Robert Carlson, Pierre Drossart, Linda Elkins-Tanton, Richard M. Goody, Richard Hartle, Jeffrey Kargel, Walter S. Kiefer, Vladimir Kranopolsky, Emmanuel Lellouch, Paula Martin, Christopher McKay, Roger Phillips, David Stevenson, Jeff Taylor, and Tielong Zhang. Finally, we are grateful for the support of the Chapman Conference and this monograph from NASA, ESA and the University of Colorado's Laboratory for Atmospheric and Space Physics.

This book is dedicated to all intrepid Venus explorers, past and future.

Larry W. Esposito
Ellen R. Stofan
Thomas E. Cravens

Exploring Venus as a Terrestrial Planet
Geophysical Monograph Series 176
Copyright 2007 by the American Geophysical Union.
10.1029/176GM01

Exploring Venus: Major Scientific Issues and Directions

Larry W. Esposito

Laboratory for Atmospheric and Space Physics, University of Colorado, Boulder, Colorado, USA

Ellen R. Stofan

Proxemy Research, Rectortown, Virginia, USA

Thomas E. Cravens

University of Kansas, Lawrence, Kansas, USA

INTRODUCTION

Venus has been a prime target of space exploration since the launch of Venera-1 in 1961. In 1962, Mariner 2 determined that the surface of Venus is hot, providing the first confirmation of its immense greenhouse effect. Venus has now been visited by numerous flybys, orbiters, atmospheric probes, landers, and balloons! Magellan's radar pierced the planet-encircling clouds to provide a global map of the Venus surface. Table 1 lists the chronology of Venus missions. Despite the numerous missions, the Venus environment provides a difficult target, and many significant questions remain unanswered. The state of current knowledge, the open questions, and ways to address them are discussed in the following chapters.

HIGHLIGHTS OF CURRENT ISSUES

The atmosphere, surface, and interior of Venus, and the relationships between them, were the subjects of focused scientific questions that were discussed at the Chapman conference. Conference talks highlighted the unique nature of Venus, where the interior, surface, and atmospheric history are closely coupled. Many conference participants noted that the deepest part of the Venus atmosphere, the lowest 12km or so, is still mostly unexplored. The extremes of this

Exploring Venus as a Terrestrial Planet
Geophysical Monograph Series 176
Copyright 2007 by the American Geophysical Union.
10.1029/176GM02

harsh environment caused a failure of the in-situ measurements of the Pioneer Venus probes. The super-rotation of the Venus atmosphere and the high velocity winds at the cloud-tops are still unexplained. The mineralogy of the surface is unknown, although Venera measurements are consistent with a variety of basalts. The paucity of craters shows the surface age to be only about 750 million years, but the Beta-Atla-Themis region may be even younger; in fact, the surface is likely to be still active. An early magnetic field (Venus has no detectible field now) could have left remnant magnetic signatures in the surface rocks. The climate history of Venus is sensitive to clouds and the rate of volcanic activity in the past. If Venus had an early ocean, it may have persisted for billions of years. Thus, Venus might have possessed conditions suitable for life when life arose on Earth, and for a substantial part of its history.

OPEN QUESTIONS: GEOLOGY

While the Magellan mission returned a nearly complete topographic and radar image dataset for Venus, significant questions remain. Analysis of the data indicate that Venus does not have an Earth-like system of plate tectonics, probably due to the lack of water, but the mechanisms and history of interior heat loss and resulting surface evolution are still greatly debated. Radar images show a surface dominated by volcanic processes, with limited amounts of extensional and compressional tectonics. Volcanic features range in size from small shields ~ 1 km across to flood basalt plains that cover 10,000 km^2. Two types of features have been identi-

Table 1. Space missions to Venus (updated from Marov and Grinspoon, 1998)

Spacecraft	Type of vehicle	Date of launch	Arrival date
Venera-1	Fly-by	2/12/61	5/19/61
Mariner-2	Fly-by	8/27/62	12/14/62
Venara-2	Fly-by	11/12/65	2/27/66
Venera-3	Entry probe	11/16/65	3/1/66
Venera-4	Descent vehicle and fly-by	6/12/67	10/18/67
Mariner-5	Fly-by	6/14/67	10/19/67
Venera-5	Descent vehicle and fly-by	1/5/69	5/16/69
Venera-6	Descent vehicle and fly-by	1/10/69	5/17/69
Venera-7	Soft lander and fly-by	8/17/70	12/15/70
Venera-8	Soft lander and fly-by	3/27/72	7/22/72
Mariner-10	Fly-by	11/3/73	2/5/74
Venera-9	Soft lander and orbiter	6/8/75	10/22/75
Venera-10	Soft lander and orbiter	6/4/75	10/25/75
Pioneer Venus Orbiter	Orbiter	5/20/78	12/4/78
Pioneer Venus bus & probes	One large and three small probes and orbiter bus	8/8/78	12/9/78
Venera-11	Lander and fly-by	9/9/78	12/25/78
Venera-12	Lander and fly-by	9/14/78	12/21/78
Venera-13	Lander and fly-by	10/30/81	3/1/82
Venera-14	Lander and fly-by	11/4/81	3/5/82
Venera-15	Orbiter	6/2/83	10/10/83
Venera-16	Orbiter	6/7/83	10/14/83
Vega-1	Lander and fly-by	12/15/84	6/11/85
Vega-2	Lander and fly-by	12/21/84	6/15/85
Magellan	Orbiter	5/4/89	8/10/90
Venus Express	Orbiter	9/11/05	11/4/06
Venus Climate Orbiter	Orbiter	(expected 2009)	

fied that are thought to form over hot plumes: hotspot rises such as Beta Regio that likely form over plumes originating at the core-mantle boundary like on Earth; and coronae, which likely form over shallower thermal instabilities. The complexly deformed plateau highlands, called "tessera" terrain, appear to be older than the plains that surround them, but the uniform crater population provides no evidence that their surfaces are significantly older. Theories for the origin of tessera include possible formation over both upwellings or downwellings. The fact that the plateaus are elevated has led to the speculation that they could be composed of more silicic rocks, and that should provide critical information about the evolution of volatiles on Venus.

The impact crater population indicates a relatively uniform crater retention age of ~750 my, resulting in a debate about the sequence and intensity of geologic activity over that time period (see chapters by Basilevsky and McGill, and Smrekar et al.). Geologic mapping of the surface is ongoing, but suffers from the lack of absolute age information. High resolution topography and visible images to provide "ground truth" for the Magellan data would also aid our understanding of the evolution of the surface. Detailed geochemical data, in particular mineralogy, would provide the means to better understand the evolution of the crust, and thus provide critical information on interior evolution, including the role and abundance of volatiles. And while gravity data has provided some ability to quantitatively model interior structure, without seismic data our ability to constrain the nature of the Venus interior is limited. See the chapters by Treiman, by Basilevsky and McGill, and by Smrekar et al.

OPEN QUESTIONS: ATMOSPHERE

Several of the main features of the Venus atmosphere remain unexplained. The dynamics of the super-rotation of the lower atmosphere transitions into a sub-solar to solar circulation in the upper atmosphere. Although this is similar to the atmosphere of Titan (the only other slowly rotating world with substantial atmosphere), we have no adequate explanation. The expected Hadley cell (with upwelling at the equator, poleward flow aloft, descending flow at high latitude, and equatorward flow near the surface) has not been observed, except for indications of poleward wind in some cloud-tracked winds. Possible Hadley cells may be related to the intriguing feature of the polar circulation, which includes the polar vortex, polar dipole, cold collar, and reversed equator-pole temperature gradient. See the chapter by Schubert et al.

Solar and thermal radiation play a dominant role in many chemical and dynamical processes that define the Venus climate. Venus maintains a large greenhouse effect because of the abundance of carbon dioxide and other atmospheric absorbers, despite the lower atmosphere receiving less solar energy than the Earth. Half of the solar flux received by the planet is absorbed in the cloud top region and only 2.6% of the solar radiation reaches the surface, leading to a hazy surface environment resembling a "cloudy day in Moscow." The greenhouse mechanism is the prime cause of the current and past Venus climate, while the particular distribution of radiative energy drives the remarkable super-rotation of the entire atmosphere. Chemical interactions between the gas and aerosols include non-linear feedbacks that are part of the complex climate system. What is the exact energy balance and how does it vary with location and Venus history? How efficient are the thermodynamic cycles of the Venus heat engine? How does Venus relate to studies of global warming on Earth and climate change on the terrestrial planets? A key requirement is for in-situ measurements that characterize the cloud aerosols and identify the unknown UV absorber and its role in the radiative balance. See the chapter by Titov, et al.

The study of Venus is key to understanding the terrestrial planets circling our Sun and those around other stars. The rich Venus chemistry leads to its planet-encircling clouds. Chemical interactions include sulfur, nitrogen, hydrogen and oxygen. They are driven by solar forcing and possibly by lightning. The photochemistry of the upper atmosphere gives way to thermal equilibrium chemistry in the lower atmosphere and to surface-atmosphere chemical interactions. Volcanic emissions likely re-supply sulfur that maintains the sulfuric acid clouds. Surprisingly few details of the primary chemical cycles have been verified by direct measurements. The size distribution, shape and composition of the majority of the cloud mass are still undetermined. A combination of direct exploration of the Venus atmosphere, high-resolution spectroscopy and laboratory investigations could markedly advance our understanding of its chemistry. See the chapter by Mills, et al.

OPEN QUESTIONS: SOLAR WIND AND ATMOSPHERIC ESCAPE

The intrinsic magnetic field of Venus is extremely weak, which allows the solar wind to interact very directly with the upper atmosphere and ionosphere of this planet. The boundary between the solar wind and the colder ionospheric plasma is called the ionopause and the location of this ionopause depends on solar wind conditions (e.g., higher solar wind dynamic pressures lead to lower altitude ionopause heights) and also depends on the solar extreme ultraviolet (EUV) and soft x-ray radiation flux. Higher solar fluxes result in a hotter, more extensive upper atmosphere and also lead to larger ionospheric plasma densities. An important component of the exosphere is the hot oxygen corona, mainly populated by the hot oxygen atoms created by the dissociative recombination of ionospheric O_2^+ ions.

One important consequence of a direct solar wind interaction is that exospheric species that find themselves above the ionosphere, once ionized, are lost to the planet as they are swept off with the solar wind flow. Atmospheric loss due to thermal (Jeans) escape and photochemically induced escape can also take place. Loss of oxygen is mainly associated with the ions created from the hot oxygen corona. Estimates of present-day atmospheric escape rates, determined using presently available spacecraft data, have very large uncertainties, but these rates appear to be very modest with respect to loss rates needed for Venus to lose an ocean of water over geological time. However, our knowledge of the evolution of the EUV solar flux is now improving due to Chandra and FUSE observations and it seems that the solar EUV flux might have been much greater in an earlier epoch, thus leading to much larger atmospheric loss rates.

We have also learned that coronal mass ejections (CME), which are associated with very high solar wind dynamic pressures, lead to very low ionopause heights and evidently greatly enhance atmospheric escape rates. On a different topic, the magnetic fields induced during CME events evidently penetrate below the ionosphere. Suitable future observations of this phenomenon could allow us to probe the electrical conductivity of the interior of Venus.

A better understanding of the solar wind interaction with Venus during CME events and their statistical occurrence rate will be essential for quantifying the long-term atmospheric escape due to CMEs. The Aspera-4 and magnetometer experiments onboard the Venus Express (in orbit only for one year at this time) will be able to make valuable new contributions to our understanding of the solar wind interaction with Venus, particularly during CME events. The Aspera-4 experiment includes sensors for measurement of energetic electrons, ions, and neutral atoms and can thus assess pick-up ion fluxes as well as photoionization rates due to the solar EUV flux and electron impact ionization rates in the upper atmosphere. An extended Venus Express mission will be needed in order for a range of solar conditions and consequent responses of the upper atmosphere and ionosphere of Venus to be studied. See the chapter by Russell et al.

OPEN QUESTIONS: EVOLUTION

A key question is the combined evolution of the Venusian interior, surface and atmosphere. This includes the history of Venus's climate, whether it ever possessed an ocean, whether Venus was once conductive to life, and if the future Earth will resemble the current Venus environment. What does Venus tell us about the evolution of other terrestrial planets? The strange alien nature of Venus is linked to its history.

As noted by the Natural Research Council's "Decadal Survey" in 2003, direct sampling of the atmospheric constituents is key to understanding how each planet formed and evolved. In particular, the heavy noble gases xenon, krypton, and argon and their isotopes provide an accessible historical record of ancient events pertinent to planetary formation and early evolutionary processes. Radiogenic elements of helium, argon, and xenon provide dating constraints on geologic processes that over the eons have delivered materials from the deep interior to the surface. The absence of plate tectonics may be due to the planet's lack of water. It is possible that the atmospheric climate evolution (through greenhouse warming and water loss) controlled the evolution of the solid planet! Sulfuric gases and water may trace active volcanism today, so that geologic activity determines the radiative state and climate. The stability of the atmosphere is ultimately intertwined with SO_2 and its role in the sulfur cycle. In situ measurements are needed to resolve these questions. A balloon for "bobbing" between the hostile lower atmosphere and the more benign cloud regions near 55 km altitude is one possible approach. See the chapter by Baines et al.

Rocky planets are characterized by complex interactions between their surfaces, interiors and climates. The degree to which the geologic history has affected its atmospheric evolution is dependent upon the resurfacing history and magma volatile content. Climate depends on feedbacks with volcanism and atmospheric water. The present sulfur dioxide abundance may be buffered by reactions with surface rocks, and the clouds of Venus could vanish periodically. Thus, rapid and dramatic climate change is possible. How long did Venus possess an ocean, and how long after that did plate tectonics cease, if it ever operated at all? These fundamental questions could be addressed by a well-instrumented lander to the Venus highlands.

OPEN QUESTIONS: HABITABILITY AND LIFE

Understanding the divergence of Venus and Earth is central to understanding the limits of habitability in the inner regions of habitable zones around solar-type stars. Venus may once have had warm, habitable oceans. Some speculate that life may have migrated to the clouds when the surface became too hostile. The unknown UV absorber has some properties in common with a photosynthetic pigment. It is likely that a large fraction of terrestrial planets discovered around other stars in the coming years may be Venus-like in composition and history. Although life might be considered a distraction in future Venus exploration, the list of desired astrobiological observations for future missions is basically identical to that required to understand Venus and its history. Thus, the first biological investigations can proceed immediately with the next stages of Venus exploration. See the chapter by Grinspoon and Bullock.

OPEN QUESTIONS: TECHNOLOGY AND FUTURE EXPLORATION

Among the mission types suggested by mission study groups over the years as being appropriate for the exploration of Venus are orbiters, low and high altitude balloons, long-lived landers, mobile probes, and sample return missions. Each of these mission architectures emphasizes different areas of Venus science. For example, balloon missions focus on the atmosphere, while landers are designed to explore the surface and subsurface. Each of these mission architectures also presents different technological challenges that must be met in order for future missions to this planet to succeed.

The surface of Venus, with its atmospheric pressure of 92 bars and temperature of 480°C, represents an extreme environment and a real challenge for spacecraft and instrument design. Technological strategies that could potentially deal with this environment include putting all electronics and sensitive spacecraft components into an environmentally controlled chamber, "hardening" all spacecraft components so that they can withstand the extreme temperatures and pressures, and a hybrid approach that combines the first two strategies. Technological developments in thermal protection strategies will be particularly important for extending mission lifetimes. Both passive and active (i.e., refrigeration) thermal control systems will need to be investigated. Passive systems include multi-layer insulation, "next generation" insulation materials, and phase change materials. These and other outstanding technological issues relevant to the future exploration of Venus are discussed in the Cutts et al. chapter.

CONCLUSION

The AGU Chapman Conference "Exploring Venus as a Terrestrial Planet" provided an overview of current missions and research directions. It highlighted key open questions and emphasized Venus's importance in understanding terrestrial planets. The results presented in this monograph give clear directions for future Venus studies and missions.

Acknowledgements. The Chapman Conference "Exploring Venus as a Terrestrial Planet" was supported financially by NASA, ESA and LASP/University of Colorado. We thank the helpful AGU staff and especially Laura Bloom at LASP who was the main organizer. Thanks to the numerous reviewers whose comments improved the chapters significantly and to the many conference participants who contributed to productive discussions, which are leading to a better understanding of Venus.

REFERENCES

Marov, M.Ya. and Grinspoon, D.H. 1998. *The Planet Venus.* P57. New Haven: Yale University Press.

Larry W. Esposito, Laboratory for Atmospheric and Space Physics, University of Colorado, 1234 Innovation Drive, Boulder, Colorado 80303-7814, USA, 303-492-5990

Ellen R. Stofan, Proxemy Research, PO Box 338, Rectortown, Virginia 20140, USA, 540-364-0092

Thomas E. Cravens, University of Kansas, 3093 Malott Hall, Lawrence, Kansas, 66045 USA, 785-864-4739

Geochemistry of Venus' Surface:
Current Limitations as Future Opportunities

Allan H. Treiman

Lunar and Planetary Institute, Houston, Texas, USA

Geochemical data about Venus' surface materials are quite limited and of poor precision. The Venera and VEGA lander missions (sources of the available data) were engineering and scientific triumphs, but their chemical analyses of the Venus surface do not permit detailed confident interpretation, such as are routine for terrestrial analyses and MER APXS rover analyses from Mars. In particular, the Venera and VEGA analyses of major elements (by XRF) did not return abundances of Na, and their data on Mg and Al are little more than detections at the 2σ level. Their analyses for K, U, and Th (by gamma rays) are imprecise, except for one (Venera 8) with extremely high K contents (~4% K_2O) and one (Venera 9) with a non-chondritic U/Th abundance ratio. The landers did not return data on other critical trace and minor elements, like Cr and Ni. In addition, the Venera and VEGA landers sampled only materials from the Venus lowlands—they did not target sites in any of the highland areas: shield volcanos, tesserae, nor the unique plateau costruct of Ishtar Terra. These limitations on current understanding of Venus' geochemistry emphasize the huge opportunities in additional chemical analyses of Venus' surface. Currently available instruments could provide much more precise analyses for major and minor elements, even within the engineering constraints of the Venera and VEGA lander systems. Such precise analyses would be welcome for basalts of Venus' lowland plains, but would be especially desirable for the highland tesserae and for Ishtar Terra. The tesserae may well represent ancient crust that predates the most recent volcanic resurfacing event and so provide a geochemical look into Venus' distant past. Ishtar Terra may be composed (at least in part) of granitic rocks like Earth's continental crust, which required abundant water to form. So, Ishtar Terra could possibly yield evidence on whether Venus once had an ocean, and thus the possibility of life.

INTRODUCTION

This chapter focuses on the chemical composition of Venus' surface, direct analyses and indirect constraints, emphasizing how little is actually known. This incognizance

Exploring Venus as a Terrestrial Planet
Geophysical Monograph Series 176
Copyright 2007 by the American Geophysical Union.
10.1029/176GM03

of Venus' geochemistry highlights the opportunities for high-impact scientific return from future investigations, by remote sensing, *in situ* analyses, and sample return.

At this time, it is fair to say that we know little of Venus' geochemistry. What we do know is greatly limited, both with respect to spatial/geologic coverage and analytical precision. This chapter does not review, or even summarize, the many geochemical interpretations studies that have appeared since the first images of Venus' surface and the first analyses of its materials. Nor does it directly address the related questions of the mineralogy of Venus' surface, and the nature and extent of surface-atmosphere interactions on Venus—these are covered elsewhere. Rather, this chapter will recount the extent of our ignorance about Venus' geochemistry, and we will show that even limited chemical data of modestly improved precision can provide huge advances in our understanding of Venus' surface, history, petrogenesis and planetary evolution.

The principal remote-sensing instrument for Venus, radar, is insensitive to the chemistry and lithology of Venus' rocks and regolith, except for the unusual low-emissivity surfaces of Venus' highlands. Some Venusian landforms have strong geochemical implications—e.g., shield volcanos almost certainly formed from basaltic lavas. Other landforms are suggestive of geochemistry (e.g., is the Ishtar Terra highland an Earth-style continent?), while others remain enigmatic (e.g., were the pancake domes felsic or basaltic lavas).

The technological and scientific triumphs of the Russian Venera and VEGA landers yielded limited chemical analyses on seven sites on Venus, and those only from lava plains and rises. Of those seven analyses, four are only of K, Th, and U; two include most (but not all) major elements of silicate rocks; and one includes K, Th, U and most major elements. All of these analyses are difficult to interpret because of their low precision by the standards of terrestrial (and MER rover) rock analyses.

In this chapter, we will consider geochemical data and inferences in order of landform type. The starting point is volcanic plains and rises (i.e., shield volcanos), which cover > 80% of Venus' surface and include all of the *in situ* chemical analyses of the Venera and VEGA landers. Within some of these plains are small, steep-sided domical volcanos, and long fluid-cut channels. Coronae are distinct and unusual volcanic tectonic features, mostly in the lowland plains. The Venusian highlands are enigmatic, and represented mostly by the tectonic landscapes of the tesserae. Unique among them is Ishtar terra, a plateau 4 kilometers above the adjacent plains surrounded by fold/fault mountain belts. The highest elevations of the highlands are coated with an enigmatic radar-reflective material. Finally, we will briefly consider impact-related deposits, including the lava flows that emanate from many Venusian craters.

VOLCANIC PLAINS AND RISES

Most of Venus' surface is extensive volcanic plains and rises, constituting ~80% of Venus' surface. Most show landforms characteristic of fluid basaltic lava [*Bruno and Taylor*, 1995; *Crumpler et al.*, 1997]; the remainder lack obvious landforms but are also interpreted as plains of basalt lava flow.

Venera and VEGA Analyses

The Venera and VEGA spacecraft all landed on and among Venus' volcanic plains and rises (Table 1, from *Kargel et al.* [1993], *Fegley* [2004]), and chemical analyses of surface materials are given in Tables 2 and 3. Major rock-forming elements were analyzed by X-ray fluorescence (XRF), using radioactive sources for X-rays (^{238}Pu and ^{55}Fe). Fluoresced X-rays were detected in gas-discharge proportional counters [*Barsukov et al.*, 1986]. Abundances of K, Th, and U were measured independently by their own gamma-ray emissions, which were detected by a CsI(Tl) scintillator [e.g., *Surkov et al.*, 1986].

Major element abundances from the X-ray fluorescence analyzers on Venera 13, Venera 14, and VEGA 2 analyses are given in Table 2, which differs from most presentations in two respects. First, the Venera and VEGA XRF systems could not detect Na, and it is shown as such rather than as calculated values based on unverifiable analogies with Earth basalts [e.g., *Surkov et al.*, 1986]. Second, analytical uncertainties on Venera and VEGA determinations are given as 2σ values rather than the 1σ or 75% confidence limits in the original and most subsequent papers [e.g., *Barsukov et al.*, 1986a; *Kargel et al.*, 1993; *Lodders and Fegley*, 1998; *Fegley*, 2004]. The choice of 2σ is for consistency and comparison with the terrestrial geochemical literature and with data from the MER rovers' APXS instruments. It will be seen from Table 2 that, with 2σ uncertainties, the Venera and VEGA analyses become significantly less interpretable. Foremost, MgO and MnO analyses become essentially detections or upper limits at the 2σ level, and so derived properties like Mg/Fe, Mg*(molar Mg/(Mg+Fe)), and FeO/MnO are poorly constrained. Abundances of TiO_2 and K_2O become much less precise (on a proportional basis), and both become detections in the VEGA 2 analysis.

Presentation of these data with 2σ uncertainties is not meant to denigrate the analyses themselves or the efforts expended obtaining them. Truly, these data are the best available for Venus, and represent an unrivaled technical and scientific achievement. Yet, the Venera and VEGA XRF data are not of the precision usually expected in terrestrial analyses (or hoped for in Martian rover analyses), and cannot be interpreted with the same level of confidence as such terrestrial or Martian analyses.

The Venera and VEGA analyses for K, Th, and U (by gamma-ray spectrometry), are given in Table 3. In the original reports, and most subsequent investigations, these data are presented with 1σ uncertainties ("2/3 fiducial uncertainty," [*Barsukov et al.*, 1986b]). As with the XRF data, we present them with 2σ uncertainties for better comparison with terrestrial analyses. At this level, the gamma-ray

Table 1. Venus Lander Probes and Their Landing Sites

Probe	Lat°, Long°	Geography	Geology/Tectonics	Geochemistry
Venera 8	-10.70, 335.24	E Navka Planitia	Mottled plains, with small cones, pancake domes; in a possible corona.	Very high K, Th, U.
Venera 9	31.01, 291.64	NE slope Beta Regio	Aikhylu Chasma rift system, in lava flows from Devana Chasma / Beta Regio.	Low K, U; high Th.
Venera 10	15.42, 291.51	SE margin Beta Regio	Probably from small shield volcano, Samodiva Mons, to S; possibly older flows of nearby coronas.	Low K, Th, U.
Venera 13	-7.55, 303.69	Navka Planitia, E of Phoebe Regio	Dark plains, flows from rift system; near domes and possible corona.	High K basalt.
Venera 14	13.05, 310.19	Navka Planitia, E of Phoebe Regio	Landing ellipse centered on caldera of small shield volcano (Panina Patera).	Tholeiitic basalt.
VEGA 1	8.10, 175.85	Rusalka Planitia, N of Aphrodite Terra	Dark, smooth, widely fractured lava plains.	Low K, U, Th.
VEGA 2	-7.14, 177.67	Rusalka Planitia to E edge of Aphrodite Terra	Dark, smooth, widely fractured lava plains. Between Eigin and Sith coronae.	Tholeiitic basalt; low K, U, Th.

From *Kargel et al.,* [1993], *Abdrakhimov and Basilevsky* [2002], *Fegley* [2004].

analyses for Venera 10, and VEGA 1 and 2 are essentially all detections or upper limits (Table 3). Only for the unusual rock of the Venera 9 site are all the gamma-ray analyses significant at the 2σ level.

Geochemistry of Venus' Surface

Even recognizing that the Venera and VEGA chemical analyses are much less precise than typical terrestrial or Martian rover results, one can still derive some information on Venus' geochemistry. Inferences here follow those of earlier studies [e.g., *Surkov et al.,* 1984; *Hess and Head,* 1990; *Barsukov,* 1992; *Kargel et al.,* 1993; *Fegley,* 2004], although with skepticism similar to that of *Grimm and Hess* [1997].

In general, the Venera and VEGA analyses can be reasonably interpreted as representing primary rocks, or nearly isochemical alterations of them (e.g., by addition of SO_2 from the atmosphere). Scenarios involving chemical alteration are possible or even likely, i.e. mobility of Te or Fe to explain the radar reflective coatings of on the Venusian highlands [*Wood,* 1997; *Schaefer and Fegley,* 2004] , or of Ca to explain the low Ca/Al ratios in the surface materials [*Grimm and Hess,* 1997].

U-Th-K. The naturally radioactive elements uranium, thorium, and potassium, are the major sources of post-accretion heat in Venus. They are also highly incompatible in most igneous settings, and so are crucial tracers of melt production and differentiation. In nebular processes, U and Th are both highly refractory and are not readily separated by their volatility. Thus, nearly all planetary materials have U and Th abundances in the same ratio as CI chondrites. Potassium is more volatile than U or Th, and so K/U and K/Th ratios are commonly much lower than CI. In planetary differentiation,

these elements do not enter metal appreciably, and so their abundance ratios are not affected by core formation. In most silicate igneous processing, Th and U behave similarly so that the Th/U ratio remains relatively constant. In fact, K, U, and Th are readily separated only in processes at relatively low temperature, like aqueous alteration.

In the Venus basalt analyses, U and Th abundances are broadly consistent with a CI chondritic ratio, although determinations of U are relatively imprecise. The V8 analysis is exceptional in its very high Th content—6.5 ppm or 220 x CI—which (along with its high K content; Table 3) are clear signs that a portion of Venus has undergone significant chemical differentiation! Another interesting observation is that Th/U at the V9 site (the most precise of the analyses), is greater than the CI ratio at the 2σ level of uncertainty. This deviation could represent analytical inaccuracy or an overestimate of precision, or could represent a real fractionation of U from Th. In the latter case, one could reasonably invoke fractionations involving garnet [*van Westeren et al.,* 2000; *Perterman et al.,* 2003], aqueous fluids [e.g., *Stopar et al.,* 2004], or possibly ionic fluids like carbonate or sulfate magma [e.g., *Kargel et al.,* 1994; *Jones,* 1995].

The potassium contents of materials at most of the Venera and VEGA sites are relatively low—4000 ppm or ~0.4 % K_2O (Table 3)—comparable to those of common terrestrial basalts. The glaring exceptions are V8 and V13 (Table 2), which have ~4% K_2O—very high for an Earth basalts (Figure 1b, 1c) and comparable to some highly felsic (e.g., granitic), and carbonatitic igneous rocks [e.g., *DuBois et al.,* 1963; *Keller and Krafft,* 1990; *Kargel et al.,* 1993; *Fegley,* 2004]. As above, these high values require significant or extensive differentiation within Venus' crust and mantle. The K/U ratios of the Venera and VEGA samples are all

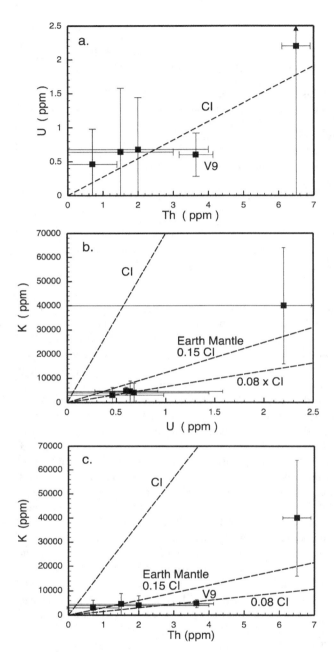

Figure 1. Abundances of uranium, thorium, and potassium in Venus basalts. Data from Table 2, uncertainty bars are 2 σ. (a). U versus Th. All analyses are consistent with a U/Th ratio like that of CI chondrites [*Anders and Grevesse, 1989*], except the Venera 9 analyses (V9). (b). K versus U. All analyses are consistent, within 2σ uncertainties, with the same K/U ratio as the Earth's mantle [*McDonough and Sun, 1995*]. (c). K versus Th. All analyses but that of Venera 9 (V9) are consistent, within 2σ uncertainties, with the K/Th ratio of the Earth's mantle [*McDonough and Sun, 1995*]. K/Th value for Mars' crust from *Taylor et al.* [2007]. Considered with Figure 1a, the V9 sample appears enriched in Th compared to K and U.

consistent within 2σ limits of that of the Earth's mantle (Figure 1b), as noted earlier [e.g., *McDonough and Sun*, 1995]. The K/Th ratios are also within 2σ limits of that of the Earth's mantle (Figure 1c), except for the V9 analysis. Considering K, Th, and U abundances together, the V9 analysis likely represents excess Th above the chondritic expectation of Th/U.

Ca-Ti-Al. Major element analyses from X-ray fluorescence provide several constraints on the origins of the Venera and VEGA basalts. A seldom-used set of constraints from major elements comes from Ca, Ti, and Al. These elements are all refractory and lithophile in planetary accretion and core formation, and are all incompatible in magma genesis from primitive mantles—Ti more incompatible, and Ca rather less so. For basalts produced by relatively high degrees of partial melting (leaving residues of olivine ± orthopyroxene), such as might be expected for the voluminous lavas of Venus, the Ca/Ti/Al of the basalts would be close to those of the source region. Ca, Al, and Ti were analyzed with some precision by the Venera and VEGA landers (unlike Na and Mg, for instances, which would be crucial in most investigations of the geneses of igneous rocks).

In a graph of Ca/Al vs. Ti/Al (Figure 2), basalts with relatively simple petrogeneses and high degrees of partial melting should have Ca/Al and Ti/Al values like those CI chondrites, again because the most abundant residual mantle minerals (olivine ± orthopyroxene) retain little of those elements. In fact, simple basalts like the eucrites meteorites have chondritic Ca/Al and Ti/Al, as do terrestrial MORB, which is consistent with their formation from high degrees of partial melting from undepleted or slightly depleted sources. The Martian basalts analyzed in Gusev crater by the MER Spirit lander also have chondritic Ca/Al and Ti/Al (Figure 2). Conversely, the Martian meteorite basalts have super-chondritic Ti/Al and most have super-chondritic Ca/Al. Lunar basalts also show superchondritic Ca/Al and TI/Al, which represent source mantle depletion in Al by early formation of the Moon's anorthositic crust. The lunar basalts' Ti/Al ratio, from chondritic to >100 times chondritic, is inferred represent melting of ilmenite-rich cumulates in the lunar mantle [e.g., *Shearer et al.*, 1999]. Terrestrial alkaline rocks of oceanic associations and ocean island basalts (OIB) extend from the MORB group near CI, to high Ti/Al ratios at moderately sub-chondritic Ca/Al. Enrichment in Ti without concomitant enrichment in Ca and Al may indicate mantle metasomatic processes (like those responsible for MARID materials [e.g., *Dawson and Smith*, 1977; *Sweeney et al.*, 1993]), while the sub-chondritic Ca/Al likely reflects their high alkali content. The three major element analyses of Table 2 are given in

Table 2. Major Element Compositions (XRF) of Venus Surface Materials and Others

	Venera 13	2σ	Venera 14	2σ	Vega 2	2σ	MER-RAT Humphrey	2σ	WSU BCR-P	2σ
SiO_2	45.1	± 6.0	48.7	± 7.2	45.6	± 6.4	46.3	± 0.4	55.18	± 0.06
TiO_2	1.6	± 0.9	1.25	± 0.8	0.2	± 0.2	0.58	± 0.06	2.28	± 0.02
Al_2O_3	15.8	± 6.0	17.9	± 5.2	16.0	± 3.6	10.78	± 0.15	13.62	± 0.04
Cr_2O_3	n.d.	.	n.d	.	n.d		0.68	± 0.03	n.d.	
FeO	9.3	± 4.4	8.8	± 3.6	7.7	± 2.2	18.6	± 0.12	12.73	± 0.04
MnO	0.2	± 0.2	0.16	± 0.16	0.14	± 0.24	0.41	± 0.01	0.184	± 0.002
MgO	11.4	± 12.4	8.1	± 6.6	11.5	± 7.4	9.49	± 0.12	3.52	± 0.12
CaO	7.1	± 2.0	10.3	± 2.4	7.5	± 1.4	8.19	± 0.06	6.99	± 0.04
Na_2O	n.d.		n.d.		n.d.		2.8	± 0.2	3.38	± 0.10
K_2O	4.0	± 1.2	0.2	± 0.14	0.1	± 0.16	0.13	± 0.05	1.74	± 0.01
SO_3	1.6	± 2.0	0.35	± 0.6	1.9	± 1.2	1.09	± 0.03	n.d.	
P_2O_5	n.d.		n.d	.	n.d		0.57	± 0.07	0.378	± 0.002
Cl	< 0.3		< 0.4		< 0.3		0.32	± 0.01	n.d	
Sum	96.1		95.8		90.6		99.9		100.00	

Venera and VEGA analyses from compilations of *Kargel et al.* [1993] and *Fegley* [2004], after comparison with original data (see these references for original sources). Uncertainties are based on the 1σ values tabulated by these authors
'MER-RAT Humphrey' is by the APXS instrument (MER Spirit Rover, Gusev Crater Mars) of an olivine-phyric basalt [*Gellert et al.*, 2006].
WSU BCR-P is a high-quality, replicate analyses of a standard basalt (Columbia River basalts), typical of the best obtainable analyses on Earth [*Johnson et al.*, 1999].

Figure 2, although Ti in the VEGA 2 analysis is at detection limit.

All these Venus basalts could, within 2σ uncertainties, have chondritic Ca/Al; it is more likely they have sub-chondritic Ca/Al (Figure 2). This is comparable to most terrestrial alkaline rocks, in which Al is paired to great extent with Na and K rather than Ca. Perhaps more relevant to Venus is that partial melts of dry eclogite (garnet pyroxenite) with MORB-like compositions also have subchondritic Ca/Al [*Pertermann and Hisrchmann*, 2003; *Perterman et al.*, 2003; see *Klemme et al.*, 2002]. It is also possible that the subchondritic Ca/Al ratio represents loss of Ca during alteration and weathering [*Grimm and Hess*, 1997]. A plausible scenario for this loss builds on the experiments of *Treiman and Allen* [1994], in which basalt glass exposed to CO_2 gas at ~500°C became coated with thin layers of calcite and magnetite. In the real Venus atmosphere, anhydrite ($CaSO_4$) would replace calcite and could be abraded from the rock and transported elsewhere by wind. Like so much else in Venus geochemistry, this scenario is speculative.

The Venera analyses appear to have super-chondritic Ti/Al, and so (in Ca-Al-Ti) are more similar to terrestrial alkaline basalt than MORB. This suggestion that V13 is an alkaline basalt fits closely with its K_2O content of 4.0% and the chemical similarity between it and terrestrial leucitite lava (among the most K-rich and silica-poor known; [*Barsukov*, 1992; *Kargel et al.*, 1993; *Fegley*, 2004]). However, the V14 and V2 basalts have low K_2O and are commonly compared with terrestrial MORB [e.g., *Surkov et al.* 1984; *Kargel et al.* 1993; *Fegley*, 2004]. These compositions are broadly similar to basaltic partial melts of anhydrous MORB-like eclogite in their superchondritic Ti/Al and low K_2O [*Pertermann and Hisrchmann*, 2003]; it should be noted that these experimental partial melts have Na_2O near 3.5%, while the common extrapolation is for the Venus basalts to have only ~2% Na_2O [*Surkov et al.*, 1986; *Kargel et al.*, 1993; *Fegley*, 2004]. The VEGA 2 analysis is further unusual in hav-

Table 3. Radioactive Element Abundances in Venus Surface Materials and Others

ppm	Venera 8	2σ	Venera 9	2σ	Venera 10	2σ	VEGA 1	2σ	VEGA 2	2σ	MER-RAT Humphrey	2σ
K	40,000 ± 24,000		4700 ± 1600		3000 ± 3200		4500 ± 4400		4000 ± 4000		1300 ± 500	
U	2.2 ± 2.4		0.60 ± 0.32		0.46 ± 0.52		0.64 ± 0.94		0.68 ± 0.76		-	-
Th	6.5 ± 0.4		3.65 ± 0.48		0.70 ± 0.74		1.5 ± 2.4		2.0 ± 2.0		-	-

Venera and VEGA analyses from compilations of *Kargel et al.* [1993] and *Fegley* [2004], after comparison with original data (see these references for original sources). Uncertainties are based on the 1σ values tabulated by these authors.
'MER-RAT Humphrey' is by the APXS instrument (MER Spirit Rover, Gusev Crater Mars) of an olivine-phyric basalt [*Gellert et al.*, 2006].

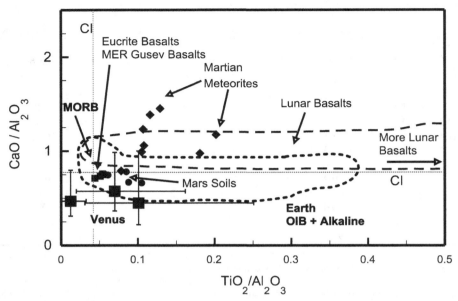

Figure 2. Abundance ratios CaO/Al_2O_3 and TiO_2/Al_2O_3 for solar system basalts and CI chondrites. Data for Venus basalts from Table 2; uncertainty bars are 2σ. Data for other basalts from literature sources including *BVSP* [1981] and *Gellert et al.* [2006]. See discussion in text.

ing subchondritic Ti/Al, even though the Ti value is an upper limit at the 2σ level.

FeO. The FeO content of a basaltic magma derived from a peridotitic (olivine-rich) mantle and the FeO content of that mantle are comparable [see *Longhi et al.*, 1992; *Robinson and Taylor*, 2001]. FeO contents of evolved magmas are higher than the mantle value, so that the FeO content of the Venera and VEGA basalts is a rough upper limit to the FeO content of a peridotitic mantle source. The Venera and VEGA analyses average about 8% FeO, which is comparable to those of most primitive Earth MORBs. This similarity suggests that the mantle source of the Venera basalt, if peridotitic, has a comparable FeO content as the Earth's peridotite mantle.

Mg/Fe. For a basalt, the molar Mg/Fe ratio of a basalt and the parameter $Mg^* = 100 \cdot Mg/(Mg+Fe)$ are crucial markers for the degree of fractionation of a basalt, and of the extent of a mantle's Fe loss to core formation. Thus, it is unfortunate that the Venera and VEGA analyses for Mg are so imprecise (Table 2). At the 2σ level, the MgO value for the V13 rock is an upper limit, and that for V14 is uncertain to ~80%. The most precise data are for the V2 sample, from which one calculates $Mg^*=73^{+13}_{-21}$. This nominal Mg^* is only slightly different from those of primitive Earth basalts ($Mg^*=68$), which represent equilibrium with terrestrial mantle olivine of Fo_{91}. If the V2 basalt does have $Mg^*=73$, it could suggest

formation from a depleted mantle source, which would be consistent with its sub-chondritic Ti/Al ratio. Or, its high Mg^* could suggest that Venus' whole mantle is more magnesian than the Earth's, which could possibly imply that Venus was originally less oxidized than the Earth and that more of its original iron was metallic and sequestered into its core.

FeO/MnO. The Fe/Mn ratio of basalts and their constituent minerals, commonly cited given as FeO/MnO, is a useful constraint on the extent of core formation in a differentiated planet. This arises because Fe and Mn have similar volatility during nebular condensation and similar behavior during silicate differentiation. However, Mn does not enter Fe-rich metal during core formation, so that the Fe/Mn ratio acts as a tracer for the extent of core formation and metal separation in a planetary body

Thus, it is unfortunate that the Venera and VEGA analyses for Mn are so imprecise (Table 2). Formally, the V13, V14 and V2 analyses for Mn are upper limits only at the 2σ level, and restrict FeO/MnO to being greater than ~25. This limit is not diagnostic, and includes CI chondrites (FeO/MnO=95) and most differentiated planets, including the Earth, the moon, Mars, and Vesta. The FeO/MnO ratio could conceivably be more precise than either analysis by itself, if some of the uncertainties in FeO and MnO are correlated. Taking the nominal values in the analyses of the Venus surface materials, FeO/MnO averages about 50, which is similar to those of Earth basalts at ~60 [*Kargel et al.*, 1993] and of Martian and eucrite basalts at ~35 [e.g., *Drake et al.*, 1989; *Treiman et al.*, 2000].

Summary. From the limited and partial chemical analyses of Venus surface materials, the volcanic plains and rises appear to be built of basalts, which are broadly similar to those found on Earth (e.g., in Mg*, FeO/MnO, Ti abundance, Ca/Al ratio). Many of the Venus basalts show signs of strong differentiation, either by enrichment (as in the high K contents of the V8 and V13 samples), depletion (as in the high low Ti/Al and high Mg* of the V2 sample), or by alteration (the superchondritic Th/U of the V9 sample). Considered with 2σ uncertainties, it is not clear that the Venus basalts can be correlated with particular types or tectonic settings as defined on Earth.

Perhaps more evident from these analyses is how many issues remain unresolved. Few analyses are available, and those are imprecise and lack critical major and trace elements. With so few analyses, it is inevitable that vast interesting portions of Venus remain unsampled. In particular, we lack any data from Venus' highlands. In addition, many of the lander sites are complex, and it remains unclear which of several units there were actually analyzed.

The Venera and VEGA missions together produced only three major element analyses, and only one of those was accompanied by analyses for U and Th. The major element analyses did not include Na, and obtained Mg and Al at very low precision. At four other sites, only one has analyses of radioactive elements, and many of those are only upper limits at the 2σ level. All of those analyses, major and radioactive elements are imprecise by terrestrial and modern spacecraft standards (Tables 2, 3).

With so few analyses for a whole Earth-sized planet, it is inevitable that most geomorphic and geologic regions remain unsampled. All of Venera and VEGA landers targeted sites in a few subsets of the volcanic plains. The Venera 9, 10, 13, and 14 landers sampled a fairly small region—the eastern flanks of the large shield volcanos of Phoebe and Beta Regio: 290–310°E, -10–30°N. And the VEGA 1 and 2 landers both were between Sith corona and Sapas Mons: ~175°E, -10–10°N. Thus, there are no *in situ* data on large volcanic edifices, rift zones, tesserae, fold/fault belts, or the Ishtar Terra highlands. And even within the volcanic plains, nearly all areas and geologic structures remain unstudied.

The landing sites of the Venera and VEGA spacecraft are uncertain to a radius of ~300 km, so for most sites it is not clear which geologic terrane was analyzed [*Kargel et al.*, 1993]. Several of the landing sites have simple, uniform geology and geomorphology so that it seems fairly clear which unit was analyzed. The V10 and V1 landing sites are on relatively homogeneous flat volcanic plains (Figs. 7, 8 of *Kargel et al.* [1993]), the V2 site is almost certainly on flows emanating from Sith corona (Figs. 7, 8 of *Kargel et al.* [1993]), and the V14 landing area is nearly all on a small shield volcano/caldera complex (Fig. 3 of *Kargel et al.* [1993]). However, the geochemically unusual analyses (V8, V9, and V13) are all from complex sites. The V8 landing site (very high K and Th) is a mottled plain (in radar imagery) with several varieties of lava flows, many small shield volcanos, and a pancake dome (Fig. 1 of *Kargel et al.* [1993]). These features may be associated with an incipient corona structure [*Basilevsky et al.* 1992]. The V9 landing site (super-chondritic Th/U) is from a complex, intensely faulted plains region near the Rhea Mons volcano of Beta Regio (Fig. 6 of *Kargel et al.* [1993]). The site appears to include several different sorts of lava flows, and several small volcanic constructs. The V13 site (basalt with 4% K$_2$O) includes a broad rifted zone, and a portion of a small corona and several steep-sided domes lie near the nominal landing region.

It is, of course, tempting to associate these incomplete and imprecise chemical analyses with particular tectonic settings, especially those of the Earth. These associations are perilous, as few (if any) of the plate tectonic regimes on Earth are present on Venus. For instance, Venus basalts may be chemically like MORB, but Venus has no mid-ocean ridges or divergent plate boundaries—in fact, its tectonic processes are the subjects of significant debate. In this case, it is more reasonable to describe the basalts in chemical terms (e.g., mildly depleted tholeiite) or by inferred mantle source region (e.g., partially depleted spinel peridotite) rather than the tectonic setting (e.g., MORB). The only tectonic association we discern in the Venera and VEGA data sets is that the two high-K analyses (V8 and V13) are from sites near coronae. With so few analyses, this association is highly speculative.

Steep-Sided Domes

Across the relatively flat, extensive volcanic plains of Venus are scattered steep-sided domes and 'pancakes', which are also interpreted as volcanic constructs [*McKenzie et al.*, 1992; *Pavri et al.*, 1992; *Head et al.*, 1992]. Included here are modified domes, like 'ticks', which probably formed by mechanical collapse of the margins of domes with more regular plan forms [*Guest et al.*, 1992; *Bulmer and Guest*, 1996]. The shapes and steep sides of these domes appear imply to a magma that is much more viscous than the fluid basalts invoked for the extensive volcanic plains and rises of Venus. This apparent high viscosity has been taken to imply a widely different chemical composition, or some difference in physical state.

Steep-sided domes on Earth are typical of silica-rich lavas, dacitic and granitic in composition, and Venus' domes may be comparable [e.g., *Fink et al.*, 1993; *Fink and Bridges*, 1995]. The Venera 8 chemical analyses, rich in K, Th, and U, was first interpreted as representing this sort of evolved silicic magma [*Nikolaeva*, 1990], and the site is near a pancake dome [*Kargel et al.*, 1993]. The V8 chemical analysis is also consistent with more mafic magma types, and the V8 lander returned no photos of its surroundings. Many steep-sided domes are found near the summits of large volcanos and volcanic rises [*Crumpler et al.*, 1997], which would be consistent with their formation as late silicic differentiates (e.g., the Puu Waawaa trachyte dome on Hualala'i volcano, Hawai'i [*Cousens et al.*, 2003]).

The apparent high viscosities of the dome and pancake lavas have been explained in two other ways. First, an abundance of crystals (or xenocrysts) will greatly increase the viscosity of basaltic magmas [*Sakimoto and Zuber*, 1995], both by the mechanical effect of being a two-phase fluid and by the chemical effect that the magma remaining among the crystals (in a closed system) will be more evolved and silicic. Similarly, an abundance of bubbles in the lava (i.e., a foam) will have a mechanical effect similar to abundant crystals and will produce a high effective viscosity [*Pavri et al.*, 1992; *Crumpler et al.*, 1997]. Through experimental and theoretical modeling of lava dome shapes, *Fink and Griffiths* [1998] inferred that the Venus domes formed most likely from andesitic or crystal-rich basaltic andesite lavas.

At this point, even the assumption of a high effective viscosity is in doubt, as *Bridges* [1995] and *Clague et al.* [2000] have shown that pancake-shaped domes can form directly from low-viscosity basaltic magma. They report flat-topped volcanic cones, quite similar to pancake domes, on Earth as products of continuously overflowing submarine ponds of fluid, tholeiitic basalt lava. This specific mechanism may not be relevant to Venus, but Bridges and Clague provide a scenario in which a landform suggestive of very-high viscosity lava can form from slow eruption of fluid lava.

Thus, the chemical compositions of Venus' steep-sided domes and pancakes remain essentially unconstrained. These domes may not be important in the broad scheme of Venus' evolution and volcanism, but yet may provide crucial clues to its magma genesis and differentiation pathways and thus the development of its atmosphere.

Long Channels (Canali)

Most landforms of the volcanic plains and rises are reasonably ascribed to basaltic lavas [e.g., *Crumpler et al.*, 1997]. However, in several places the basaltic plains are incised by long meandering channels, designated canali [*Baker et al.*,

1992, 1997]—the fluid that cut the canali is not known. Canali are long channels, from 500 to 6800 km in length, sinuous and meandering, incised into volcanic plains (and rarely tesserae). Their widths are typically 3 km, ranging up to 10 km, and are constant over the lengths of each canale. Shorter channels are comparable to the lunar sinuous rilles, and can reasonably be ascribed to the flow of basaltic lava. But the longer canali, with their constant widths, are enigmatic.

The challenge presented by canali is to understand what sort of fluid could flow, at Venus' surface, for the distance and time needed to produce channels 500 to 6800 kilometers long. *Baker et al.* [1992] showed that the canali required a low-viscosity fluid that was either very hot or had a melting temperature close to that of the Venus surface. Basaltic and komatiitic lavas seem to be precluded by their rapid cooling rates on Venus [*Baker et al.*, 1992; *Komatsu et al.*, 1992; *Gregg and Greeley*, 1993]. The rapid cooling could be reduced by formation of an insulating cover, partially as a roof (although the channels are too wide to be spanned to form a lava tube), and partially by solid plates floating on and carried by the flow [*Gregg and Greeley*, 1993]. On the other hand subsurface flows of basaltic lava could conceivably produce long channels [*Lang and Hansen*, 2006], but it is not clear if sub-surface flows could produce the observed meanders of the canali.

Possible liquids other than basalts include liquid sulfur, carbonate-rich lavas (carbonatites), and even water if the channels formed when Venus' surface were considerably cooler than at present. Liquid sulfur is a plausible product of reaction between oxidized atmospheric sulfur and reducing silicate rocks, but its low vapor pressure in the Venus atmosphere may imply that it would evaporate too rapidly to produce long channels [*Komatsu et al.*, 1992]. Carbonate-rich and carbonate-sulfate lavas were considered by *Kargel et al.* [1994], and shown to have appropriately low viscosities, erosive potentials, and melting temperatures [*Treiman and Schedl*, 1983; *Treiman*, 1995]. They also showed that carbonate-sulfate liquids could plausibly form in the shallow crust by burial of basalt rock that was weathered at the surface.

Although these canali are a local feature of limited areal extent, they may be critically important for understanding Venus' subsurface. For instance, a Venus crust permeated with carbonate-sulfate magma will have vastly different rheological and electrical properties than a fluid-absent crust. And, as shown here, our understanding of these canali and their fluids is rudimentary and speculative.

VENUS' HIGHLANDS

Besides the volcanic lowlands and rises (volcanos), Venus' surface is marked by two distinct sorts of uplands: highland plateaus, and the unique high plain and surrounding mountains of Ishtar Terra. The geochemistries of the uplands are unknown. At this time, no lander spacecraft has visited either sort of upland, and most of our understanding of their geochemistry is limited to inferences from radar geomorphology and gravity anomalies. The exception to our complete ignorance is that portions of the uplands show an unusually low emissivity in radar wavelengths, which constrains (but not uniquely) the chemistry of materials in the topmost few mm of the surface.

Highland Plateaus

Highland plateaus of Venus stand a few kilometers above the surrounding lowland plains, and comprise ~8% of the planet's surface. Commonly, plateau surfaces intensely deformed by multiple generations of faults, yielding a surface morphology called tessera [*Barsukov* et al., 1986c; *Hansen and Willis*, 1996]. Edges of the plateaus are commonly embayed and partially engulfed by lava flows (basalt) from the adjacent volcanic lowlands and rises; inside some plateau are regions which appear to have been resurfaced by lava flows, and a few tesserae cut by canali. No spacecraft have landed on the plateaus, although the nominal landing site of V10 includes some tessera terrane [Fig. 7 of *Kargel et al.*, 1993].

The chemical compositions and rock types of the tesserae are essentially unknown; the available constraints come from geomorphology and gravity/topography. Volatile-bearing minerals, created in some earlier cooler epoch, could persist over long durations on the highlands [*Johnson and Fegley*, 2000, 2003a, 2003b], but are undetectable in available data. The geomorphology of tesserae is complex and disrupted extensively by fault surfaces, but small undeformed islands are scattered among the faults. *Ivanov* [2001] found that these islands showed the same surface morphologies as the lowland volcanic plains and inferred that tesserae developed from older plains material and are thus basaltic.

Geophysical data (gravity and topography) for the plateaus and tessera are consistent with a basaltic composition and density, but do not exclude less dense materials. Magellan gravity data suggest that the topography of the tesserae and highland plateaus are isostatically compensated, corresponding to a crust of 20-40 km of basaltic rock [e.g., *Simons et al.*, 1997; *Anderson and Smrekar*, 2006]. Geophysical data do not, however, exclude the possibility of thinner, less dense material under the tesserae.

Ishtar Terra

The Ishtar Terra upland is unique on Venus, and bears a strong geomorphic resemblance to a continent on Earth. The center of Ishtar Terra is the relatively flat, little deformed Lakshmi Planum, which stands 3.5 km above mean planetary radius (which is only slightly higher than the average elevation of the volcanic lowlands). For comparison, the average height of a continent on Earth stands ~5 km above the average ocean depth. Lakshmi Planum is bordered on most sides by curved mountain belts (Maxwell, Freya, and Danu), as are some continents on Earth (e.g., North America). The mountain belts around Ishtar are compressional, in which rock masses appear to have been thrust onto and over Lakshmi Planum [*Kaula et al.*, 1992, 1997]. On Earth, similar mountains are common now and in the geologic record as foreland thrust belts [*Ansan et al.*, 1996] (like the Himalayas and the Rocky Mountain front range), where one continental block is forced over another. Early work likened Ishtar Terra to the Tibetan plateau [*Kiefer and Hager*, 1991], which formed in a continent-continent collision.

The geomorphic similarity of Ishtar and a terrestrial continent has led to speculation that Ishtar might (like a continent) be composed, at least in part, of low-density rock like granite (*sensu lato*) [*Jull and Arkani-Hamed*, 1995]. On the other hand, the high-standing topography and gravity anomalies might represent a thick Fe-depleted (reduced density) lower crust and upper mantle [*Hansen and Phillips*, 1995]. These ideas are based on, and tested through, modeling the gravity and topography of Ishtar, but the tests have been frustratingly inconclusive, as it is difficult (or impossible) to untangle the effects of density and thickness variations in the crust, density variations in the mantle, flexural support, and dynamic support. The most recent inferences admit the possibility, but not the certainty, of granitic (low-density) rock beneath the mountain belts around Lakshmi Planum [*Kaula et al.*, 1997; *Hansen et al.*, 1997].

Low-Emissivity Deposits

The highest elevations on Venus, including several tesserae and the mountain ranges surrounding Ishtar Terra (especially Maxwell Montes), appear significantly brighter in Magellan and Venera radar imagery than do lower slopes. This difference is ascribed to the material properties of the highlands, in having a lower radar emissivity (low-ε), rather than being significantly rougher at radar wavelengths [*Pettengill et al.*, 1982, 1988]. It is generally agreed that the low-ε materials are alterations of, or coatings on, underlying rock materials; the altitude dependence of the low-ε materials implies a significant involvement of atmospheric transport or reaction. Beyond that, there is disagreement about the nature and significance of the low-ε material.

Earlier models of the low-ε material emphasized normal rock loaded with grains of strong dialectrics: pyrite, FeS_2

[*Klose et al.,* 1992; *Wood,* 1994]; perovskite, $CaTiO_3$ [*Fegley et al.,* 1992]; or exotic ferro-electric substances [*Shepard et al.,* 1994]. The bistatic radar experiment of *Pettingill et al.* [1996] on the low-ε material of Maxwell Montes showed that its radar electrical properties are not consistent with the loaded dialectic model, but are more consistent with a thin coating of a semiconductor material. Again, several choices were proposed, all involving vapor transport through the atmosphere and deposition in the highlands. Frosts of volatile metals or chalcogenides [*Brackett et al.,* 1995] are possible, with metallic tellurium being the favored material [*Pettengill et al.,* 1997; *Schaefer and Fegley,* 2004; *Fegley,* 2004]. However, *Wood* [1997] argued that coatings of pyrite or magnetite (Fe_3O_4) would provide the requisite electrical properties, and are more reasonable geochemically than metallic tellurium. Preliminary results from the Venus Express spacecraft radar seem inconsistent with a widespread conductor (like tellurium) and more consistent with a loaded dialectic material [*Simpson et al.,* 2007]. Obviously, this controversy is unresolved, and may not be resolved without high-precision analyses of trace components (e.g., Te) in Venus's atmosphere, *in situ* elemental and mineralogical data in areas of low-ε, or sample return.

CONCLUSIONS

The chemical compositions of Venus' crustal materials are poorly known. The incredible engineering and scientific successes of the Venera and VEGA landers yielded individual chemical analyses from a few sites. Unfortunately, with so few landing sites and so many interesting and diverse terrane types, many important volcanic and tectonic remain unsampled, as (most likely) do many important rock composition. Also unfortunately, the chemical analyses have very large uncertainties that preclude their interpretation beyond broad generalities.

Even with these huge caveats, the Venera and VEGA analyses show that Venus is geochemically diverse. Its basalts include varieties both depleted in incompatible elements (e.g., K, U, Th) and enormously enriched in them. This variety among the few analyses shows significant magmatic or mantle differentiation. In addition, one analysis may show a non-chondritic Th/U ratio, which would not be attainable under silicate magmatic differentiation and could indicate the chemical action of other fluids (e.g., water or carbonatite).

However, the available analyses have barely touched the surface of Venus' geochemistry. We have no information on the chemical compositions of Venus' highlands, or on its major shield volcanos, and only a few analyses of the basalts in its extensive lowland volcanic plains. This lack of data on the plains is exacerbated by the available data itself.

The Venera and VEGA analyses represent a wide range of basalts, and who knows what other varieties are present?

Other constraints on the chemistry of Venus' surface are limited and ambiguous. The radar-reflective (low emissivity) highlands are probably coated by thin layers of semiconductor material, as shown by bistatic polarized radar [*Pettingill et al.,* 1996]. The chemistry of this coating remains a mystery. Some small volcanic domes and flows have forms and sizes that suggest silica-rich lavas, but basalt lava (rich in phenocrysts or bubbles) would have similar shapes. And, part of the Ishtar Terra plateau may be underlain by silica-rich rocks (e.g., granite), but other explanations are available.

So, Venus is now a geochemical terra incognita. We know less of Venus' geochemistry than those of many asteroids, intact or disrupted [e.g., *Keil,* 1989; *Drake,* 2001; *McCoy et al.,* 2002; *Floss et al.,* 2003]! Any new data on Venus' geochemistry would be greatly welcome!

Remote Sensing Opportunities

Geochemistry by orbital remote sensing is extraordinarily difficult on Venus because of its thick, relatively opaque atmosphere. The atmosphere and its clouds do not permit surface analyses by either gamma rays nor thermal emission spectra, as have proven so useful for Mars [e.g., *Feldman et al.,* 2004; *Christensen et al.,* 2005]. However, a few orbital observations may be useful in surface chemistry.

First, the Magellan bistatic radar experiment proved very useful in constraining the nature of the radar-reflective, low-emissivity highlands [*Pettengill et al.,* 1996], touched a miniscule portion of Venus' surface. The VeRa (Venus Radio Science Experiment) on the Venus Express spacecraft, currently in orbit at Venus, will perform additional bistatic radar experiments to cover much more of the surface [*Titov et al.,* 2006; *Häusler et al.,* 2006; *Simpson et al.,* 2007]. These experiments are expected to further constrain the electrical properties of the low-emissivity coatings on Venus's highlands, and possibly provide some constraints on the composition and structure of other regions of Venus' crust.

Second, Venus' atmosphere is not completely opaque to all wavelengths of light—there are narrow windows in the near infrared between 0.85 and 1.18 µm [*Moroz,* 2002; *Hashimoto and Sugita,* 2003]. It may be possible to retrieve surface emissivities in these windows (after correction for atmospheric effects) [*Hashimoto and Sugita,* 2003; *Marinangeli et al.,* 2004], and thereby retrieve broad constraints on surface rock types, i.e. felsic versus mafic igneous rock.

Optical remote sensing of the Venusian surface would much more rewarding from a balloon platform floating below the cloud base [e. g., *Klassen and Greeley,* 2003]. Below the clouds, the surface would be generally visible in all wavelengths, but spectroscopic measurements will be affected by multiple reflections and emissions between the clouds and the ground (downwelling and multiple radiances); yet, it should be possible to obtain useful spectra in the visible and NIR [*Moroz*, 2002].

Lander Science Opportunities

Landers with contact (or near-contact) instruments are the only way (short of sample return) to obtain detailed, precise geochemical data about the Venusian surface. The Venera and VEGA missions demonstrated conclusively that Venus landers are possible and can yield an extraordinary science return. The challenges and opportunities now are to improve on Venera and VEGA. Venus rovers or long-duration landers present huge technical challenges, and are beyond a near-term realistic appraisal. So, one can assume that the next Venus surface landers might have sizes, volumes, and thermal protection systems comparable to those of Venera and VEGA, and comparable lifetimes of only a few hours. But significant improvements are possible in instrumentation, sample selections, and lander location.

Instruments. The Venera and VEGA instruments were excellent at the time for chemical analyses by X-ray fluorescence and intrinsic gamma-ray spectrometry, and there have been significant advances in both types of analysis. For X-ray fluorescence analyses of major and minor elements, two systems designed for Mars could work nearly as well in a Venera-style spacecraft on Venus. The APXS system deployed now on Mars in the Mars Exploration Rover (MER) landers [*Rieder et al.*, 2003] is robust and has provided precise (and accurate) chemical analyses [*Gellert et al.*, 2006]. However, the MER analysis in Table 2 (with very low uncertainties) represents data collection overnight, so that a 1-hour analysis would be considerably less precise. Another XRF system is CheMin, slated to fly on the Mars Science Laboratory (MSL) '09 rovers system, which is a combined XRF and XRD analytical instrument [*Blake et al.*, 2005; *Sarrazin et al.*, 2005]. Because CHEMIN has an active X-ray tube, it can produce a higher flux of fluoresced X-rays than does APXS, and so presumably provide more precise element chemistry than APXS. Both the APXS and CheMin instruments would be usable in a Venus lander spacecraft, would provide quantitative analyses of the crucial element Na, and provide huge improvements in analytical precision on the other major elements.

Gamma-ray spectrometry has also seen advances since the CsI(Tl) detectors used on Venera and VEGA, but it is not clear that these advances are more advantageous for Venus. The constraints for gamma ray analyses on Venus are: low count rates (intrinsic radioactivity only); short count time (limited lander lifetime); difficulty (cost and mass) of supporting cryogenic detectors (thermal constraints); and the need to detect gamma rays from only three elements (K, U, and Th). Of currently available detector systems, CsI(Tl) is still the most efficient in terms of countable events per incident gamma ray. Other scintillator materials, like BGO (bismuth germante), can run warm and have better energy resolution than CsI(Tl) [*Feldman et al.*, 1999], but that energy resolution is not helpful for analysis of K, U, and Th by their own radioactive decay. A different system, direct gamma ray detection in germanium, is in use in the GRS instrument on the Mars Odyssey Spacecraft [*Boynton et al.*, 2004]. Ge detectors have very high energy resolution, but require cryogenic cooling and also detect only about 1/20 of the gamma rays that a comparable CsI(Tl) detector would. If CsI(Tl) is the gamma detector of choice for Venus, improvements over Venera and VEGA would probably have to come from larger detector volumes.

On the other hand, it is possible that other analytical techniques, besides XRF and passive gamma counting, would be useful for Venus. Laser-induced breakdown spectroscopy, LIBS [*Cremers and Radziemski*, 2006], is the laser-activated equivalent of spark-source emission spectroscopy. A LIBS instrument is baselined for the MSL'09 rover mission to Mars as part of the ChemCam package [*Wiens et al.*, 2005]. LIBS must be considered seriously for Venus, because of the very short time it requires per analysis, and its sensitivity to some uncommon but geochemically useful elements (including H, Ni, and U, [*Wiens et al.*, 2005]). The high surface pressure of Venus does affect LIBS performance, an effect which is being studied [*Arp et al.*, 2004; *Salle et al.*, 2005].

If analyses of other trace elements are deemed critical, techniques besides passive gamma counting may be required. Pulsed neutron sources [*Akkurt et al.*, 2005] may provide the capability of INAA-like analyses for prompt gamma emission spectrometry on Venus landers [*Latif et al.*, 1999; *Ebihara and Oura*, 2001]. Depending on neutron energies and permitted count times, analyses may be possible for most major elements and many trace elements, including H, B, S, Cl, Co, Ni, and Sm [e.g., *Oura et al.*, 2003].

Sample Access and Number. A significant weakness in the Venera and VEGA analyses is that they could only analyze a single sample at each landing site. All the sites showed at least two types of material—rock and regolith—so multiple analyses would be important. Also, it would be very useful to measure chemical compositions of rock surfaces and interiors, to constrain the nature and extent of surface-atmosphere interactions. This sort of study by the MER rovers has proved important for determining the 'real' compositions of the basalts at the Spirit Rover site [*Gellert et al.*, 2006]. So, it would seem very important to have the capability of doing at least two chemical analyses (by XRF), and of obtaining at least two different samples for analyses. This capability has significant implications for sample handling systems and analysis durations.

Lander Locations. As noted above, all of the in situ analyses of Venus' geochemistry are from its lowland basalt plains. Even though there is significant geochemical variability in the plains, landing on other sorts of sites would greatly enhance our understanding of the planet. Geochemistry from lavas on a major volcanic rise (a shield volcano like Sapas Mons) or a corona feature would help determine if and how they are related to mantle plumes. Geochemistry from a tessera site would help explain the origins of these upland regions, and probably something of the older history of Venus. Geochemistry from a radar-reflective (low-ε) highlands would help delimit surface-atmosphere interactions. And geochemistry of a suitable site on Ishtar Terra might reveal whether Venus has a true, Earth-style continent, with all its implications for abundant water [*Campbell and Taylor*, 1983]. Truly, most of Venus is geochemical *terra incognita*.

Lander / Balloon. An interesting twist on these *in situ* analytical schemes is involvement of a balloon. In this scenario, a sample would be collected from a balloon or tether, and would then be carried up into the atmosphere to a lower temperature and pressure. There, analytical instruments would have more time to investigate the chosen sample. This scenario would probably not include gamma-ray analyses, and sample selection would depend on where Venus' winds blew the balloon.

Sample Return Opportunities

Sample return would be the most informative spacecraft outcome for Venus geochemistry, but also would be the most ambitious, speculative, and costly. Venus sample return presents many technical challenges, including Venus' high surface temperature, the need to return through a thick atmosphere, and an abundance of corrosive gases and particulates. Yet studies of Venus sample return missions are ongoing [e.g., *Rodgers et al.*, 2000; *Sweetser et al.*, 2000, 2003], and sample return appears to be possible, if costly.

Venus sample return would allow Earth's full analytical armamentarium to be applied, and would permit the same sort of detailed informative investigations now possible on Martian (and other) meteorites—age, magma composition, mantle composition, core size, surface alteration processes and pathways, etc—see *Treiman* [2005] for a review of the understanding gained from one class of Martian meteorites. *Jones and Treiman* [1998], *MacPherson* [2001], and *Shearer and Borg* [2006] have shown the importance returned samples from Mars and the moon—returned samples from Venus would be equally important.

Acknowledgments. This work was suggested by S. Smrekar, following the February 2006 Chapman conference. Treiman is a co-I on the CHEMIN XRD/XRF instrument on the MSL'09 mission. W. Kiefer helped enormously with access to the geophysical literature, and M. Zolotov helped with the organization and scope. I owe a debt to B. Fegley Jr., who first interested me in Venus' geochemistry. Careful and thoughtful reviews were provided by G. J. Taylor, W. Kiefer, J. Filiberto, and an anonymous expert; their suggestions are greatly appreciated. Lunar and Planetary Institute contribution No. 1347.

REFERENCES

Abdrakhimov, A. M., and A. T. Basilevsky (2002), Geology of the *Venera* and *Vega* Landing-Site Regions, *Solar System Research, 36*, 136–159.

Akkurt H., J. L. Groves, J. Trombka, R. Starr, L. Evans, S. Floyd, R. Hoover, L. Lim, T. McClanahan, R. James, T. McCoy, and J. Schweitzer (2005), Pulsed neutron generator system for astrobiological and geochemical exploration of planetary bodies, *Nuclear Instruments and Methods in Physics Research B, 241*, 232-237.

Anders, E., and N. Grevesse (1989) Abundances of the elements: Meteoritic and solar, *Geochim. Cosmichim. Acta, 53*, 197-214.

Anderson, F. S., and S. E. Smrekar (2006), Global mapping of crustal and lithospheric thickness on Venus, *J. Geophys. Res., 111*, E08006, doi:10.1029/2004JE002395.

Ansan, V., P. Vergely, and P. Masson (1996), Model of formation of Ishtar Terra, Venus, *Planet. Space Sci., 44*, 817—831.

Arp, Z. A., D.A. Cremers, R. D. Harris, D. M. Oschwald, G. R. Parker, and D. M. Wayne (2004), Feasibility of generating a useful laser-induced breakdown spectroscopy plasma on rocks at high pressure: Preliminary study for a Venus mission, *Spectrochim. Acta B, 59*, 987-999.

Baker, V. R., G. Komatsu, T. J. Parker, V. C. Gulick, J. S. Kargel, and J. S. Lewis (1992), Channels and valleys on Venus - Preliminary analysis of Magellan data, *J. Geophys Res., 97*, 13421-13444.

Baker, V. R., G. Komatsu, V. C. Gulick, and T. J. Parker (1992), Channels and Valleys, In: Bougher, S.W., Hunten, D.M., Phillips, R.J. (Eds.), Venus II: Geology, Geophysics, Atmosphere, and Solar Wind Environment. Univ. of Arizona Press, Tucson, pp. 757–793.

Barsukov, V. L. (1992), Venusian Igneous Rocks, in *Venus Geology, Geochemistry, and Geophysics* (eds. V. L. Barsukov, A. T. Basilevsky, V. P. Volkov, and V. N. Zharkov). 165-176. Univ. Arizona Press, Tucson AZ.

Barsukov, V. L., Yu. A. Surkov, L. V. Dmitriev, and I. L., Khodakovskiy (1986a) Geochemical studies on Venus with the landers from the Vega 1 and Vega 2 probes. *Geochem. Intl., 23(7)*, 53-65.

Barsukov, V. L., Yu. A. Surkov, L. V. Dmitriev, and I. L. Khoda-kovskii (1986b), Geochemical studies of Venus by Vega-1 and Vega-2 landers, *Geokhimiia*, 1986, 275-288.

Barsukov, V. L., et al., (1986c), The geology and geomorphology of the Venus surface is revealed by the radar images obtained by Veneras 15 and 16, *Proc. Lunar Planet. Sci. Conf. 16th, J. Geophys. Res. 91*, D378-D398.

BVSP (Basaltic Volcanism Study Project) (1981), *Basaltic Volcanism on the Terrestrial Planets*, Pergamon, NY, 1286 p.

Basilevsky, A. T., M. A. Ivanov, and O. V. Nikolaeva (1992) Geology of the Venera 8 landing site region from Magelland data: Morphological and geochemical considerations, *J. Geophys. Res., 97*, 16315-16335.

Blake, D. F., P. Sarrazin, D. L. Bish, S. J. Chipera, D. T. Vaniman, S. Feldman, and S. Collins (2005), CHEMIN: A definitive mineralogy instrument in the analytical laboratory of the Mars Science Laboratory (MSL '09), *Lunar Planet. Sci., XXXVI*, Abstract #1608. CD-ROM, Lunar and Planetary Institute, Houston.

Boynton, W. V., W. C. Feldman, I. G. Mitrofanov, L. G. Evans, R. C. Reedy, S. W. Squyres, R. Starr, J. I. Trombka, C. d'Uston, J. R. Arnol[0], P. A. J. Englert, A. E. Metzger, H. Wänke, J. Brückner, D. M. Drake, C. Shinohara, C. Fellows, D. K. Hamara, K. Harshman, K. Kerry, C. Turner, M. Ward, H. Barthe, K. R. Fuller, S. A. Storms, G. W. Thornton, J. L. Longmire, M. L. Litvak and A. K. Ton'chev (2004), The Mars Odyssey Gamma-Ray Spectrometer Instrument Suite, *Space Sci. Rev., 110*, 37-83.

Brackett, R.A., Fegley Jr., B., Arvidson, R.E., 1995. Volatile transport on Venus and implications for surface geochemistry and geology. *J. Geophys. Res., 100*, 1553–1563.

Bridges, N. T. (1995), Submarine analogs to Venusian pancake domes, *Geophys. Res. Lett., 22*, 2718-2784.

Bruno, B. C., and G. J. Taylor (1995), Morphologic investigation of Venusian lavas, *Geophys. Res. Lett., 22*, 1897-1900.

Bulmer, M. H., and J. E. Guest (1996), Modified volcanic domes and associated debris aprons on Venus, in *Volcano Instability on the Earth and Other Planets*, Geol. Soc. Spec. Publ., 110 , eds. W. J. McGuire, A. P. Jones and J. Neuberg, pp. 349-371, Geol. Soc. Amer.

Campbell, I. H., and S. R. Taylor (1983), No water, no granite - No oceans, no continents, *Geophys. Res. Lett., 10*, 1061-1064.

Christensen, P. R., H. Y. McSween, Jr, J. L. Bandfield, S. W. Ruff, A. D. Rogers, V. E. Hamilton, N. Gorelick, M. B. Wyatt, B. M. Jakosky, H. H. Kieffer, M. C. Malin and J. E. Moersch (2005), Evidence for magmatic evolution and diversity on Mars from infrared observations, *Nature, 436*, 504-509, doi:10.1038/nature03639.

Clague, D. A., J. G. Moore, and J. R. Reynolds (2000), Formation of submarine flat-topped volcanic cones in Hawai'i, *Bull. Volcanol., 62*, 214-233.

Cousens, B. L., D. A. Clague, and W. D. Sharp (2003), Chronology, chemistry, and origin of trachytes from Hualalai Volcano, Hawaii, *Geochem. Geophys. Geosyst., 4(9)*, 1078, doi:10.1029/2003GC000560.

Cremers, D. A., and L. J. Radziemski (2006), *Handbook of laser-induced Breakdown Spectroscopy*, John Wiley and Sons, New York.

Crumpler, L. S., J. C. Aubele, D. A. Senske, S. T. Keddie, K. P. Magee, and J. W. Head (1997), Volcanoes and centers of volcanism on Venus. In: Bougher, S.W., Hunten, D.M., Phillips, R.J. (Eds.), Venus II: Geology, Geophysics, Atmosphere, and Solar Wind Environment. Univ. of Arizona Press, Tucson, pp. 697–756.

Dawson, J. B., and J. V. Smith (1977), The MARID (mica-amphibole-rutile-ilmenite-diopside) suite of xenoliths in kimberlite, *Geochim. Cosmichim. Acta, 41*, 309-310. .

Drake, M. J. (2001) The eucrite / Vesta story, Meteorit. Planet. Sci., 36, 501-513.

Drake, M. J., H. E. Newsom, and C. J. Capobianco (1989), V, Cr, and Mn in the Earth, Moon, EPB, and SPB and the origin of the Moon: Experimental studies, Geochim. Cosmochim. Acta, 53, 2101-2111.

DuBois, C. G. B, J. Furst, N. J. Guest, and D. J. Jennings (1963) Fresh natro-carbonatite lava from Oldoinyo L'Engai, Nature, 197, 445-446.

Ebihara, M., and Y. Oura (2001), Applicability of prompt gamma-ray analysis to the initial analysis of the extraterrestrial materials for chemical composition, *Earth Planets Space*, 53, 1039–1045.

Fegley, B., Jr. (2004), Venus, Chapter 21. pp. 487-507. In *Meteorites, Comets, and Planets* (e.d. A.M. Davis) Vol. 1 *Treatise on Geochemistry* (Turekian, K.K. and Holland, H.D., eds.) Elsevier-Pergamon, Oxford.

Fegley, B. Jr., A. H. Treiman, and V. L. Sharpton (1992), Venus surface mineralogy: observational and theoretical constraints, *Proc. Lunar Planet. Sci. Conf. 22nd*, 3–19, Lunar and Planetary Institute, Houston.

Feldman, W. C., B. L. Barraclough, K. R. Fuller, D. J. Lawrence, S. Maurice, M. C. Miller, T. H. Prettyman, and A. B. Binder (1999), The Lunar Prospector gamma-ray and neutron spectrometers, *Nuclear Instruments and Methods in Physics Research A, 422*, 562-566.

Feldman, W. C., T. H. Prettyman, S. Maurice, J. J. Plaut, D. L. Bish, D. T. Vaniman, and M. T. Mellon (2004), Global distribution of near-surface hydrogen on Mars, *J. Geophys. Res., 109*, E09006, doi:10.1029/2003JE002160.

Fink, J. H., and N. T. Bridges (1995), Effects of eruption history and cooling rate on lava dome growth, *Bull. Volcanol, 57*, 229-239.

Fink, J. H., and R. W. Griffiths (1998), Morphology, eruption rates, and rheology of lava domes: Insights from laboratory models, *J. Geophys. Res., 103*, 527-54x. 97JB02838.

Fink, J. H., N. T. Bridges, and R. E. Grimm (1993) Shapes of venusian 'pancake' domes imply episodic emplacement and silicic composition, *Geophys. Res. Lett., 20*, 261-264.

Floss, C., G. Crozaz, G. McKay, T. Mikouchi, and M. Killgore (2003), Petrogenesis of angrites, *Geochim. Cosmochim. Acta, 67*, 4775– 4789.

Gellert, R., R. Rieder, J. Brückner, B. C. Clark, G. Dreibus, G. Klingelhöfer, G. Lugmair, D. W. Ming, H. Wänke, A. Yen, J. Zipfel, and S. W. Squyres (2006), Alpha Particle X-ray Spectrometer (APXS): Results from Gusev crater and calibration report, *J. Geophys. Res., 111*, E02S05, doi:10.1029/2005JE002555

Gregg T. K. P. and R. Greeley (1993), Formation of Venusian canali: Considerations of lava types and their thermal behaviors, *J. Geophys. Res., 98*, 10873-10882.

Grimm, R. E., and P. C. Hess (1997), The crust of Venus, In: Bougher, S.W., Hunten, D.M., Phillips, R.J. (Eds.), Venus II: Geology, Geophysics, Atmosphere, and Solar Wind Environment. Univ. of Arizona Press, Tucson, pp. 1205–1244.

Guest, J. E., M. H. Bulmer, J. C. Aubele, K. K. Beratan, R. Greeley, J. W. Head, G. A. Michaels, C. M. Weitz, and C. R. Wiles (1992), Small volcanic edifices and volcanism in the plains of Venus, *J. Geophys. Res., 97*, 15,949-15,966.

Hansen, V. L., and R. J. Phillips (1995), Formation of Ishtar Terra, Venus: Surface and gravity constraints, Geology, 23, 292-296.

Hansen, V. L., and J. J. Willis (1996), Structural analysis of a sampling of tesserae: Implications for Venus geodynamics, Icarus, 123, 296-312.

Hansen, V. L., J. J. Willis, and W. B. Banerdt (1997), Tectonic Overview and Synthesis, In: Bougher, S.W., Hunten, D.M., Phillips, R.J. (Eds.), Venus II: Geology, Geophysics, Atmosphere, and Solar Wind Environment. Univ. of Arizona Press, Tucson, pp. 797–844.

Hashimoto, G. L., and S. Sugita (2003), On observing the compositional variability of the surface of Venus using nightside near-infrared thermal radiation, *J. Geophys. Res., 108,* 5109, doi:10.1029/2003JE002082.

Häusler, B., M. Pätzold, G. L. Tyler, R. A. Simpson, M. K. Bird, V. Dehant, J.-P. Barriot, W. Eidel, S. Remus, J. Selle, S. Tellmann, and T. Imamura (2006), radio science investigations by VeRa onboard the Venus Express spacecraft, *Planet. Space Sci. 26*, proof online.

Head, J. W., L. S. Crumpler, J. C. Aubele, J. E. Guest, and R. S. Saunders (1992), Venus volcanism - Classification of volcanic features and structures, associations, and global distribution from Magellan data, *J. Geophys. Res., 97*, 13,153-13,197.

Hess, P. C., and J. W. Head (1990), Derivation of primary magmas and melting of crustal materials on Venus: Some preliminary petrogenetic considerations, *Earth Moon Planets, 50/51*, 57-80.

Ivanov, M. A. (2001), Morphology of the tessera terrain on Venus: Implications for the composition of tessera material, *Solar System Res., 35*, 1–1780.

Johnson, D. M., P. R. Hooper, R. M. and Conrey (1999), XRF analyses of rocks and minerals for major and trace elements on a single low-dilution Li-tetraborate fused bead. *Advances X-ray Analyses, 41*, 843-867.

Johnson, N. M. and B. Fegley Jr. (2000), Water on Venus: New insights from tremolite decomposition, *Icarus* 146, 301-306.

Johnson, N. M. and B. Fegley Jr. (2003a), Tremolite decomposition on Venus II. Products, kinetics, and mechanism, *Icarus* 164, 317-333.

Johnson, N. M. and B. Fegley Jr. (2003b), Longevity of fluorine-bearing tremolite on Venus, *Icarus* 165, 340-348.

Jones, J. H. (1995), Experimental trace element partitioning, 73-104 in T. H. Ahrens (ed.) *Rock Physics and Phase Relations: A Handbook of Physical Constants,* Amer. Geophys. Un., Washington DC.

Jones, J. H., and A. H. Treiman (1998), Bringing Mars home - Opportunities and challenges presented by the Mars sample return mission. *Lunar Planet. Info. Bull., 85*, 12-17.

Jull. M. G., and J. Arkani-Hamed, (1995), The implications of basalt in the formation and evolution of mountains on Venus, *Phys. Earth Planet. Interiors, 89*, 163-175.

Kargel, J. S., G. Komatsu, V. R. Baker, and R. G. Strom (1993), The volcanology of Venera and VEGA landing sites and the geochemistry of Venus, *Icarus, 103*, 235-275.

Kargel, J. S., B. Fegley Jr., and A. H. Treiman (1994), Carbonate-sulfate volcanism on Venus?, *Icarus, 112*, 219-252.

Kaula, W. M., D. L. Bindschadler, R. E. Grimm, V. L. Hansen, K. M. Roberts, and S. E. Smrekar (1992), Styles of deformation in Ishtar Terra and their implications, *J. Geophys. Res., 97*, 16,085-16,120

Kaula, W. M., A. Lenardic, D. L. Bindschadler, and J. Arkani-Hamed (1997), Ishtar Terra, In: Bougher, S.W., Hunten, D.M., Phillips, R.J. (Eds.), Venus II: Geology, Geophysics, Atmosphere, and Solar Wind Environment. Univ. of Arizona Press, Tucson, pp. 879–900.

Keil, K. (1989), Enstatite meteorites and their parent bodies, Meteoritics, 24, 195-208.

Keller, J., and M. Krafft (1990) Effusive natrocarbonatite activity of Oldoinyo L'engai, June 1988, *Bull. Volcanol., 52*, 629-645.

Kiefer, W. S., and B. H. Hager (1991), Mantle downwelling and crustal convergence - A model for Ishtar Terra, Venus, *J. Geophys. Res., 96*, 20,967-20,980.

Klassen, K. P., and R. Greeley (2003), VEVA Discovery mission to Venus: exploration of volcanoes and atmosphere, *Acta Astronautica, 52*, 151-158.

Klemme, S., J. D. Blundy, and B. J. Wood (2002), Experimental constraints on major and trace element partitioning during partial melting of eclogite, *Geochim. Cosmochim. Acta, 66*, 3109-3123.

Klose, K. B., J. A. Wood, and A. Hashimoto (1992) Mineral equilibria and the high radar reflectivity of Venus mountaintops, *J. Geophys. Res., 97*, 16353-16369.

Lang N. P., and V. L. Hansen (2006), Venusian channels formation as a subsurface process, *J. Geophys. Res., 111*, E04001, doi: 10.1029/2005/E002629.

Latif S. A., Y. Oura, M. Ebihara, G. W. Kallemeyn, H. Nakahara, C. Yonezawa, T. Matsue, and H. Sawahata (1999), Prompt gamma-ray analysis (PGA) of meteorite samples with emphasis on the determination of Si, *J. Radioanal. Nucl. Chem., 239*, 577-580.

Lodders, K., and B. Fegley Jr. (1998), *The Planetary Scientist's Companion*, Oxford University Press, N.Y., 371 pp.

Longhi, J., E. Knittle, J. R. Holloway, and H. Wänke (1992), The bulk composition, mineralogy and internal structure of Mars, p. 184-208 in *Mars* (eds. H. H. Kieffer, B. M. Jakosky, C. W. Snyder, and M. S. Matthews), Univ. Ariz. Press.

MacPherson, G. (2001), The First Returned Martian Samples: Science Opportunities *Science Planning for Exploring Mars, Part 3*, Jet Propulsion Laboratory Pub. 01-7.

Marinangeli L. L., K. Baines, R. Garcia, P. Drossart, G. Piccioni, J. Benkhoff, J. Helbert, Y. Langevin, G.G. Ori , G. Komatsu, and

I. C. Pope (2004), Venus surface investigation using VIRTIS onboard the ESA/Venus Express Mission (abstract), *Lunar Planet. Sci., XXXV*, Abstract #1363.pdf.

McCoy, T. J., M. S. Robinson, L. R. Nittler, and T. H. Burbine (2002), The Near Earth Asteroid Rendezvous Mission to Asteroid 433 Eros: A Milestone in the Study of Asteroids and their Relationship to Meteorites, *Chemie der Erde, 62*, 89-121.

McDonough, W., and S. S. Sun (1995), The composition of the Earth, *Chem. Geol., 120*, 223-253.

McKenzie, D., P. G. Ford, F. Liu, and G. H. Pettengill (1992), Pancakelike domes on Venus, *J. Geophys. Res., 97*, 15,967-15,976.

Moroz, V. I. (2002), Estimates of visibility of the surface of Venus from descent probes and balloons, *Planet. Space Sci., 50*, 287-297.

Nikolaeva, O. V. (1990), Geochemistry of the Venera 8 material demonstrates the presence of continental crust on Venus, *Earth Moon Planets, 50/51*, 328-341.

Oura, Y., N. Shirai, and M. Ebihara (2003), Chemical composition of Yamato (Y) 000593 and Y000749: Neutron induced prompt gamma-ray analysis study, *Antarctic Meteorite Research (NIPR), 16*, 80-93.

Pavri, B., J. W. Head III, W. B. Klose, and L. Wilson (1992), Steep-sided domes on Venus - Characteristics, geologic setting, and eruption conditions from Magellan data, *J. Geophys. Res., 97*, 13,445-13,478.

Simpson, R. A., G. L. Tyler, B. Häusler, and M. Pätzold (2007), Search for anomalous surface properties at Maxwell Montes with Venus Express bistatic radar, Lunar Planet. Sci. XXX-VIII, Abstract #2240. CD-ROM, Lunar and Planetary Institute, Houston.

Sweetser, T., J. Cameron, G.-S. Chen, J. Cutts, R. Gershmann, M. S. Gilmore, J. L. Hall, V. Kerzhanovich, A. McRonald, E. Nilsen, W. Petrick, D. Rodgers B. Wilcox, A. Yavrouian, W. Zimmerman, and the JPL Advanced Project Design Team (2000), Venus surface sample return: A weighty high-pressure challenge, *Adv. Astronaut. Sci., 103*, 831-844, (Proc. AAS/AIAA Astrodynamics Conf., Aug. 16-19, 1999, Girdwood, Alaska,) 1631.

Sweetser, T., C. Peterson, E. Nilsen, and B. Gershman (2003), Venus sample return missions - A range of science, a range of costs, *Acta Astronautica, 2*, 165-1721631.

Pertermann, M., and M. M. Hirschmann, (2003), Anhydrous partial melting experiments on MORB-like eclogite: Phase relations, phase compositions, and mineral-melt partitioning of major elements at 2—3 GPa, *J. Petrol., 44*, 2173-2201.

Pertermann, M., M. M. Hirschmann, K. Hametner, D. Günther, and M. W. Schmidt (2003), High field strength element/rare earth element fractionation during partial melting in the presence of garnet: Implications for identification of mantle heterogeneities, *Geochem. Geophys. Geosyst., 5*, 10.1029/2003GC000638.

Pettengill, G. H., P. G. Ford, and B. D. Chapman (1988), Venus: Surface electromagnetic properties, *J. Geophys. Res., 93*, 14881-14892.

Pettengill, G. H., P. G. Ford, and R. A. Simpson (1996), Electrical properties of the Venus surface from bistatic radar observations, *Science, 272*, 1628–1631.

Pettengill, G. H., B. A. Campbell, D. B. Campbell, and R. A. Simpson (1997), Surface scattering and dielectric properties, In: Bougher, S.W., Hunten, D.M., Phillips, R.J. (Eds.), Venus II: Geology, Geophysics, Atmosphere, and Solar Wind Environment. Univ. of Arizona Press, Tucson, pp. 591–636.

Rieder, R., R. Gellert, J. Brückner, G. Klingelhöfer, G. Dreibus, A. Yen, and S. W. Squyres (2003), The new Athena alpha particle X-ray spectrometer for the Mars Exploration Rovers, *J. Geophys. Res., 108*, ROV 7-1, CiteID 8066, DOI 10.1029/2003JE002150.

Robinson, M. S., and G. J. Taylor (2001), Ferrous oxide in Mercury's crust and mantle, *Meteorit. Planet. Sci., 36*, 841-847.

Rodgers, D., M. Gilmore, T. Sweetser, J. Cameron, G.-S. Chen, J. Cutts, R. Gershmann, J. L. Hall, V. Kerzhanovich, A. McRonald, E. Nilsen, W. Petrick, C. Sauer, B. Wilcox, A. Yavrouian, W. Zimmerman, and the JPL Advanced Project Design Team (2000), Venus sample return: A hot topic, *Proc. IEEE Aerospace Conf., 7*, 473-4831631.

Sallé, B., D. A. Cremers, S. Maurice, and R. C. Wiens (2005), Laser-Induced Breakdown Spectroscopy for Space Exploration Applications: Influence of the Ambient Pressure on the Calibration Curves Prepared from Soil and Clay Samples, *Spectrochim. Acta B, 60*, 479-490.

Sarrazin, P., D. Blake, S. Feldman, S. Chipera, D. Vaniman, and D. Bish (2005), Field deployment of a portable X-ray diffraction/X-ray flourescence instrument on Mars analog terrain, *Powder Diffraction, 20*, 128-133.

Schaefer, L., and B. Fegley Jr. (2004), Heavy metal frost on Venus, *Icarus, 168*, 215–2191631.

Shearer, C. K., and L. E. Borg (2006), Big returns on small samples: Lessons learned from the analysis of small lunar samples and implications for the future scientific exploration of the Moon, *Chemie der Erde, 66*, 163-185.

Shearer, C. K., and J. J. Papike (1999), Magmatic evolution of the Moon, *Amer. Mineral., 84*, 1469-1494.

Shepard, M. K. R., E. Arvidson, R. A. Brackett, and B. Fegley Jr. (1994), A ferroelectric model for the low emissivity highlands on Venus, *Geophys. Res. Lett., 21*, 469-472.

Stopar, J. D., G. J. Taylor, and W. Boynton (2004), Aqueous alteration pathways for K, Th, and U on Mars, *Lunar Planet. Sci., XXXV*, Abstract #1429. Lunar and Planetary Institute, Houston.

Surkov, Yu. A., V. L. Barsukov, L. P. Moskalyeva, V. P. Kharyukova, and A. L. Kemurdhzian (1984), New data on the composition, structure and properties of Venus rock obtained by Venera 13 and Venera 14, *Proc. Lunar Planet Sci. Conf. 14th, J. Geophys. Res., 89 Suppl.*, B393-B402.

Surkov, Yu. A., L. P. Moskalyeva, V. P. Kharyukova, A. D. Dudin, G. G. Smirnov, and S. Ye. Zaitseva (1986), Venus rock compositions at the Vega 2 landing site, *Proc. Lunar Planet Sci. Conf. 17th, J. Geophys. Res., 91 Suppl.*, E215-E218.

Sweeney, R. J., A. B. Thompson, and P. Ulmer (1993), Phase relations of a natural MARID composition and implications for MARID genesis, lithospheric melting and mantle metasomatism, *Contrib. Mineral. Petrol., 115*, 225-241.

Sweetser, T., J. Cameron, G.-S. Chen, J. Cutts, R. Gershmann, M. S. Gilmore, J. L. Hall, V. Kerzhanovich, A. McRonald, E. Nilsen, W. Petrick, D. Rodgers B. Wilcox, A. Yavrouian, W. Zimmerman, and the JPL Advanced Project Design Team (2000), Venus surface sample return: A weighty high-pressure challenge, *Adv. Astronaut. Sci., 103*, 831-844, (Proc. AAS/AIAA Astrodynamics Conf., Aug. 16-19, 1999, Girdwood, Alaska,) 1631.

Sweetser, T., C. Peterson, E. Nilsen, and B. Gershman (2003), Venus sample return missions - A range of science, a range of costs, *Acta Astronautica, 2*, 165-1721631.

Taylor, G. J., W. Boynton, J. Brückner, H. Wänke, G. Dreibus, K. Kerry, J. Keller, R. Reedy, L. Evans, R. Starr, S. Squyres, S. Karunatillake, O. Gasnault, S. Maurice, C. d'Uston, P. Englert, J. Dohm, V. Baker, D. Hamara, D. Janes, A. Sprague, K. Kim, and D. Drake (2007), Bulk composition and early differentiation of Mars, *J. Geophys. Res., 111,* E03S10, doi:10.1029/2005JE002645.

Titov, D. V., H. Svedhem, D. McCoy, J. -P. Lebreton, S. Barabash, J. -L. Bertaux, P. Drossart, V. Formisano, B. Haeusler, O. I. Korablev, W. Markiewicz, D. Neveance, M. Petzold, G. Piccioni, T. L. Zhang, F. W. Taylor, E. Lellouch, D. Koschny, O. Witasse, M. Warhaut, A. Acomazzo, J. Rodrigues-Cannabal, J. Fabrega, T. Schirmann, A. Clochet, and M. Coradini (2006), Venus Express: Scientific goals, instrumentation, and scenario of the mission, *Cosmic Research, 44*, 334-348. Doi 10.1134/S0010952506040071.

Treiman, A. H. (1995), Ca-rich carbonate magmas: A regular solution model, with applications to carbonatite magma-vapor equilibria and carbonate lavas on Venus, *Amer. Mineral, 80*, 115–130.

Treiman, A. H. (2005), The nakhlite Martian meteorites: Augite-rich igneous rock from Mars, *Chemie der Erde, 65*, 203-270.

Treiman, A. H., and C. C. Allen (1994), Chemical weathering on Venus: Preliminary results on the interaction of basalt and CO_2, *Lunar Planet Sci., XXV*, 1415-1416.

Treiman, A. H., and A. Schedl (1983), Properties of carbonatite magma and processes in carbonatite magma chambers, *J. Geol., 91*, 437-447.

Treiman, A. H., J.D. Gleason, and D. D. Bogard (2000), The SNC meteorites are from Mars, *Planet. Space Sci., 48*, 1213-1230.

van Westeren, W., J. D. Blundy, and B. J. Wood (2001), High field strength element/rare earth element fractionation during partial melting in the presence of garnet: Implications for identification of mantle heterogeneities, *Geochem. Geophys. Geosyst., 2*, 10.1029/2000GC000133.

Wiens, R., S. Maurice, N. Bridges, B. Clark, D. Cremers, K. Herkenhoff, L. Kirkland, N. Mangold, G. Manhés, P. Mauchien, C. McKay, H. Newsom, F. Poitrasson, V. Sautter, C. d'Uston, D. Vaniman, and S. Shipp (2005), CHEMCAM science objectives for the Mars Science Laboratory (MSL) rover, *Lunar Planet. Sci., XXXVI*, Abstract #1580. Lunar and Planetary Institute, Houston.

Wood, J. A. (1997), Rock weathering on the surface of Venus. In: Bougher, S.W., Hunten, D.M., Phillips, R.J. (Eds.), Venus II: Geology, Geophysics, Atmosphere, and Solar Wind Environment. Univ. Arizona Press, Tucson, pp. 637–664.

Allan H. Treiman, Lunar and Planetary Institute, 3600 Bay Area Boulevard, Houston, Texas, USA

Surface Evolution of Venus

Alexander T. Basilevsky

Vernadsky Institute of Geochemistry and Analytical Chemistry,
Russian Academy of Sciences, Moscow, Russia
Department of Geological Sciences, Brown University, Providence, Rhode Island, USA

George E. McGill

Department of Geosciences, University of Massachusetts, Amherst, Massachusetts, USA

This chapter contains short descriptions of material units and structures observed on the surface of Venus as well as an abbreviated history of discoveries, that led to the current knowledge of this planet's geology. It is shown that observed units and structures are broadly similar and commonly exhibit similar age sequences in different regions of the planet, although there is debate as to what degree these sequences can be integrated into any single global stratigraphic model. There is a broad consensus concerning the recent general geodynamic style of Venus (no plate tectonics), on the dominating role of basaltic volcanism in the observed crust-forming processes, and on the significant roles of both compressional and extensional tectonic deformation. Also not under debate is that we see the morphologic record of only the last ~1 b.y. (or less) of the history of this planet and that during this time period the role of exogenic resurfacing was very minor. Several important unresolved questions of Venus geology are formulated and suggestions for future missions, that could lead to resolving them are given.

INTRODUCTION

The bulk properties (diameter, mean density) of Venus and Earth do not differ very much. In contrast, the surface environments are very different. The atmosphere of Venus is 96.5% CO_2 and at mean elevation this atmosphere has a pressure of 95 bars, and a temperature of 737K. Both pressure and temperature vary significantly with elevation. Venus is entirely shrouded in clouds, and the surface temperature varies only slightly as a function of latitude. Winds near the surface are thus very gentle; capable of transporting fine materials but not very effective as agents of erosion. Liquid water is obviously not possible on the surface, and thus the water-related processes that dominate surface modification on Earth are not active on Venus. Venus exhibits a bewildering array of structural features, in large part because the extreme slowness of erosion and burial results in preservation of structures for geologically long intervals. In an Earth-like surface environment most of these structures would be eroded or buried.

This chapter is intended to provide a description of the surface geology of Venus as background to a brief discussion of our understanding of the evolution of Venus' crust. This chapter also will provide context for the chapters devoted to the description of the interior and atmosphere of Venus. We briefly describe here the major geological processes responsible for the current state of the surface, as deduced from the available observations, and then consider the issue of surface evolution. We begin with a short review of the history of the study of Venus, noting the major discoveries achieved

Exploring Venus as a Terrestrial Planet
Geophysical Monograph Series 176
Copyright 2007 by the American Geophysical Union.
10.1029/176GM04

at different stages of exploration. Then we consider the current views of the surface structure and evolution. Finally, we summarize the most important findings to date in Venus geology and surface evolution, formulate the important questions to be resolved in future studies, and make suggestions for future measurements and missions necessary to address these issues. The global map of Venus presented in a shaded relief version colored according to altitude levels (Plate 1) provides a guide to the major physiographic components of the planet. More detailed physiographic maps with feature and terrain names are available on: http://planetarynames. wr.usgs.gov/vgrid.html. Figures in the text are mostly portions of Magellan synthetic aperture radar (SAR) images. These and other Magellan images with the maximum available resolution can be found on: http://pdsmaps.wr.usgs.gov/ PDS/public/explorer/html/fmapadvc.htm. Because chapter length for this monograph is strictly limited, we are unable to provide an extended reference list, and thus cite only representative examples of appropriate works.

HISTORY OF VENUS EXPLORATION

Before the era of space science the lack of visible surface features on Venus when viewed through a telescope suggested the presence of dense clouds. Based on incorrect assumption that the clouds were made of water vapor it was believed by some that the surface environment of Venus is similar to warm and humid areas on Earth. In the 1960's the radio-telescope observations of Venus provided evidence of its very high surface temperature and slow retrograde rotation [e.g., *Cruikshank*, 1983]. At approximately the same time, Mariner 2, the first successful mission to Venus, measured on flyby a radio brightness consistent with a hot planet surface [*Barath et al.*, 1964]. Then several missions to Venus obtained increasing evidence of a planet unlike the Earth, culminating in 1970 with the Venera 7 mission, the first successful landing on the planet's surface. Early Earth-based radar observations led to the discovery of large (~1,000–2,000 km across) radar-bright surface features (e.g., Alpha) of unknown origin [e.g., *Goldstein*, 1965; *Goldstein and Ramsey*, 1972].

Then a series of Venera and Vega landers acquired key information on the surface: the gamma ray (GRS) and X-ray fluorescence spectrometry (XRFS) showed that contents of potassium, thorium and uranium, as well as a number of petrogenic elements in the surface material (beneath the landers down to a few decimeters depth in the case of GRS analysis, and a few cm^3 sample taken by the drilling device in the case of XRFS analysis), are approximately the same as in terrestrial mafic rocks [e.g., *Barsukov*, 1992; *Surkov*, 1997]. TV panoramas showed finely layered rocks and soil in

local lows (Figure 1). According to a number of other *in-situ* observations these rocks were easily crushable and thus obviously porous [*Florensky et al.*, 1978; *Basilevsky et al.*, 1985]. Recently, *Basilevsky et al.* (2004) suggested a hypothesis that this layered material could be deposited from the atmosphere and be composed of partly indurated sediment of the fine fraction of ejecta of upwind impact craters. This hypothesis implies that the source of the sampled material at the Venera sites could be derived from the kilometers-deep subsurface and not necessarily be representative of what we see in Magellan-scale images in the vicinity of the lander as it was considered in earlier publications [e.g., *Kargel et al.*, 1993; *Weitz and Basilevsky*, 1993].

Very important data were acquired by the Pioneer Venus mission flown in 1978 consisting of four probes and an orbiter. Two results were of particular importance to geology. One was the discovery by one of the probes of the so-called deuterium anomaly, a D/H ratio larger by a factor of 150 than it is in Earth's oceans [*Donahue et al.*, 1982]. This implies significant hydrogen escape from the planet, suggesting that early in its history Venus could have lost a large amount of water [e.g., *Donahue and Russel*, 1997]. Some models even suggest the presence of oceans in the early history of Venus [e.g., *Kasting et al.*, 1984; *Kasting*, 1988; *Grinspoon and Bullock*, 2003].

Another result of the Pioneer Venus mission important for understanding Venus geology was radar mapping by its orbiter, which provided a nearly global picture of the planet's topography, gravity and surface roughness characteristics (e.g., *Masursky et al.*, 1980). It was shown that the planet's surface is dominated by plains with elevations close to the mean planetary radius (Plate 1). Part of the plains ("lowlands") were correctly interpreted as relatively young basaltic plains, while the "rolling plains" were considered as "ancient" terrain of possibly "granitic" composition. The older age of the rolling plains was deduced from the observation of numerous circular features then considered as possible lava-filled impact basins. As later shown by observations from Venera 15/16 and Magellan, these circular features are so-called coronae (see below) whose abundance does not indicate surface age. The "uplands" were generally correctly interpreted as a result of tectonic and volcanic crustal thickening [e.g., *Masursky et al.*, 1980]. Analysis of the global topography led to the discovery of belts of rifts resembling continental rifts on Earth [*McGill et al.*, 1981; Schaber, 1982].

The next data set of geologic importance was acquired by the Venera 15/16 twin mission (1983–84), which provided synthetic aperture radar (SAR) images with 1–2 km resolution of about 25% of the planet's surface (Figure 2) [*Kotelnikov et al.*, 1989]. These images, like those mentioned

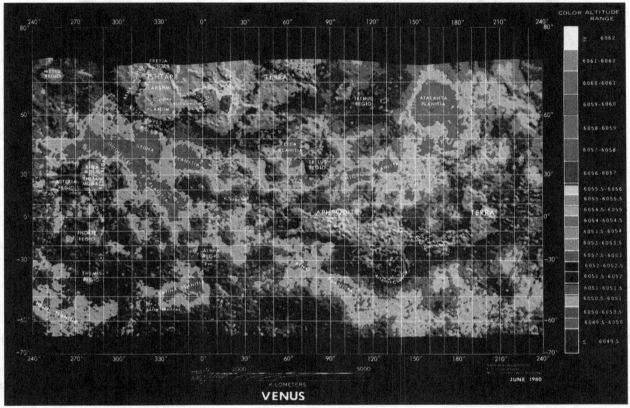

COLOR ALTITUDE RANGE

≥	6062
	6061–6062
	6060–6061
	6059–6060
	6058–6059
	6057–6058
	6056–6057
	6055.5–6056
	6055–6055.5
	6054.5–6055
	6054–6054.5
	6053.5–6054
	6053–6053.5
	6052.5–6053
	6052–6052.5
	6051.5–6052
	6051–6051.5
	6050.5–6051
	6050–6050.5
	6049.5–6050
≤	6049.5

JUNE 1980

VENUS

Plate 1. Global hypsometric map of Venus based on the Pioneer Venus Orbiter altimetry. Although its coverage at high latitudes is less complete than the later Magellan hypsometruc map, it shows almost all major features of the Venusian surface.

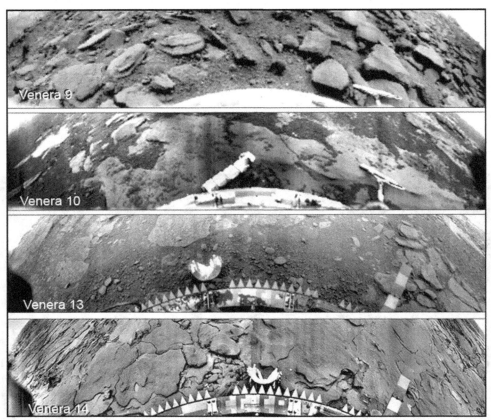

Figure 1. TV panoramas taken at the Venera 9, 10, 13 and 14 landing sites. Finely layered rocks and darker soil are seen. The layered rocks are crushable and porous [*Florensky et al.*, 1978; *Basilevsky et al.*, 1985] and could be ejecta from the upwind craters deposited from the atmosphere. The teeth seen on the spacecraft in the center-lower parts of the Venera 13 and 14 panoramas are 5 cm apart.

below that were acquired by the Magellan mission, provided pictures of the surface very similar to photographic and TV images. The SAR images show the "illumination-shading" effect when slopes looking toward the illumination source (to the radar in the case of SAR images or to the Sun in the case of photographic and TV images) look brighter while the slopes looking in the opposite direction look darker than the horizontal surfaces. Similar to photographic and TV images, SAR images also show differences in the surface brightness not related to the illumination-shading effect but related to the surface reflectivity. In the case of photographic and TV images the brightness variation are due to optical reflectivity (albedo), but in the case of SAR images the variations are due to radar reflectivity which is a combined function of surface roughness (the rougher, the brighter) and surface electromagnetic properties (metallic and semimetallic surfaces look bright). There are some other differences between the visible range and radar images not important for understanding figures of this chapter.

The Venera 15/16 SAR image resolution was sufficient to directly observe many key elements of Venus geology [*Barsukov et al.*, 1986; *Basilevsky et al.*, 1986]. Dominance of volcanic plains was confirmed, as was the presence of indisputable large volcanic constructs. First identified as significant components of Venus geology in this mission were ridge belts, tesserae and coronae (see below). A population of impact craters was found that suggested a mean surface age for the studied area to be 0.5 to 1 b.y. [*Ivanov et al.*, 1986]. These discoveries suggested a planetary geology that is very different from that of Earth. The Magellan mission allowed extending these Venera 15/16 findings to the entire planetary surface.

The Magellan spacecraft arrived in the Venus vicinity in 1990 and orbited the planet through 1994. The mission provided a nearly global radar survey (SAR images of 100–200 m resolution), measurements of altimetry and radiophysical properties, and derived measurements of gravity anomalies [Figure 3; *Saunders et al.*, 1992]. Magellan data showed that

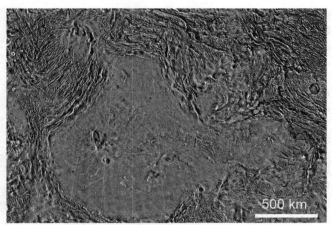

Figure 2. The Venera 15/16 photomap of the volcanic Lakshmi Planum plateau and the surrounding high mountain belts presumably formed due to intensive compressional deformation. Lava flows radiating from the volcanic calderas Colette (left) and Sakajawea (right) are seen on the plateau.

the global geology of Venus is generally similar to that found by Venera 15/16, but the better resolution of Magellan images provided more detail, and thus much more information on mechanisms of formation for materials and structures, and on time relations between them. The descriptions of surface features and units that follow are primarily based on Magellan data analysis.

CURRENT VIEW OF SURFACE FEATURES AND UNITS

The accumulated results of the space missions and Earth-based observations coupled with laboratory and theoretical work suggest that the surface of Venus is dominated by volcanic landforms, mostly plains, that are slightly, moderately or highly deformed by various tectonic structures. Plains materials compose up to 80% of the surface of Venus. The most widespread of these are called "regional plains", which occupy 50–60% of the surface and form a background on which are seen material units and structures either superposed on regional plains, or forming inliers of different sizes embayed by the regional plains. The pre-regional plains inliers include tessera terrain as well as moderately to highly deformed varieties of plains. Below we describe some of the main material units and structures.

In this chapter we use the units as they were identified and used by *Basilevsky and Head* [1998; 2000] for global consideration of Venus geology. Other researchers, describing and/or mapping individual regions identified units appropriate for these regions and based on their approach to this problem. These local units in some publications are close

to the global ones [e.g., *Ivanov and Head*, 2000a] while in other ones may be partially close and partially different or significantly different [e.g., *Brian et al.*, 2005; *Hansen and DeShon*, 2000a]. For clarity, describing units we mention these different names. We describe units generally in the order "from older to younger" deduced by many researchers in different regions of the planet. The question how these local stratigraphic columns, reflecting time sequences, can or can not be integrated into a global stratigraphic column is considered in the Discussion part of this chapter.

Tessera terrain. This terrain forms "islands" and "continents", occupying in total ~8% of the surface [*Ivanov and Head*, 1996]. Tessera exposures stand topographically above the surrounding regional plains. The surface of tessera terrain is very rough, being dissected by numerous criss-crossing ridges and grooves a few km wide and tens of km long (Figure 4). In the place where it was first identified [*Barsukov et al.*, 1996] the tessera tectonic fabric resembles that of the tile roof that later led to the terrain name (tessera = tile in Greek). Ridges are evidently formed by compressional tectonic deformation while many grooves are extensional structures [e.g., *Sukhanov*, 1992; *Ivanov and Head*, 1996; *Hansen et al.*, 1997]. The surface of tessera terrain looks bright on the Magellan images, due to high meter-decameter roughness [*Ford et al.*, 1993] that is at least in part a result

Figure 3. The Magellan SAR mosaic of the hemisphere centered at 0° latitude and 180° longitude. Bright wispy linear zones are rifts. Junction of several rifts in the center-right is Atla Regio. In the north is seen a network of ridge belts. In the south, survey gaps are seen.

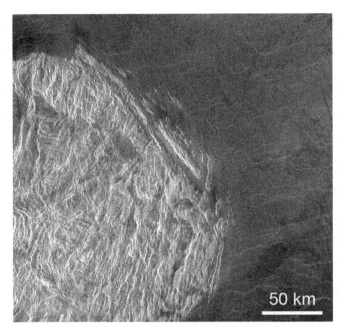

50 km

Figure 4. NE part of Alpha Regio tessera (bright) embayed by regional plains (dark) that implies that tessera terrain is older than regional plains. Fragment of Magellan image centered at 22°S, 10°E.

of tectonic deformation. This deformation implies episode(s) of intensive tectonic activity, which formed what we see now as tessera terrain. The composition of tessera material is unknown. Some researchers suggest it has a basaltic composition; others believe that tessera may be made of more feldspathic material resembling to some degree anorthosites on the Moon or granites on Earth [*Nikolaeva et al.*, 1992]. Surrounding plains of apparent volcanic origin embay the tessera [e.g., *Basilevsky and Head*, 1995, 1998, 2000, *Bleamaster and Hansen*, 2005; *Ivanov and Head*, 2001]. In some rare cases, when tessera is in contact with moderately and highly deformed plains, it can be seen that materials composing these plains embay tessera terrain. Because tessera terrain is observed in practically all regions of Venus, this intensive tectonism could be of global or close to global scale, although the data in hand are insufficient to determine if the deformation resulting in the formation of tessera terrain occurred at similar times globally or else at different times in different places.

In the western part of Ishtar Terra, in close association with tessera terrain, are seen *Mountain belts*, which surround the volcanic plateau Lakshmi [e.g., *Barsukov et al.*, 1986; *Basilevsky et al.*, 1986; *Crumpler et al.*, 1986; *Pronin*, 1992; *Solomon et al.*, 1992]. They consist of clusters of parallel (within the given belt) ridges (see Figure 2). The mountain belts are among the topographically highest features on Venus. The summit of the highest of them, Maxwell Montes,

stands more than 11 km above the mean planetary radius. Parallel ridging and high altitudes of the mountain belts are considered as evidence that they formed due to horizontal compression. These mountain ranges merge laterally into tessera terrain so maybe these two are close in age. As in the case of tessera terrain, the composition of the mountain range material is not known.

The top parts of the mountain ranges (and of some other highlands of Venus as well) have extremely high radar reflectivity, appearing very bright on the Magellan images. This brightening appears above some critical altitude, although the exact value varies somewhat in different parts of the planet. It was suggested by some researchers that above this altitude, surface material has undergone a specific weathering in which the iron in silicates, such as pyroxenes and olivines, is segregated into minerals having a high electric conductivity (iron oxides or sulfides) [*Klose et al.*, 1992; *Wood*, 1997]. Other researchers however suggested that it could be a temperature-controlled deposition of some heavy metals and/or their sulfides and sulfosalts, whose components sublimed from the surface rocks of the topographically lower and thus hotter regions [*Pettengill et al.*, 1997; *Schaefer and Fegley*, 2004].

Densely fractured plains are also called by many mappers *densely lineated* [*Campbell and Campbell*, 2002; *Ivanov and Head*, 2004a] or *lineated plains* [e.g., *McGill*, 2000]. They are seen in many areas of Venus as relatively small (tens of km to 100–200 km across) "islands" of densely fractured material (that is basis for the unit name) standing a few hundred meters above the surrounding regional plains (Figure 5). If one ignores the fractures, the terrain forming these islands is plains [e.g., *Basilevsky and Head*, 1998, 2000; *Brian et al.*, 2005]. These islands are generally considered to consist of plains-forming volcanics, probably of basaltic composition, that have been deformed by closely spaced fractures. The structural pattern of the fractures is generally subparallel within a given island. The islands of densely fractured plains are obviously embayed by the regional plains [e.g., *Basilevsky and Head*, 1998, 2000; *Brian et al.*, 2005; *Campbell and Campbell*, 2002; *McGill*, 2000]. In rather rare cases, where they are in contact with tessera terrain, their material seems to embay tessera (Figure 5). A terrain of this type is also observed in many coronae (see below), where radial and concentric structural patterns of dense fracturing are typical. Global abundance of outcrops of the densely fractured plains material is about 3 to 5% [*Basilevsky and Head*, 2000a].

Ridge belts. Most ridge belts are long, relatively narrow areas that are somewhat elevated above surrounding regional plains (Figure 6). They contain plains-like materials that have been deformed into individual ridges that generally

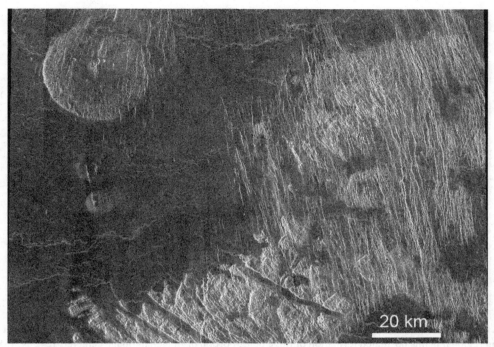

Figure 5. Densely fractured plains (image right, cut by NNW set of faults) embaying tessera (bottom, cut by multiple sets of faults) indicating that emplacement of the material of the densely fractured plains postdated formation of tessera terrain. Both densely fractured plains and tessera are embayed by even younger regional plains (dark areas). A steep-sided volcanic dome is seen in the upper left. Its age relations with the surrounding regional plains are ambiguous. Portion of Magellan image centered at 46°N, 360°E.

parallel the long dimensions of the belts. They occur as continuous belts or as patches that appear to be partially flooded by lavas of the surrounding regional plains. The material composing ridge belts is commonly slightly rougher than the material of regional plains, and it probably also consists of basaltic lavas. Folding into relatively broad (3–5 km) ridges arranged into parallel bands is typical of these belts and most likely resulted from horizontal compression. Ridge belts differ from the mountain belts described above in their much lower altitudes (hundreds of meters) and probably in the density of deformational structures. In most observed cases, regional plains embay the belts [e.g., *Basilevsky and Head*, 1998, 2000; *Bridges and McGill*, 2002; *Ivanov and Head*, 2004a] although in some places their age relations are ambiguous [e.g., *Rosenberg and McGill*, 2001]. Global abundance of ridge belts is about 3 to 5% [*Basilevsky and Head*, 2000a].

Regional plains. Although studied in detail in the initial Magellan images [*Guest et al.*, 1992; *Head et al.*, 1992], regional plains were first identified as a specific unit by *Basilevsky and Head* [1995] under the name of *Plains with wrinkle ridges*. In that work "Plains with wrinkle ridges" included what is now called *Shield plains*. In later studies

"Plains with wrinkle ridges" and "Shield plains" began to be mapped separately [e.g., *Aubele*, 1996; *Ivanov and Head*, 2001]. Many researchers prefer to call the first one "Ridged plains" [e.g., *Bridges and McGill*, 2002; *Campbell and Campbell*, 2002] or "Regional plains" [e.g., *Brian et al.*, 2005; *McGill*, 2000] as we do here. Regional plains occupy 50–60% of the surface [*Basilevsky and Head*, 2000a]. They have a rather smooth (at the scale of observation) surface (Figure 7), often with flow-like features, which are inferred to be solidified lava flows [*Guest et al.*, 1992; *Head et al.*, 1992]. These plains are deformed by a network of relatively narrow (1–2 km) gently-sloping ridges tens to hundreds of km in length, called "wrinkle ridges". Wrinkle ridges are the result of "wrinkling" of the surface by moderate horizontal compression [e.g., *McGill*, 1993]. Large areas occupied by the flow-like features (100–200 km long flows are common) and very gentle slopes, along which the flows were emplaced, indicate high-yield eruptions of low viscosity, probably basaltic, lava which formed plains that were subsequently deformed by wrinkle ridges. The suggestion of basaltic composition of the lava is supported by the analysis of the Magellan images for the landing ellipses of the Venera/Vega landers. This analysis showed that the landing sites,

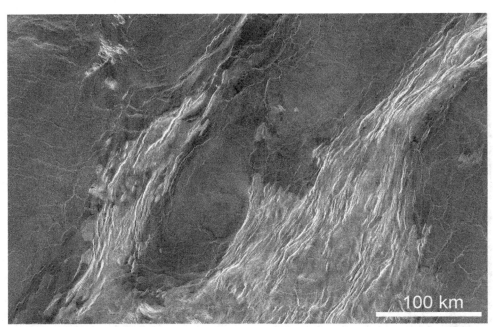

Figure 6. Part of Vedma Dorsa ridge belt (light) embayed by regional plains (dark) suggesting that emplacement of lavas composing regional plains postdated formation of ridge belts. Portion of Magellan image centered at 42°N, 155°E.

where the basaltic composition was measured [Venera 9, 10, 13, Vega 1, 2; *Barsukov*, 1992; *Surkov*, 1997], are dominated by regional plains [*Abdrakhimov and Basilevsky*, 2002]. But if the layered rocks seen in the Venera TV panoramas are sediments of the fine fraction of ejecta from upwind impact craters, as suggested above, these analyses may partly represent other units underlying regional plains.

Within these plains are seen sinuous channels of 2–5 km width and lengths of hundreds of kilometers [*Baker et al.*, 1997]. One of the channels, Baltis Vallis, is 6,800 km long, a distance that is about 1/6 of the circumference of Venus (Figure 7). It is not yet clear how these channels formed, but the channel morphology implies erosion by some liquid. The most popular view is that the channels resulted from thermal and/or mechanical erosion by flowing high-temperature lava, perhaps komatiitic lava. Similar channels (although not so long) are known on the Moon and even smaller analogs are seen on the flanks of some terrestrial volcanoes. The major enigma of the channels on Venus is their great length. Calculations show that the temperature of the surface and near-surface atmosphere is, despite its very high temperature relative to Earth, still cold compared to basaltic or komatiitic lava. Thus komatiitic lava in particular should solidify rather quickly, inhibiting channel formation. Mechanical erosion by liquids having significantly lower melting temperature, such as sulfur or alkaline carbonatite melt, are considered as possible alternatives [e.g., *Kargel et al.*, 1994].

Shield plains. These generally form fields of small shields a few hundred kilometers across. Their surface is peppered with numerous gentle-sloping volcanic shields 5 to 15 km in diameter (Figure 8). Coalescing flanks of the shields form most of the plains surface. Different morphologic analyses show that individual shields can predate regional plains, can be coeval with them, or can postdate them. The relative proportions of these pre, syn, and post varieties are a subject of debate. *Addington* [2001] and *Hansen* [2005] concluded that most post-date regional plains, whereas *Ivanov and Head* [2004b] concluded that pre-regional-plains varieties dominate over the syn- and post-regional-plains examples. Most likely the main reason for this disagreement stems from different criteria used for these studies, and different areas of shield plains sampled. Global abundance of shield plains was estimated by *Basilevsky and Head* [2000a] as close to 10–15%.

The gentle slopes of the shields imply that they are made of low viscosity, probably basaltic lava. The Venera 8 lander, whose landing ellipse is dominated by shield plains, found contents of potassium, uranium and thorium (beneath the lander down to a few decimeters depth) significantly higher than those typical for the majority of terrestrial basalts. This led to the suggestion that some parts of the shield plains are composed of alkaline basalts, or even geochemically more evolved rocks, which would have higher viscosities than normal basalt [*Basilevsky*, 1997].

Lobate plains and flows. These are observed in two partly overlapping varieties. The first one (lobate plains) typically forms fields of individual flows not directly associated with volcanic edifices. The flows are of variable radar brightness, suggesting variability in their surface texture (Figure 9 left). The flows are tens to 200–300 km long and from a few to a few tens of kilometers wide. They typically overlay regional plains and embay other units described above. More than 200 flow fields were observed on the surface of Venus, each with an area larger than 50,000 km² [*Crumpler et al.*, 1997]. Their sources are often associated with rift zones (see below). Some of these young lava flows occur in association with coronae. The great length of the flows on very gentle-sloping surfaces suggests low viscosity of the lava. This, in turn, is usually considered as evidence of basaltic composition.

Lobate flows are also seen associated with *volcanic constructs* composing the second subunit of the "lobate plains and flows" unit (Figure 9 right). More than a hundred constructs larger than 100 km in diameter and about 300 constructs 20–100 km in diameter are observed on Venus [*Crumpler et al.*, 1997; *Magee and Head*, 2001]. The youngest lavas related to these constructs are clearly superposed on regional plains. The highest volcano on Venus, Maat Mons, stands about 9 km above mean planetary radius. Lava flows radiating from Maat Mons cover an area about 800 km across. These large and intermediate sized volcanoes are morphologically very similar to basaltic shield volcanoes on Earth, although the latter are typically smaller than their counterparts on Venus. Measurements by the Venera 14 lander, whose landing ellipse is dominated by lava flows associated with the Panina Patera volcanic construct, showed a basaltic composition of the surface material [*Barsukov*, 1992; *Surkov*, 1997]. However the TV panoramas of the site show that the material analyzed at this place is represented by finely layered rocks so its source could instead be ejecta of upwind impact craters, and not the Potanina lava flows. Global abundance of the lobate plains and flows unit is estimated to be about 10% [*Crumpler et al.*, 1997; *Basilevsky and Head*, 2000a,b].

Smooth plains on Venus are typically radar dark. They occupy only a few percent of the surface [*Basilevsky and Head*, 2000a]. Some relatviely rare smooth plains have very sharp boundaries. Their morphology and frequent association with obvious volcanic landforms suggest that they are fields of lava flows with very smooth surfaces (Figure 10 left). Another variety of smooth plains has diffuse boundaries and is commonly associated with large impact craters. These are probably mantles of fine debris, the primary source of which is ejecta from impact craters (Figure 10 right).

Structural units. Significant tectonic deformation is observed in several units described above and some are so penetratively deformed (tessera and densely fractured

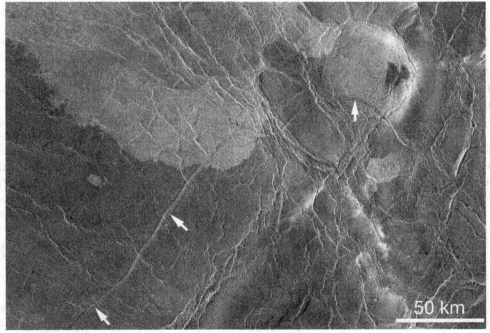

Figure 7. Two subunits of regional plains: the older (dark) eroded by the Baltis Vallis channel (arrows) and the younger (bright) which partly floods the channel suggesting that the bright subunit postdates the channel formation. Both subunits are wrinkle ridged. Portion of Magellan image centered at 48°N, 162°E.

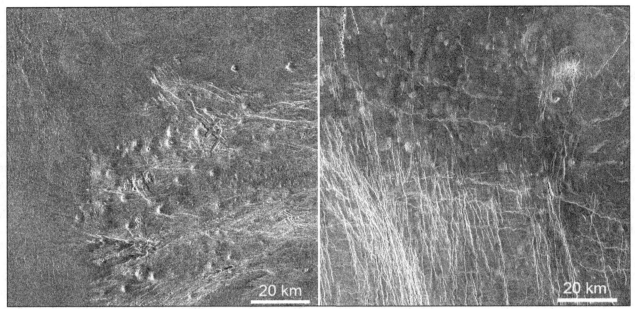

Figure 8. Two examples of shield plains. Left, shield plains predating the adjacent regional plains: the latter embay shield plains and truncate most fractures, which cut the shields; right, similar shields (upper left of the image) superposed on regional plains: the latter are cut by numerous fractures, which are truncated by the shields. These two examples show that some shields and their fields formed before the adjacent regional plains while others formed after. Magellan images centered at 47.5°S, 82°E and 19.5°N, 15.5°E correspondingly.

plains) that their original properties are not evident. But these units are embayed by younger units, so it is logical to consider them as material units. Two types of highly deformed terrain, however, show no (or partial) embayment and according to recommendations of the U.S. Geological Survey [*Tanaka*, 1994] are mapped by many researches as structural units. They are represented by two overlapping and often closely associated varieties: *rifts* and *fracture belts*. Although mapped as structural units by some workers, it is important to note that rifts and fracture belts are not material units in the same sense as are most mappable units on Venus. Their combined global abundance is about 9–10%, of which about 2/3 are young rifts and 1/3 fracture belts [*Price*, 1995; *Basilevsky and Head*, 2002b].

Young rifts on Venus are rather similar to continental rifts on Earth and form a global system up to 40,000 km long [*Masursky et al.*, 1980; *McGill et al.*, 1981; *Schaber*, 1982]. They are typically topographic troughs, whose floors may be a few of kilometers below the neighboring non-rifted terrain, whereas the rims of the troughs are commonly uplifted above the neighboring terrain. The walls and floors of the troughs are fractured, commonly very heavily (Figure 11 left and center). The anastomosing character of the fracturing and significant changes in the width of individual fractures are often typical of rifts. Young rifts cut regional plains. Young

post-regional-plains lava fields and volcanic constructs are commonly associated with rifts. They are locally superposed on some elements of rifts but locally are cut by rift faults. The general consensus is that rifts formed in the environment of tectonic extension.

Fracture belts are partly older and partly younger than regional plains [e.g., *Hansen et al.*, 1997; *Ivanov and Head*, 2001]. They are typically seen as highly fractured areas partly flooded by lavas of regional plains (Figure 11 right) and partly fracturing regional plains. It thus is possible that this relatively old rifting was more extensive than one can judge from the abundance of fracture belts that are now observed. Compared to the younger rifts, fracturing within fracture belts is more homogeneous and less anastomosing.

It is necessary to add that some researchers avoid defining and mapping structural units [e.g., *Hansen*, 2000; *Hansen and DeShon*, 2002; *Bleamaster and Hansen*, 2005]. For example, part of a very heavily tectonized rift zone of Aphrodite Terra (Figure 11 center) instead of being mapped as "rifted terrain" was mapped by *Bleamaster and Hansen* [2005] as "chasmata flow material". Their conclusion that it is material of volcanic flows was derived from the following consideration: They assumed that Venus could host only three basic rock types: igneous, metamorphic and sedimentary. Because of "absence of surface water and the paucity of eolian erosion on Venus" they rejected the suggestion of any significant presence of sedimentary rocks. Because of the absence on this planet of widespread erosion they considered surface exposures of metamorphic and intrusive igneous rocks as not likely. So they conclude that "surface rocks most likely originated as extrusive igneous rocks, that is volcanic flows" (page 2 of the pamphlet accompanying the geologic map of *Bleamaster and Hansen* [2005]).

Coronae. Several hundred volcanic-tectonic structures called "coronae" ("corona" singular) are observed on the surface of Venus. They were first discovered in the analysis of radar images of Venera 15/16 [*Barsukov et al.*, 1986]. Coronae appear to be unique to Venus. Coronae are oval to circular features typically 100 to 300 km in diameter (Figure 12), although a few are even larger. They have a tectonically deformed annulus, which generally stands a few hundred meters above the surrounding plains. The area inside the annulus is commonly lower than the surrounding plains and partly flooded with plains-forming volcanics. Aprons of young lobate volcanic flows are commonly seen radiating from many coronae. A core is seen at the center of some coronae as an elevated and tectonically deformed area. Coronae are considered to form as a result of rising of hot mantle plumes/diapirs [*Basilevsky et al.*, 1986; *Stofan et al.*, 1997]. The diapir raised the upper lithosphere and crust, and during its ascent produced magmatic melts, some of which reached the surface and formed the corona-associated lava flows. When the diapir cooled, the uplifted surface subsided producing the structure now called a "corona". Some coronae are scattered among the regional plains. Others form clusters and chains associated with rift zones.

Impact craters. In the Magellan images of Venus more than 960 impact craters from 1.5 to 270 km in diameter were identified [*Schaber et al.*, 1998; *Herrick et al.*, 1997]. Their size frequency distribution is obviously controlled by the screening effect of the massive Venus atmosphere. Most statistical tests show that the distribution of impact craters around the planet is indistinguishable from a random one [*Schaber et al.*, 1992; *Phillips et al.*, 1992; *Strom et al.*, 1994; *Kreslavsky*, 1996], although some more specific tests suggest the possibility of a partly non-random distribution (see below) [*Hauck et al.*, 1998].

The morphology of impact craters on Venus is essentially similar to that of impact craters on other planets and satellites. They are circular depressions surrounded by an elevated rim (Figure 13). The rim, and the area surrounding it, are knobby and typically radar bright due to a surface layer of relatively rough ejecta excavated from the crater. The general morphology of impact craters on Venus is correlated with their size (Figure 13A, B, C): Craters smaller than 10–20 km in diameter have an irregular floor. Larger craters have a central peak on the floor. Even larger (>50–60 km) craters have a concentrically ringed floor. A similar size dependence (although with different transitional diameters

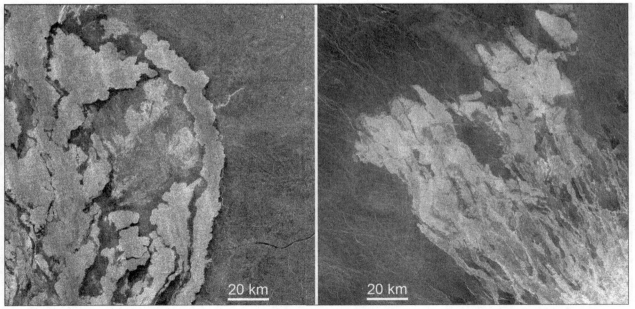

Figure 9. Two varieties of lobate flows. Left are flows of the NE part of Mylitta Fluctus lobate plains whose source is Kalaipahoa Linea rift; right are lobate flows of the NW outskirt of Sapas Mons volcano. Both varieties are superposed on regional plains. Magellan images centered at 54°S, 357°E and 10.5°N, 186.5°E correspondingly.

between types) is observed for craters on other planetary bodies. On the other bodies, however, craters of smaller size are not irregular-floored but have a relatively smooth bowl-shaped floor. This difference is due to break-up of relatively small crater-forming projectiles on their way through the massive atmosphere of Venus. The result is that in these cases the planet is impacted not by a single projectile, but a swarm of dispersed fragments.

Around many craters on Venus flow-like features are seen extending from the knobby ejecta (Figure 13D). These features (commonly called "outflows") are believed to be flows of high-temperature melt produced by the crater-forming impact. Their abundance on Venus is probably due to the higher temperature of the surface of Venus, compared to other planets and satellites, and thus the higher temperature of the upper crust. This would tend to increase the amount of impact melt produced [*Schultz*, 1992]. Some ejecta outflows show evidence that their material is rather easily redeposited by wind. This suggests that these could be formed by flows of fine-grained material suspended in the air, thus mimicking formation of turbidities flows on the continental slopes of terrestrial oceans.

Radar dark haloes are seen in association with many impact craters on Venus (Figure 13). The haloes are considered due to relatively fine-grained debris deposited as a part of the crater formation process. With increasing time, haloes degrade and disappear, so the presence and prominence of

the halo can be used as a measure of crater age [*Arvidson et. al.*, 1992; *Basilevsky and Head*, 2002a; *Basilevsky et al.*, 2003]. Some areally extensive haloes have a parabolic form with the parabola apex pointing to the east (Figure 13E). These dark parabolas are considered to have formed due to delivery of crater ejecta into the upper parts of the atmosphere and consequent settling. During this settling, the fine fraction of the ejecta is entrained by strong zonal winds, which blow at high speed towards the west. The dark parabolas associated with the youngest craters degrade with time into non-parabolic haloes. The total amount of relatively fine debris produced by impacts after formation of the regional plains and widely distributed by the wind is equivalent to a global layer a few meters thick (Garvin, 1990).

Aeolian features. In the absence of liquid water on Venus, exogenic resurfacing is dominated by aeolian processes [*Greeley et al.*, 1997]. The orientation of aeolian features is indicative of the dominant directions of near-surface winds and thus can be helpful in studies of the dynamics of the lower atmosphere. Part of the observed aeolian features could be formed by strong winds, which are thought to accompany impact cratering events [*Ivanov et al.*, 1992]. The observed aeolian features are represented by radar-dark mantles, wind streaks, yardangs and dunes. The first two types of aeolian features are rather common on Venus, while the features of the second two types, large enough to be seen on the Magellan images, are observed in only a few localities.

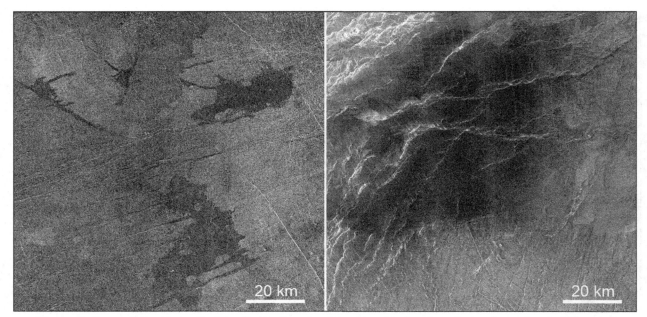

Figure 10. Two varieties of smooth plains. Left is the volcanic variety with sharp boundaries observed east of Ohogetsu corona, right is aeolian variety with diffuse boundaries SW of impact crater Mead, whose ejecta probably is the source of aeolian debris. Magellan images centered at 28°S, 84°E and 8°N, 52°E correspondingly.

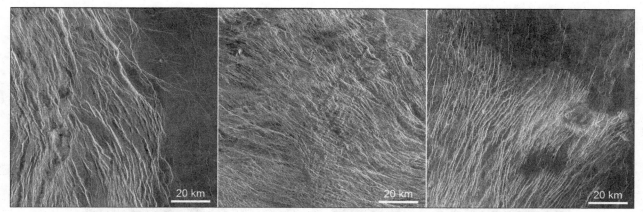

Figure 11. Examples of rifts and fracture belt. Left is fragment of the young Ganis Chasma rift composed of anastomosing faults cutting regional plains; center is fragment of the young Jana Chasma rift, saturated with densely spaced faults; right is relatively old fracture belt composed of more homogeneous and less anastomosing faults embayed by regional plains. Magellan images centered at 12°N, 198°E, 12°S, 112°E, and 32°S, 227.5°E correspondingly.

Dark mantles commonly are seen in association with impact craters, forming halos of different sizes and forms (see above). The source of the dark-mantle material is fine debris formed by crater-forming impacts. In many cases the dark mantles have lost direct contact with impact craters and occupy wind-shadow localities in local topographic lows and behind, or against, positive topographic features, becoming a variety of smooth plains (see above).

Wind streaks are the most abundant aeolian features on Venus. They vary significantly in shape (linear, fan-like, wispy), size (from a few to tens of kilometers), and radar brightness (bright, dark, mixed). Figure 14A, B shows examples of their variety. Wind streaks obviously formed as the result of accumulation and/or erosion of loose surface material due to wind turbulence behind topographic features. It is rather typical that lateral boundaries of wind streaks are diffuse.

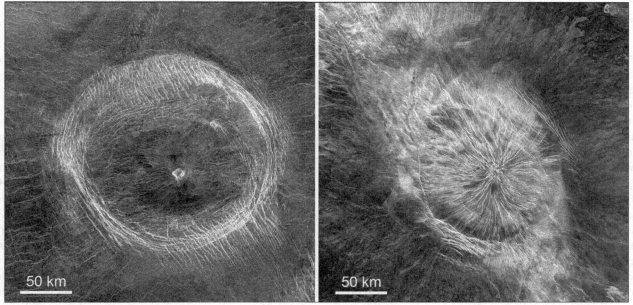

Figure 12. Two examples of coronae. Left, Thourus corona, 6.5°S, 12.9°E, D = 190 km. Its annulus is embayed by regional plains, post-regional-plains activity is very minor; right, Dhorani corona, 8°S, 12.9°E, D = 150 km, showing extended apron of post-regional-plains lava flows. These two examples show that morphologically observable activity of some coronae terminated before the emplacement of adjacent regional plains (Thourus) while activity of others continued after the regional plains emplacement (Dhorani).

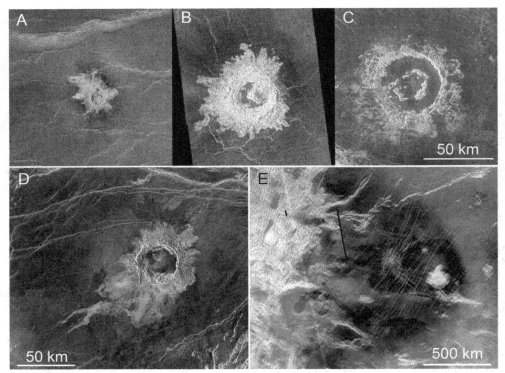

Figure 13. Examples of impact craters: A – crater Avene, 40.4°N, 149.4°E, D = 10 km, showing faint associated halo and irregular morphology of its floor; B - crater Buck, 5.7°S, 349.6°E, D = 22 km, with faint halo and a prominent central peak; C - crater Barton, 27.4°N, 20.2°E, D=50 km, having no halo and a double-ring structure; D – crater Flagstad, 54.3°S, 18.9°E, D=39 km, with prominent halo and ejecta outflow; E – crater Stuart, 30.8°S, 20.2°E, D=68 km with associated radar-dark parabola. Images A, B, and C share the same scale bar.

A field of possible *yardangs*, which are wind-erosional ridges, is observed in the vicinity of the crater Mead, the largest impact crater on Venus. They are represented by sets of parallel, linear, slightly sinuous grooves and ridges separating them (Figure 14D). Yardangs differ from wind streaks in that they have well-defined boundaries and lack a distinctive relation to topographic features, such as hills.

Two *dune fields* have been found on the Magellan images: one on plains between Fortuna and Meskhent tessera massifs (Figure 14C) and another one on plains of the northern part of Lavinia Planitia [*Greeley et al.*, 1997]. Both dune fields are in close association with large impact craters whose ejecta probably was the source of the debris involved in dune formation. The dunes are from 0.5 to 10 km long and a few hundred meters wide. Formation of dunes implies saltation of sand-sized particles, so the lack of observed dunes on Venus may indicate a deficit of debris of this size on the planet. The apparent lack of dunes could, however, be partially an observational effect. Analysis of radiophysical properties in several regions of Venus by *Bondarenko et al.* [2006] revealed a noticeable east-west asymmetry in the

radar returns which was interpreted as possible dunes/ripples of meter to decameter size range.

DISCUSSION

The material units and structures described above are observed with minor variations over most of the planet, and are included on most of the geologic quadrangle and other maps of Venus [e.g., *Bleamaster and Hansen*, 2005; *Brian et al.*, 2005; *Bridges and McGill*, 2002; *Campbell and Campbell*, 2002; *Ivanov and Head*, 2001, 2004a; *McGill*, 2000; *Rosenberg and McGill*, 2001]. Of particular interest in comparison with Earth are the belts of rift zones. These rifts differ from the Earth's mid-oceanic rifts, which are the areas of the youngest volcanism bordered by progressively older volcanics. Although generally relatively young structures, Venus rifts cut through terrains and units of different ages, resembling in this, and in a number of other characteristics, terrestrial continental rifts rather than oceanic rifts. Structures that could be considered as analogs to terrestrial zones of collision and subduction are not observed.

Thus it is concluded that geodynamics on Venus does not operate in the plate-tectonics style, but rather in some other mode [e.g., *Solomon et al.*, 1992; *Hansen et al.*, 1997; *Phillips, et al.*, 1997]. The lack of current and geologically recent plate tectonics is supported by the observation that despite the seemingly long evolution of many coronae, they do not show evidence for deformation and elongation with time suggesting that the "plate" on which they reside, has not been involved in lateral movement and intense deformation. Additional evidence against geologically recent plate tectonics on Venus is the seemingly random spatial distribution of impact craters mentioned above.

The surface geology of Venus is almost certainly dominated by basaltic volcanism forming globally extensive plains as well as abundant volcanic constructs. Except for a few unusual volcanoes with short, stubby, radar-bright flows [e.g., *Moore et al.*, 1992], only the relatively rare steep-sided domes appear to be candidates for extrusion of compositionally more evolved viscous lavas [e.g., *Pavri et al.*, 1992] although other suggestions on their nature have been published: low-eruption rate basaltic volcanoes [*Fink and Griffith*, 1998], increased content of dissolved water and difference in crystallinity [*Bridges*, 1995] or foamy basaltic lavas [*Pavri et al.*, 1992]. However, knowledge that even a minor portion of the volcanism on Venus is not basaltic is

very important for the understanding of the general petrology and geodynamical evolution of this planet, so future missions should plan measurements necessary to determine their mineralogy and petrology.

The rocks of the crust of Venus have suffered from compressional and extensional deformation. In local stratigraphic sequences and maybe on a global scale (see below) the involvement of the crustal materials in deformation generally decreased with time.

The characteristics of the most deformed and most ancient tessera terrain have yet to provide conclusive keys to the geodynamical origin of this most distinctive terrain, and several different models exist for the formation of tessera and the geodynamical evolution of the planet as a whole [e.g., *Sukhanov*, 1992; *Bindschadler et al.*, 1992; *Ivanov and Head*, 1996; *Hansen et al.*, 1997, 2000]. Future studies need to concentrate on this problem by acquiring significantly higher resolution imaging of the surface of Venus on a global or at least regional scale. This will provide a better understanding of the deformation history of tessera terrain. Further progress in geodynamic modeling and correlations of the geological information with geophysical data sets will also be very helpful.

Magellan images revealed somewhat less than 1000 impact craters on the surface of Venus. This led to a number of esti-

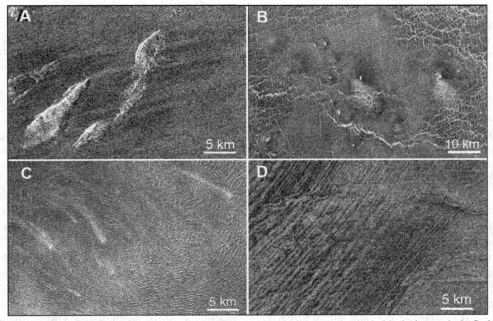

Figure 14. Examples of aeolian features: A – Depositional radar-dark wind streaks behind (downwind of) the tessera ridges protruding through the smooth plains, 0.9°S, 71.5°E. B – Erosional radar-bright wind streaks behind (downwind of) small volcanic shields, 22.3°N, 332.1°E. C - Part of the Fortuna-Meskhent dune field. Dune long axis orientation varies from E-W to NE-SW. Orientation of radar-bright wind streaks implies winds blowing towards the NW, 68°N, 90.5°E. D - Yardangs to the SE of the crater Mead, 9°N, 60°E.

mates of the mean surface age that are generally consistent with each other. The latest estimate suggests that the mean surface age of the planet is about 750 m.y., but any values between 300 m.y. and 1 b.y. cannot be excluded [*McKinnon et al.*, 1997]. This means that we see the geological record of only the latest 10 to 20% of the history of Venus. Some material representing earlier time periods could be present as inclusions in tessera terrain. This implies that higher resolution imaging of tessera (meters to tens of meters per pixel), and analysis of its composition, should be high priority goals for future missions to Venus. Some progress in understanding the mineral composition of tesserae may be expected from night-time near-IR observations planned for the Venus Express mission.

Another message from the past of this planet is the deuterium anomaly, which suggests that Venus in early times was richer in water (e.g., Donahue and Pollack, 1983). There are models suggesting that Venus could have had an ocean for several hundred million years or possibly even longer [e.g., *Kasting et al.*, 1984; *Kasting*, 1988; *Grinspoon and Bullock*, 2003]. If so, this could have affected the rheology and therefore the dynamics of the upper mantle and crust as well as magma formation and differentiation processes.

As noted by *Solomon et al.* [1992] in the early analysis of the Magellan data, in many different locations on Venus occur similar sequences of materials and structures. Basilevsky and Head [1995] mapped 36 1000 x 1000 km areas randomly distributed around the planet. Their results confirmed the observations of *Solomon et al.* [1992]. Extending that work, *Basilevsky et al.* [2000] and *Ivanov and Head* [2001] mapped the northern 25% of the planet as well as a continuous geotraverse circling the planet at ~30⁰ N. latitude. The results were interpreted to support the existence of distinctive "phases" in the crustal evolution of Venus (Figure 15) during the time span represented by the currently exposed materials and structures [see summary in *Basilevsky and Head*, 2002b].

Following many of these early analyses, *Guest and Stofan* [1999 labeled this interpretation as a "directional" model and suggested the contrasting "non-directional" model, according to which "coronae, rifts, wrinkle ridges, small and large edifices, and large flow fields have each formed throughout the portion of Venus' history revealed by presently exposed rock units" (their page 55). These ideas are currently being debated.

One of the findings of earlier works [e.g., *Basilevsky and Head*, 1998, 2000; *Ivanov and Head*, 2001] is that regional plains occupy a central position of the regional and global stratigraphic columns and that these plains appear to have a broadly similar age. Variations in the radar backscatter of the regional plains are common, implying that these plains consist of many different flows from several different pos-

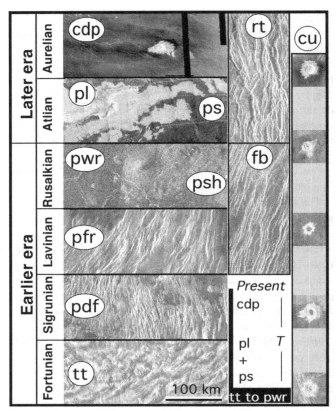

Figure 15. Diagram showing a sequence of material and structural units by Basilevsky and Head (2002). It shows material units from the oldest tessera terrain (tt), through densely fractured plains (pdf), ridge belts (here designated as pfr), regional (pwr) and shield (psh) plains, lobate flows (pl) and smooth plains (ps) to the youngest craters with dark parabolas (cdp). It shows also change in the fracturing style from older rifts (fb) to younger ones (rt), change in degree of preservation of crater associated dark haloes and approximate time span of the of the emplacement of material units (T is mean global age of the planet). Whether this sequence formed at different times in different places, or formed at the same time globally, is currently an issue of some debate.

sible sources [e.g., *Stofan et al.*, 2005]. Nevertheless, the regional plains as a whole are characterized by an impact crater distribution that was found to be not distinguishable from random [e.g., *Schaber et al.*, 1992; *Phillips et al.*, 1992]. More recently, *Hauck et al.* [1998], concluded that although the hypothesis of complete spatial randomness of the crater distribution cannot be rejected, the modeling suggests a possible spread in the ages of plains that might exceed half the mean global surface age. Unfortunately, plains used by Hauck et al. included materials that. based on geologic evidence. are younger or older than regional plains, and thus their work does not constrain the age of regional plains [see for details *Basilevsky and Head*, 2006].

Can widespread units such as the regional plains thought to be globally equivalent stratigraphic units by some [e.g., *Basilevsky and Head*, 1996; *Ivanov and Head*, 2001] be correlated using the existing crater data? *Campbell et al.* [1999] argued that determining the relative ages of separate areas within the plains can not be statistically robust because of the small number of craters available. Thus, using crater statistics alone, *Campbell et al.* [1999] would conclude that the relative ages of separated plains units cannot be confidently established.

A number of studies have been undertaken to determine the time sequences of different material units and structures, for example, on tessera terrain [e.g., *Ivanov and Head*, 1996; *Hansen et al.*, 1997; *Gilmore and Head*, 2000; *Hansen et al.*, 2000], or on the relative ages of regional plains and shield plains [e.g., *Basilevsky and Head*, 1996; *Hauck et al.*, 1998; *Campbell*, 1999; *Addington*, 2000; *Ivanov and Head*, 2001, 2004b; *McGill*, 1993, 2004; *Hansen*, 2005]. Although not resulted in general consensus, these studies have led to the descriptions of many important details, and also to greater clarity of the different researchers' positions.

We would like to conclude the discussion by a comparison with greenstone belts on Earth [e.g., *De Witt and Ashwal*, 1997]. These belts are common in Archean and earliest Proterozoic terrains, but rare or absent in younger terrains. Thus at the largest time scale, the history of greenstone belts is directional. However, based on radiometric dating, greenstone belt formation within the Archean and earliest Proterozoic clearly occurred in different places at different times. Perhaps we can think of some aspects of Venus crustal history in the same way: characterized by a prominent trend for the larger time intervals, and by repetitive components at shorter time intervals.

KEY QUESTIONS TO BE RESOLVED IN FUTURE STUDIES

As evident from the above description and discussion, analysis of the results from many missions and from laboratory and theoretical studies has led to significant progress in understanding major issues of Venus geology. However, many important questions still wait to be resolved in future studies. Below is a short list of some of them:

• What was happening on Venus between its accretion and the formation of tessera terrain?
• Did Venus once have an ocean?
• Did plate tectonics ever occur on Venus?
• Is the geological history "directional" or "non-directional" or some combination of these models?
• Is tessera terrain composed of thickened basaltic crust or of a different low-density material?

• What was the mechanism of tessera formation (deformation resulted from upwelling or downwelling) and how long was the tessera-forming phase?
• How did the folded mountain ranges surrounding Lakshmi Planum form?
• What caused the deformation typical of densely fractured plains, ridge belts and wrinkle ridges; in particular, how can we explain the stresses necessary to form areally extensive coherent arrays of wrinkle ridges?
• Is Venus still volcanically and tectonically active?
• Are coronae manifestations of mantle plumes? Are some of them still active?
• What is the origin of layered rocks seen in the Venera panoramas?

Although the Venus Express mission is designed mostly to gain data for atmospheric and plasma science, it may shed some light on the compositions of tesserae and mountain belts. This could be possible if measuring the near infrared emissivity from these terrains is successful. Progress in atmospheric dynamics and chemical composition may lead to better understanding of the surface geological processes. But serious progress in resolving the above mentioned questions demands new missions, including orbiter(s), balloons, landers and sample return.

An increase of surface imaging resolution is needed to see more detail (meters to tens of meters per pixel) and thus to better understand the nature and age relations of different material units and structures, and to link the Magellan global imagery with *in-situ* lander observations. Higher resolution SAR radar imagery, as well as images taken by future landers on their descent or by balloons, may be the preferred approach.

New landings are needed to obtain geochemical analyses on the Venus surface with the capability to broaden the range of determined elements (and possibly detect isotopes) with higher accuracy than in the past. Landing sites also should target terrains not analyzed in the past (tesserae, mountain belts, steep-sided domes). Long-term observations on the surface involving seismic measurements are crucial for understanding the internal structure and the level of present endogenous activity of Venus. For more distant perspective we should also plan sample return missions to Venus that could lead to significant progress in understanding petrogenesis on that planet and that could determine material ages now only estimated using impact crater counts.

CONCLUSIONS

The above descriptions and discussion show that although many important problems of Venus geology are still not resolved (for example, scarcity of impact craters on the planet surface complicates integration of the observed sequences of material and structural units into any single stratigraphic model) it is already clear that among other terrestrial planets Venus holds a unique place. Its relatively young surface, formed by a number of volcanic and tectonic processes, certainly distinguishes it from the smaller bodies: the Moon, Mercury and Mars. This is obviously the result of the net energy budget: in contrast to Venus the small bodies reached a stage of thick and rigid lithosphere relatively early in their histories [e.g., *Toksoz et al.*, 1978; *Solomon and Head*, 1982; *Basilevsky and Kreslavsky*, 1992; *Basilevsky*, 1994]. More enigmatic is an obvious difference between the geologies of Venus and Earth. Although if one ignores the exogenic processes and resulted materials, both planets show abundant volcanism and tectonism. However, they are different in geodynamic styles. Venus lacks evidence of active plate tectonics in the morphologically surviving part of its history (the last 1 b.y. or less). This may be due to the lack of water on its surface, which is considered by some researchers as implying very dry crust and upper lithosphere. This in turn may be responsible for high stiffness of the planet's interior precluding plate tectonics. The key difference in geochemistry is in abundance on Earth of granitic materials and lack of evidence of its presence (at least in significant amount) on Venus. This may also be due to the lack of plate tectonics (and water) on Venus because the granitic crust of Earth is considered to be a result of the plate-tectonic recycling with important role of water in the fractional melting and crystallization differentiation. All this shows that the principles controlling the key geologic processes and geodynamic styles on the planetary bodies are still not fully understood.

Acknowledgements. Comments from two anonymous reviewers significantly improved this manuscript. ATB has benefited from extensive discussions of venusian geology with James W. Head and Mikhail Ivanov. The contributions of GEM were partially supported by NASA Planetary Geology and Geophysics grant NNG06GG73G.

REFERENCES

Abdrakhimov, A. M. and A. T. Basilevsky, Geology of the Venera and Vega landing-site regions, *Sol. Syst. Res.*, 36, 136–159, 2002.

Addington, E. A., A Stratigraphic study of small volcano clusters on Venus, *Icarus*, 149, 16–36, 2001.

Arvidson, R. E., R. Greeley, M. C. Malin, R. S. Saunders, N. Izenberg, J. J. Plautt, S. D. Wall, and C. M. Weitz, Surface modification of Venus as inferred from Magellan observations of plains, *J. Geophys. Res.*, 97, 13,303–13,317, 1992.

Aubele, J. C., Akkruva small shield plains: definition of a significant regional plains unit on Venus, *Lunar Planet. Sci., XXVII*, 49–50, 1996.

Barath, F. T., A. H. Barrett, J. Copeland, D. C. Jones, and A. E. Lilley, Mariner 2 microwave radiometer experiment and results, *Astron. J.*, 69, 49–58, 1964.

Barsukov V. L., Venusian igneous rocks, *in Venus Geology, Geochemistry and Geophysics. Research Results from the USSR* edited by V. Barsukov, A. T. Basilevsky, V. P. Volkov, and V. N. Zharkov, pp. 165–176, Univ. of Arizona Press, Tucson, 1992.

Barsukov, V. L., Basilevsky, A. T., Burba, G. A., and 27 coauthors, The geology and geomorphology of the Venus surface as revealed by the radar images obtained by Veneras 15 and 16, *J. Geophys. Res.*, 91, B4, D378–D398, 1986.

Basilevsky A. T., Age of rifting and associated volcanism in Atla Regio, Venus. *Geophys. Res. Lett.*, 20, 883–886, 1993.

Basilevsky, A. T., Factors controlling volcanism and tectonism in solar system solid bodies. *Earth, Moon, and Planets*, 67, Nos 1–3, 47–49, doi: 10.1007/BF00613289, 1994.

Basilevsky, A. T., Venera 8 landing site geology revisited, *J. Geophys. Res.*, 102, 9257–926, 1997.

Basilevsky, A. T., and J. W. Head, Global stratigraphy of Venus: Analysis of a random sample of thirty-six test areas, *Earth, Moon, Planets*, 66, 285–336, 1995.

Basilevsky, A. T., and J. W. Head, Evidence for rapid and widespread emplacement of volcanic plains on Venus: Stratigraphic studies in the Baltis Vallis region, *Geophys. Res. Lett.*, 23, 1497–1500, 1996.

Basilevsky, A. T., and J. W. Head, The geologic history of Venus: A stratigraphic view, *J. Geophys. Res.*, 103, 8531–8544, 1998.

Basilevsky, A. T., and J. W. Head, Geologic units on Venus: Evidence for their global correlation, *Planet. Space. Sci.*, 48, 75–111, 2000a.

Basilevsky, A. T. and J. W. Head, Rifts and large volcanoes on Venus: Global assessment of their age relations with regional plains, *J. Geophys. Res.*, 105, 24,583–24,611, 2000b.

Basilevsky, A. T., and J. W. Head, Venus: Analysis of the degree of impact crater deposit degradation and assessment of its use for dating geological units and features, *J. Geophys. Res.*, 107, 5061, doi:10.1029/2001JE001584, 2002a.

Basilevsky, A. T., and J. W. Head, Venus: Timing and rates of geologic activity, *Geology*, 30, 1015–1018, 2002b.

Basilevsky, A. T., and J. W. Head, Impact crater air fall deposits on the surface of Venus: Areal distribution, estimated thickness, recognition in surface panoramas, and implications for provenance of sampled surface materials, *J. Geophys. Res.*, 109, E12003, doi:10.1029/2004JE002307, 2004.

Basilevsky, A. T. and J. W. Head, Impact craters on regional plains on Venus: Age relations with wrinkle ridges and implications for the geological evolution of Venus, *J. Geoph. Res.*, 111, Issue E3, CiteID E03006, 2006.

Basilevsky, A.T., and M. A. Kreslavsky, Volcanism and tectonics on the planets and satellites of the Solar system: Dependence on the body size and period of rotation around the central body, *Sol. Syst. Res.*, 32, 183–190, 1992.

Basilevsky, A. T., G. A. Burba, M. A. Ivanov, N. N. Bobina, V. P. Shashkina, and J. W. Head, Analysis of the geologic structure and compilation of the geologic map of the northern part of planet Venus, *Solar System Res.,* 34, 349–378, 2000.

Basilevsky, A. T., J. W. Head, and A. M. Abdrakhimov, Impact crater air fall deposits on the surface of Venus: Areal distribution, estimated thickness, recognition in surface panoramas, and implications for provenance of sampled surface materials, *J. Geophys. Res.,* 109, E12003, doi:10.1029/2004JE002307, 2004.

Basilevsky, A. T., J. W. Head, and I. V. Setyaeva, Venus: Estimation of age of impact craters on the basis of degree of preservation of associated radar-dark deposits, *Geophys. Res. Lett.,* 30, 1950, doi:10.1029/2003GL017504, 2003.

Basilevsky, A. T., R. O. Kuzmin, O. V. Nikolaeva, A. A. Pronin, V. S. Avduevsky, G. R. Uspensky, Z. P. Cheremukhina, and V. M. Ladygin, The surface of Venus as revealed by the Venera landings: Part II, *Geol. Soc. Am. Bull.,* 96, 137– 144, 1985.

Basilevsky, A. T., A. A. Pronin, L. B. Ronca, V. P. Kryuchkov,. A. L. Sukhanov, M. S. Markov, Styles of tectonic deformations on Venus: Analysis of Venera 15 and 16 data, *J. Geophys. Res.,* 91, B4, D399–D411, 1986.

Bindschadler, D. L., G. Schubert, W. Kaula, Coldspots and hotspots: Global tectonics and mantle dynamics of Venus, *J. Geophys. Res.,* 97, 13,495–13,532, 1992.

Bleamaster, L. S., and V. L. Hansen, Geologic map of the Ovda Regio Quadrangle (V-35), Venus, U.S. Geological Survey. *Geologic Investigation Series Map I-2808,* 2005.

Bondarenko, N. V., M. A. Kreslavsky, and J. W. Head, North-south roughness anisotropy on Venus from the Magellan radar altimeter: Correlation with geology, *J. Geophys. Res.,* 111, E06S12, doi: 10.1029/2005JE002599, 2006.

Brian, A. W., E. R. Stofan, and J. E. Guest, Geologic map of the Taussig Quadrangle (V-39), Venus, U.S. Geological Survey. *Geologic Investigation Series Map I-2813,* 2005.

Bridges, N. T., Submarine analogs to Venusian pancake domes, *Geophys. Res. Lett.,* 22, 2781–2784, 1995.

Bridges, N. T., and G. E. McGill, Geologic map of the Kaiwan Fluctus Quadrangle (V-44), Venus, U.S. Geological Survey. *Geologic Investigation Series Map I-2747,* 2002.

Campbell, B. A., Surface formation rates and impact crater densities on Venus, *J. Geophys. Res.,* 104, 21,951–21,955, 1999.

Campbell, B. A., and P. G. Campbell, Geologic map of the Bell Regio Quadrangle (V-9), Venus, U.S. Geological Survey. *Geologic Investigation Series Map I-2743,* 2002.

Cruikshank, D. P., The development of studies of Venus, *in Venus,* edited by D. M. Hunten, L Colin, T. M. Donahue, and V. I. Moroz, pp. 1–9. Univ. of Ariz. Press, Tucson, 1983.

Crumpler, L. S., J. W. Head, and D. B. Campbell, Orogenic belts on Venus, *Geology,* 14, 1031–1034, 1986.

Crumpler, L. S., J. C. Aubele, D. A. Senske, S. T. Keddie, K. P. Magee, and J. W. Head, Volcanoes and centers of volcanism on Venus, *in Venus II. Geology, Geophysics, Atmosphere, and Solar Wind Environment,* edited by S. W. Bougher, et al., pp. 757–793, Univ. of Arizona Press, Tucson, 1997.

De Witt, M. J., and L. D. Aswal, *Greenstone Belts* (Oxford Monographs on Geology and Geophysics), pp. 840, Oxford University Press, USA, 1997.

Donahue, T. M., and J. B. Pollack, Origin and evolution of the atmosphere of Venus, *in Venus,* edited by D. M. Hunten, L. Collin, T. M. Donahue and V. I. Moroz, pp. 1003–1036, The Univ. of Arizona Press. Tucson.

Donahue, T. M., J. H. Hoffman, R. R. Hodges, and A. J. Watson, Venus was wet: A measurement of the ratio of D to H, *Science,* 216, 630–633, 1982.

Fink, J. H., and R. W. Griffiths, Morphology, eruption rates, and rheology of lava domes—Insights from laboratory models, *J. Geophys. Res.,* 103, 537–545, 1998.

Florensky, C. P., L. B. Ronca, A. T. Basilevsky, G. A. Burba, O. V. Nikolaeva, A. A. Pronin, A. M. Trakhtmann, V. P. Volkov, and V. V. Zasetsky, The surface of Venus as revealed by Soviet Venera 9 and 10, *Geol. Soc. Am. Bull.,* 88, 1537–1545, 1977.

Ford, J. P., J. J. Plaut, C. M. Weitz, T. G. Farr, D. A. Senske, E. R. Stofan, G. Michaels, T. J. Parker, *Guide to Magellan Image Interpretation,* pp. 148, JPL Publication 93–24, 1993.

Garvin, J. B., The global budget of impact-derived sediments on Venus. *Earth, Moon and Planets,* 50–51, 175–190, 1990.

Gilmore, M. S., and J. W. Head, Sequential deformation of plains at the margins of Alpha Regio, Venus: Implications for tessera formation, *Meteoritics & Planet. Sci.,* 35, 667–687, 2000.

Goldstein, R. M., Preliminary Venus radar results, *J. Res. Natl. Bur. Stand,* Sect. D., 69D, 1623–1625, 1965.

Goldstein, R. M., and H. C. Rumsey, A radar image of Venus, *Icarus,* 17, 699–703, 1972.

Greeley, R., K. C. Bender, R. S. Saunders, G. Schubert, and R. A. Simpson, Aeolian processes and features of Venus, *in Venus II. Geology, Geophysics, Atmosphere, and Solar Wind Environment,* edited by S. W. Bougher, et al., pp. 547–589, Univ. of Arizona Press, Tucson, 1997.

Grinspoon, D. H., and M. A. Bullock, Did Venus experience one great transition or two? *Bull. Amer. Astron. Soc.,* 35. 44–03, 2003.

Guest, J. E., and E. R. Stofan, A new view of the stratigraphic history of Venus, *Icarus,* 139, 55–66, 1999.

Guest, J. E., M. H. Bulmer, J. Aubele, K. Beratan, R. Greeley, J. W. Head, G. Michaels, C. Weitz, and C. Wiles, Small volcanic edifices and volcanism in the plains of Venus, *J. Geophys. Res.,* 97, 15,949–15,966, 2002.

Hansen, V. L., Geologic mapping of tectonic planets: *Earth. Planet. Sci. Lett.,* 176, 527–542, 2000.

Hansen, V. L., Venus's shield terrain, *Bull. Geol. Soc. Amer.,* 117, 808–822; doi: 10.1130/B256060.1, 2005.

Hansen, V. L., and H.R. DeShon, Geologic map of the Diana Chasma Quadrangle (V-37), Venus, U.S. Geological Survey. *Geologic Investigation Series Map I-2752,* 2002.

Hansen, V. L., J. J. Willis, W. B. Banerdt, Tectonic overview and synthesis, *in Venus II. Geology, Geophysics, Atmosphere, and Solar Wind Environment,* edited by S. W. Bougher, et al., pp. 797–844, Univ. of Arizona Press, Tucson, 1997.

Hansen, V. L., R. J. Phillips, J. J. Willis, R. R. Ghent, Structures in tessera terrain, Venus: Issues and answers, *J. Geophys. Res.*, 105, 4135–4152, 2000.

Hauck, S. A., R. J. Phillips, M. H. Price, Venus: Crater distribution and plains resurfacing models, *J. Geophys. Res.*, 103, 13,635–13,642, 1998.

Head, J. W., L. S. Crumpler, J. C. Aubele, J. E. Guest, R. S. Saunders, Venus volcanism: Classification of volcanic features and structures, associations, and global distribution from Magellan data, *J. Geophys. Res.*, 97, 13,153–13,197, 1992.

Herrick, R. R., V. L. Sharpton, M. C. Malin, S. N. Lyons, and K. Feely, Morphology and morphometry of impact craters, *in Venus II. Geology, Geophysics, Atmosphere, and Solar Wind Environment*, edited by S. W. Bougher, et al., pp. 1015–1046, Univ. of Arizona Press, Tucson, 1997.

Ivanov, B. A., A. T. Basilevski, V. P. Kriuchkov, I. M. Chernaia, Impact craters of Venus—Analysis of Venera 15 and 16 data, *J. Geophys. Res.*, 91, D413–D430, 1986.

Ivanov, B. A., I. V. Nemchinov, V. A. Svetsov, A. A. Provalov, and V. M. Khazins, Impact cratering on Venus: Physical and mechanical midels, *J. Geophys. Res.*, 97, 16,167-16,181, 1992.

Ivanov, M. A. and J. W. Head, Tessera terrain on Venus: A survey of the global distribution, characteristics, and relation to surrounding units from Magellan data, *J. Geophys. Res.*, 101, 14861–14908, 1996.

Ivanov, M. A. and J. W. Head, Geology of Venus: Mapping of a global geotraverse at 30°N latitude, *J. Geophys. Res.*, 106, 17515–17566, 2001.

Ivanov, M. A., and J. W. Head, Geologic map of the Atalanta Planitia Quadrangle (V-4), Venus, U.S. Geological Survey. *Geologic Investigation Series Map I-2792*, 2004a.

Ivanov, M. A., and J. W. Head Stratigraphy of small shield volcanoes on Venus: Criteria for determining stratigraphic relationships and assessment of relative age and temporal abundance, *J. Geophys. Res.*, 109, CiteID E10001, doi: 10.1029/2004JE002252, 2004b

Kargel, J. S., R. L. Kirk, B. Fegley, A. H.Treiman, Carbonatite-sulfate volcanism on Venus, *Icarus*, 112, 219–252, 1994.

Kargel, J. S., G. Komatsu, V. R. Baker, R. G. Strom The volcanology of Venera and VEGA landing sites and the geochemistry of Venus, *Icarus*, 103, 253–275, 1993.

Kasting, J. F., J. B. Pollack, and T. P. Ackerman, Response of Earth's atmosphere to increases in solar flux and implications for loss of water from Venus, *Icarus*, 57, 335–355, 1984.

Kasting, J. F., Runaway and moist greenhouse atmospheres and the evolution of Earth and Venus, *Icarus*, 74, 472–494, 1988.

Klose, K. B., J. A. Wood, A. Hashimoto, Mineral equilibria and the high radar reflectivity of Venus mountaintops, *J. Geophys. Res.*, 97, 16,353–16,369, 1992.

Kotelnikov V. A., V. R. Yashchenko, A. F. Zolotov and 14 others eds. *Atlas of Venus surface*, pp 328, The Main Department of Geodesy and Cartography at the Councel of Ministers of the USSR, Moscow, 1989 (in Russian).

Kreslavsky, M. A., The duration of Venusian plains formation derived from data on impact craters, *Sol. Syst. Res.*, 30, 466–473, 1996.

Masursky H., E. Eliason, P. G. Ford, G. E McGill., G. H. Pettengil, G. G Schaber., G Schubert. Pioneer Venus radar results—Geology from images and altimetry, *J. Geophys. Res.*, 85, 8232–8260, 1980.

Magee, K. P., and J. W. Head, Large flow fields on Venus: Implications for plumes, rift associations, and resurfacing, *in Mantle Plumes: Their Identification Through Time*, edited by R. E. Ernst and K. L. Buchan, *Spec. Pap. Geol. Soc. Am.*, 352, 81–101, 2001.

McGill, G. E., Geology and geophysics,. *Nature*, 296, 14–15, 1982.

McGill, G. E., Wrinkle ridges, stress domains, and kinematics of Venusian plains, *Geophys. Res. Lett.*, 20, 2407–2410, 1993.

McGill, G. E., Geologic map of the Sapho Patera Quadrangle (V-20), Venus, U.S. Geological Survey. *Geologic Investigation Series Map I-2637*, 2000.

McGill, G. E., Tectonic and stratigraphic implications of the relative ages of Venusian plains and wrinkle ridges, *Icarus*, 172, 603–612, 2004.

McGill G. E., S. L Steenstrup, C. Barton, and P. G. Ford, Continental rifting and the origin of Beta Regio, Venus, *Geophys. Res. Lett.*, 8, 737–740, 1981.

McKinnon, W. B., K. J. Zahnle, B. A. Ivanov, and H. J. Melosh, Cratering on Venus: models and observations, *in Venus II. Geology, Geophysics, Atmosphere, and Solar Wind Environment*, edited by S. W. Bougher, et al., pp. 969–1014, Univ. of Arizona Press, Tucson, 1997.

Moore, H. J., J. J. Plaut, P. M. Schenk, and J. W. Head, An unusual volcano on Venus, *J. Geophys. Res.*, 97, 13,479-13,493, 1992.

Nikolaeva, O. V., M. A. Ivanov, and V. K. Borozdin, Evidence on the crustal dichotomy, *in Venus Geology, Geochemistry and Geophysics. Research Results from the USSR* edited by V. L. Barsukov, A. T. Basilevsky, V. P. Volkov, and V. N. Zharkov, pp. 129–139, Univ. of Arizona Press, Tucson, 1992.

Pavri, B., J. W. Head, B. Klose, L. Wilson, Steep-sided domes on Venus: Characteristics, geologic setting, and eruption conditions from Magellan data, *J. Geophys. Res.*, 97, 13,479–13,493, 1992.

Pettengill, G. H., B. A. Campbell, D. B. Campbell, and R. A. Simpson, Surface scattering and dielectric properties, *in Venus II: Geology, Geophysics, Atmosphere, and Solar Wind Environment*, edited by S. W. Bougher, D. M. Hunten and R. J. Phillips, pp. 527–546, Univ. of Arizona Press, Tucson, 1997.

Phillips, R. J., C. L. Johnson, S. J. Mackwell, P. Morgan, D. T. Sandwell, and M. Zuber, Lithospheric mechanisms and dynamics of Venus, *in: Venus II: Geology, Geophysics, Atmosphere, and Solar Wind Environment*, edited by S. W. Bougher, D. M. Hunten and R. J. Phillips, pp. 1163–1204, Univ. of Arizona Press, Tucson, 1997.

Phillips, R. J., R. F. Raubertas, R. E. Arvidson, I. C. Sarkar, R. R. Herrick, N. Izenberg, and R. E. Grimm, Impact craters and Venus resurfacing history, *J. Geophys. Res.*, 97, 15,923–15,948, 1992.

Price, M., *Tectonic and volcanic map of Venus,* Dept. of Geol. Sci., Princeton Univ., Princeton, N. J. 1995.

Pronin, A. A., The Lakshmi phenomenon, *in Venus Geology, Geochemistry and Geophysics. Research Results from the USSR,* edited by V. L. Barsukov, A. T. Basilevsky, V. P. Volkov, and V. N. Zharkov, pp. 68–81, Univ. of Arizona Press, Tucson, 1992.

Rosenberg, E., and G. E. McGill, Geologic map of the Pandrosos Dorsa Quadrangle (V-5), Venus, U.S. Geological Survey. *Geologic Investigation Series Map I-2721,* 2001.

Saunders, R. S., A. J. Spear, P. C. Allin and 24 coauthors, Magellan mission summary, *J. Geophys. Res., 97,* 13,067–13,090, 1992.

Schaber, G. G., Venus—Limited extension and volcanism along zones of lithospheric weakness. *Geophys. Res. Lett.,* 9, 499–502, 1982.

Schaber, G. G., R. G. Strom, H. J. Moore, and 7 co-authors, Geology and distribution of impact craters on Venus: What are they telling us? *J. Geophys. Res., 97,* 13,257–13,301, 1992.

Schaber, G. G., R. L. Kirk, R. G. Strom, Data base of impact craters on Venus based on analysis of Magellan radar images and altimetry data, *USGS Open-File Report 98–104,* 19 p. 1998.

Schaefer, L., and B. Fegley, Heavy metal frost on Venus, *Icarus,* 168, 215–219, 2004.

Schultz, P. H., Atmospheric effects on ejecta emplacement and crater formation on Venus, *J. Geophys. Res., 97,* 16,183–16,248, 1992.

Solomon, S. C., and J. W. Head, Mechanisms for lithospheric heat transport on Venus: Implications for tectonic style and volcanism, *J. Geophys. Res., 87,* 9236–9246, 1982.

Solomon, S. C., S. C. Smrekar, D. L. Bindschadler, and 8 coauthors, Venus tectonics: An overview of Magellan obsrvations, *J. Geophys. Res., 97,* 13,199–13,255, 1992.

Stofan E. R., V. E. Hamilton, D. M. Janes, and S. E. Smrekar, Coronae of Venus: Morphology and origin, *in Venus II. Geology, Geophysics, Atmosphere, and Solar Wind Environment,* edited by S. W. Bougher, et al., pp. 931–965, Univ. of Arizona Press, Tucson, 1997.

Stofan, E., R., A. W. Brian, and J. E. Guest, Resurfacing styles and rates on Venus: assessment of 18 venusian quadrangles, *Icarus,* 173, 312–321, 2005.

Strom, R. G., G. G. Schaber, and D. D. Dawson, The global resurfacing of Venus, *J. Geophys. Res.,* 99, 10,899–10,926, 1994.

Sukhanov, A. L., Tesserae, *in Venus Geology, Geochemistry and Geophysics. Research Results from the USSR,* edited by V. Barsukov, A. T. Basilevsky, V. P. Volkov, and V. N. Zharkov, pp. 82–95, Univ. of Arizona Press, Tucson, 1992.

Surkov, Y. A., *Exploration of Terrestrial Planets from Spacecraft: Instrumentation, Investigation, Interpretation,* 2nd ed., 446 pp., John Wiley, Hoboken, N. J., 1997.

Tanaka, K.L. (compiler), *Venus geologic mappers' handbook,* second edition, U.S. Geological Survey Open-File Report 94-438, 50 pp., 1994

Toksoz, M. N., A. T. Hsui, and D. H. Johnson, Thermal evolution of the terrestrial planets, *The Moon and the Planets,* 18, 281–320, 1978.

Weitz C. M., and A. T. Basilevsky, Magellan observations of the Venera and Vega landing site regions. *J. Geophys. Res.,* 98, 17069–17097, 1993.

Wood, J. A., Rock weathering on the surface of Venus, in *Venus II. Geology, Geophysics, Atmosphere, and Solar Wind Environment,* edited by S. W. Bougher, et al., pp. 637–664, Univ. of Arizona Press, Tucson, 1997.

Tectonic and Thermal Evolution of Venus and the Role of Volatiles: Implications for Understanding the Terrestrial Planets

Suzanne E. Smrekar[1], Linda Elkins-Tanton[2], Johannes J. Leitner[3],
Adrian Lenardic[4], Steve Mackwell[5], Louis Moresi[6],
Christophe Sotin[7], and Ellen R. Stofan[8]

Venus is similar to Earth in size and bulk composition. The dramatic differences between the two planets indicate that planetary size alone does not control geologic evolution. Earth's geology is dominated by plate tectonics, or active lid convection. The crater retention age of Venus demonstrates that the planet has been very geologically active in the last b.y., but there is no evidence of active plate tectonics. Instead the surface is dominated by diverse volcanism and tectonism, and highlands interpreted as mantle upwellings and downwellings. Venus' surface is much hotter and drier than Earth's, suggesting that the interior may be drier as well. The volatile content has a major effect on both the rheology of the interior and predicted melt products. The low water content of the interior may create a stronger lithosphere and preclude asthenosphere formation, leading to the development of a stagnant rather than an active lid convective regime. A stable, stagnant lid fosters development of phase changes and other density anomalies that can lead to delamination and associated opportunities for melting. Additionally, the thicker boundary layer formed in a stagnant lid regime may increase the number of expected plumes. Mars also is in a stagnant lid regime, probably due to early heat loss rather than volatile loss. Here we discuss what is known about interior volatile content and explore the implications of a dry interior for volcanism, tectonism, and interior convection as a working hypothesis to explain the profound differences in the evolution of Venus, Earth, and Mars.

[1] Jet Propulsion Laboratory, California Institute of Technology, Pasadena, California, USA.

[2] Massachusetts Institute of Technology, Cambridge, Massachusetts, USA.

[3] University of Vienna, Vienna, Austria.

[4] Rice University, Houston, Texas, USA.

[5] Lunar and Planetary Institute, Houston, Texas, USA.

[6] Monash University, Victoria, Australia.

[7] University of Nante, Nante, France.

[8] Proxemy Research, Rectortown, Virginia, USA.

Exploring Venus as a Terrestrial Planet
Geophysical Monograph Series 176
Copyright 2007 by the American Geophysical Union.
10.1029/176GM05

1. INTRODUCTION

The present data, principally from the imaging radar and altimeter on the Magellan mission, suggest that there is no plate tectonics on Venus (Solomon et al., 1992; Hansen and Phillips, 1993), thus inhibiting the recycling of the volatiles from the surface to the mantle. By comparison with Earth, the fate of water on Venus is interesting not only for understanding the atmospheric and climate evolution and possible implications for life, but also for investigating the consequences of the lack of water on tectonics and mantle dynamics. As we search for terrestrial-like planets in other star systems, it is essential to understand what causes planets to diverge down different evolutionary paths, particularly with respect to volatiles. In this paper we examine the role of water (and other volatiles) in the overall evolution of Venus and compare its evolution to that of Earth and Mars.

Venus is often called Earth's twin due to its similarity in size, bulk density, and location in the solar system. However, greenhouses gasses in the atmosphere result in a surface temperature of ~470°C and an atmospheric pressure of 90 bars. No surface features carry the morphology of past water erosion. Presently Venus' atmosphere contains only 30 ppm of water (Donahue et al., 1997). Venus' D/H ratio of 150 times that of Earth indicates that the planet lost significant water relative to its initial atmosphere (Donahue et al., 1997; Grinspoon, 1993). Indirect evidence also suggests that its interior may be drier than that of Earth. For example, analysis of the gravity and topography data imply that there is no low viscosity asthenosphere (Phillips, 1986; Kiefer and Hager, 1991; Smrekar and Phillips, 1991), which is generally believed to be a result of water weakening (e.g. Karato and Jung,1998). Clearly oceans and water-driven erosion dominate the surface of Earth, and volatiles are recycled into the interior via subduction. Mars had abundant water early in its history, as evidenced by both water erosion (Carr, 1995) and rocks deposited or altered in the presence of water (Bibring et al., 2005; Murchie et al., 2007). Today Mars has polar caps containing water ice and the atmosphere contains about 0.03% water (Owens, 1992), ten times that of Venus (Grinspoon, 1993). Small-scale (~ 1 km) Martian gullies possibly created by water seepage may have formed less than a million years ago (Malin and Edgett, 2000). Malin et al.'s (2006) dramatic discovery showed that at least 2 gullies have been active within the last 5–7 years. However it is possible that these recent gullies formed via dry flow (McEwen et al. 2007). Latitude dependent mantling deposits (Head et al., 2003) may also be a result of water or other volatile processes. But whether or not these processes reflect deeper reservoirs of water or represent interaction with atmospheric water alone is not clear. Analyses of different Martian meteorites provide disparate results regarding the amount of water in the interior of Mars (e.g., Longhi, 1991;McSween et al., 2001). In particular, some analyses of Martian meteorites indicate water contents of up to 1–2% in magma source regions (Johnson et al. 1991; McSween and Harvey,1993).

An early theory suggested that size was the dominant factor controlling the geologic evolution of the terrestrial planets (Kaula, 1975). The larger the planet, the longer it takes to cool due to the greater surface area to volume ratio. To first order, this theory is valid for the terrestrial planets, with the surface of Mercury and the Moon being the oldest, Mars the next oldest, and Earth and Venus being similarly youthful. Yet the geologic history of each planet is distinctly different. Despite the similarities between Venus and Earth,

Venus lacks the organized system of plate tectonics that controls geologic processes on Earth. Recent modeling that takes into account the effects of the volatiles, in addition to temperature, on rheology shows that the mechanism of heat loss to the surface is strongly controlled by the relative strength of the lithosphere and mantle convective stresses (e.g., Solomatov and Moresi, 1997, 2000). Although duration of geologic activity appears to be a function of size, the very different evolutionary paths that Venus and Earth have followed may be due to the history of volatiles on the two planets. All of the terrestrial planets, only Earth has unquestionably retained abundant water in both its atmosphere and interior.

In this chapter we focus on the role of water and other volatiles in the tectonic and thermal evolution of terrestrial planets, beginning with an overview of the tectonic history of the Mars, Earth, and Venus. We do not discuss Mercury since it is significantly smaller than the other terrestrial planets and does not appear to have developed prolonged tectonic activity. We then outline what is known about the volatile content of the interior of the three planets. We review recent advances in the understanding of the implications of volatile content on rheology, interior convection, and geochemistry. Together, these effects have a profound influence on the overall planetary evolution, rivaling the effects of temperature following the earliest stage of evolution. Finally we describe new data sets that would provide a step forward in constraining our knowledge of interior processes on Venus.

2. BACKGROUND: TECTONIC HISTORY OF EARTH, MARS, AND VENUS

The topography of Earth and Venus reflects their gross differences in tectonic and convective style (Plate 1). The same may be true for Mars (Plate 1), although the origin of the dichotomy between the northern plains and southern highlands is debated, and large basins dating from early bombardment are also prevalent (Plate 1). In this section we briefly describe the major tectonic features of each planet, and theories for their origin.

Earth is the only body in the solar system known to have plate tectonics, with geologic deformation and volcanism largely driven by this process. New oceanic crust forms in areas where mantle material rises in long, linear sections, or spreading centers (Plate 1), with the decompression melting of mantle material generating basaltic melt. The melt, melt residuum, and cooled mantle forms new oceanic lithosphere that spreads apart to form ocean basins. Oceanic crust eventually subducts and is recycled back into the mantle, carrying volatiles back into the interior. The ocean floor is lower than the continents due to the greater density of basalt relative

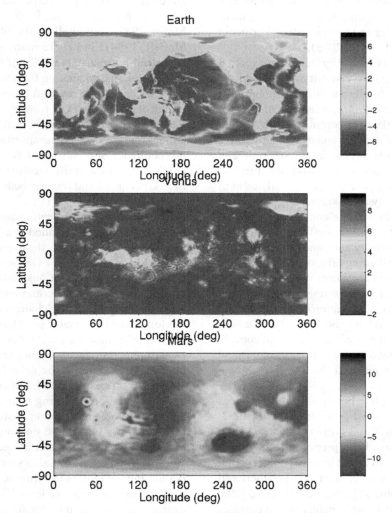

Plate 1. Topographic elevation (km) for Earth, Venus and Mars. On Earth, long, semi-linear oceanic ridges and continental mountain belts mark the location of plate boundaries. Venus has several distinct highland regions, but no linear zones characteristic of plate tectonics. The topography of Mars is dominated by the northern lowlands – southern highlands dichotomy boundary and the huge Tharsis volcanic rise.

to the silicious continental crust and the on average thinner crust (Plate 1). The continental crust has accumulated over much of Earth's history, forming as the subducted basaltic crust has been remelted. The formation of such a secondary crust may require water to produce the high silica content (Campbell and Taylor, 1983). The continental crust is comparatively stable, but may locally become gravitationally unstable and delaminate, introducing volatiles into the mantle (Kay and Kay, 1993; Ducea and Saleeby, 1998; Farmer et al., 2002; Elkins-Tanton and Grove, 2003).

The motion of the plates is fueled by mantle convection (Schubert et al., 2001). The low viscosity asthenosphere may be required to facilitate plate motion (e.g. *Cathles, 1975; Hager and Richards, 1989; Mitrovica, 1996;* Tackley 2000a,b; Richards et al., 2001). The origin of Earth's asthenosphere beneath oceanic lithosphere and tectonically active continental lithosphere is not agreed upon. Stixrude and Lithgow-Bertelloni (2004) argue that sub-solidus mineralogical responses to increasing pressure may be sufficient to explain the seismic low velocity zone without resort to partial melting or near-saturation by volatile components. Karato and Jung (1998) suggest that sub-solidus effects of water may also explain both the low viscosity zone and the seismic low velocity zone, due to the weakening effect of water on bond strengths in olivine and other constituent minerals.

About 90% of Earth's heat is lost through subduction of the plates back into the interior, formation of new crust, and conduction though the lithosphere (Sclater, 1980; Davies, 1998). Mantle plumes are also part of Earth's overall convective pattern, forming 'hotspots' such as the Hawaiian island chain. Such plumes account for up to 10% of Earth's heat loss (Davies, 1992a).

There is no evidence for plate boundaries in the topography and radar data for Venus (e.g. Solomon et al., 1992; Hansen and Phillips, 1993). The 3 large rift zones (Parga, Hecate, and Dali/Diana) have been proposed to be a possible exception. Some large rift zones have been interpreted as subduction zones (McKenzie et al., 1992; Schubert and Sandwell, 1995). However others (Hansen and Phillips, 1993; Hamilton and Stofan, 1996) argued against this hypothesis for a number of reasons, including the switch in subduction direction along the rift, the rift branches, the abundance of volcanoes through out the rift zones, and the continuity of some features across the rift. Despite the absence of terrestrial-style plate tectonics, there is abundant geologic activity, and widespread volcanism and deformation (see Plate 1). There are 5 highly deformed plateau highlands and 9 large volcanic rises (Solomon et al., 1992; Phillips and Hansen, 1994; Stofan and Smrekar, 2005) that range from 1000 km to 3000 km in diameter. The plateaus are believed to form via either mantle upwelling (Bindschadler

and Head, 1991; Kidder and Phillips, 1996; Phillips and Hansen, 1998; Hansen et al., 1999) or downwelling (Grimm and Phillips, 1991; Kiefer and Hager, 1991; Bindschadler et al., 1992; Lenardic et al. 1995). Ishtar Terra is the largest plateau highland, and is unique in that it is surrounded on three sides by wide expanses of linear mountain belts. The large volcanic rises are interpreted as sites of mantle upwellings (e.g., Masursky et al., 1980; McGill, 1983; Phillips and Malin, 1983; Campbell et al., 1984; Senske et al., 1992, McGill, 1994), and share many of the characteristics of terrestrial hotspots. Over 85% of the surface is made up of volcanic plains, emanating from sources that include volcanic edifices, coronae and rift fractures (Banerdt et al., 1997; Stofan et al., 2005). Data from the Venera and Vega landers indicated a basaltic composition at three of the landing sites, although more evolved compositions are suggested by the morphologies of a number of volcanic features (e.g., Head et al., 1992). Low strain tectonic features such as wrinkle ridges, polygons, and fracture belts are common in the plains (e.g., Banerdt et al., 1997).

The average crater retention age of the surface is on the order of 700 Ma. (McKinnon et al., 1997), but given uncertainties in impactor flux, the average age could be as little as 300 Ma or as much as 1 Ga. Studies of the crater population on Venus are contentious. Initial studies concluded that only a small number of craters were modified by tectonic and volcanic processes (Schaber et al., 1992). The apparent randomness of the crater population, along with the large number of unmodified craters pointed towards a model of catastrophic resurfacing event around 700 my ago, followed by a major decline in the resurfacing rate (Strom et al., 1994; Turcotte et al., 1999). Other, non-random models of crater population, resulting from a wide spread in surface age, fit the observed crater distribution at least as well (Hauck et al., 1998), and on-going small-scale resurfacing is consistent with the crater population (Phillips et al., 1992). Campbell (1999) also showed that attempts to use populations of craters to estimate crater ages on a specific geologic unit is not possible, unless one assumes that every example of a given unit forms rapidly at the same time. With further in depth analysis, recent studies of the crater population have found a much larger number of modified craters (Herrick, 2006; Herrick and Forsberg-Taylor, 2003 Bond and Warner, 2006). Given the small number of craters on Venus (~1000), very large regions are needed to give a valid age date. Within this constraint, there do not appear to be areas of significantly different age (Campbell, 1999). However bringing in addition information, in the form of modification of impact crater halos indicts that the surface may be divided into 3 general areas with differing ages (Phillips and Izenberg, 1995).

The crater population on Venus indicates that the planet experienced major resurfacing in the last By, but the rate at which this resurfacing event occurred is a subject of much debate (e.g., Basilevsky et al., 1997; Stofan et al., 2005). Numerous models have been proposed to explain global resurfacing followed by a lower rate of geologic activity. One hypothesis is that the rate of geologic activity has decreased over time due to cooling of the lithosphere and mantle. As the lithosphere cools, it could become unstable and founder, allowing Venus to resurface. Parmentier and Hess (1992) proposed that foundering is caused by an increase in both the chemical and thermal density of the lithosphere and a thick layer of mantle residuum. Turcotte et al. (1999) proposed that episodic subduction cannot fully account for Venus' heat loss, and that vigorous tectonic activity is needed as well. Others have suggested that geologic activity decreased gradually over time (Solomon, 1993), due to an increase in lithospheric viscosity resulting from cooling (Grimm, 1994; Arkani-Hamed, 1993, 1994). Cooling of the mantle has also been proposed to increase lithospheric buoyancy and inhibit plate tectonics (Phillips and Malin, 1983; Herrick, 1994).

Perhaps the leading hypothesis to understand Venus' geologic evolution is that the lack of water on Venus resulted in a sufficiently strong lithosphere to allow a stagnant lid convective regime to develop, in which the convective stresses are too weak to overcome the yield strength of the lithosphere. Numerous studies have explored the model of stagnant lid convection for Venus and other planets (Solomatov and Moresi, 1996, 1997, 2000; Reese et al., 1999, Moresi and Solomatov, 2000). These convection models simulate the effect of the lithosphere on the style of convection. When the yield strength of the lithosphere is sufficiently large, neither surface plates nor subduction can occur (Richards et al., 2001). Phillips and Hansen (1998) discuss the idea of the transition from thin (mobile) lid to stagnant lid because the convective stress imparted to the lithosphere drops below its yield stress. However, one advantage of the hypothesis that the transition to a stagnant lid regime is caused by a lack of water in the lithosphere or the mantle or both is that it offers an explanation of how Venus and Earth developed different tectonic styles. Further, the transition to a stagnant lid explains the reduced rate of geologic activity following resurfacing. One consequence of a stagnant lid regime is that the strong, thick lithosphere acts to insolate the mantle, raising the mantle temperature. Below we explore both the possible causes and the implications of a stagnant lid regime in detail. In particular we explore the effects of a stagnant lid and low volatiles on tectonic, volcanic and convective style.

With a diameter roughly half that of Earth and Venus, it is believed that Mars lost its interior heat and engine for major geologic activity relatively quickly. However, the recent determination of its k_2 Love number (Yoder et al., 2003) suggests that the Martian core is liquid and that the lower mantle temperature must be larger than the melting point of iron alloys. It is thus possible that the Martian mantle is still convecting with a thick stagnant lid that limits the cooling rate of its interior. Isotopic evidence suggests that the majority of the crust on Mars formed within 25–50 My of solar system formation (Solomon et al., 2005 and references therein). The core formed even earlier, within ~25 My (Klein et al., 2004; Foley, et al., 2005). Recent investigations of the phase relations within the Martian mantle suggest that the early atmosphere formed when accreted water was degassed during formation of the crust following accretion of hydrated minerals (Medard and Grove, 2006). The magnetized crust on Mars (Connerney et al., 1999) indicates that Mars once had an active dynamo, possibly due to a plate tectonic phase (Nimmo and Stevenson, 2000) or core formation (Stevenson, 2001; Williams and Nimmo, 2004).

The surface is dominated by two main features that give Mars its unique geologic signature: the global dichotomy and the Tharsis rise (Figure 1). The dichotomy is a topographic elevation change of 1–6 km that separates the smooth northern lowlands from the heavily cratered southern highlands. Approximately a third of the boundary has a steep topographic slope, with offsets of 4–6 km (Frey et al., 1998; 2002). Another third has a gentler slope. The massive Tharsis volcanic rise straddles the final third. Numerous hypotheses for the formation of the dichotomy have been proposed, including degree one mantle convection (Wise et al., 1979; Breuer et al., 1997, 1998; Zhong and Zuber, 2001), plate tectonics (Sleep, 1994; Lenardic et al., 2004), and impact basin formation (Wilhelms and Squyres, 1984; Frey and Schultz, 1988). The recent impact crater age dating of the northern lowlands indicates that the basement age is similar to that of the highlands, despite the younger volcanic or sedimentary fill covering the lowlands (Frey, 2006; Watters et al., 2006). The Tharsis rise is believed to have formed above a very long-lived mantle upwelling (e.g. Harder and Christensen, 1996; Kiefer et al., 1996; Roberts and Zhong, 2004; Ke and Solomatov, 2006). An alternative hypothesis is that a major impact initiated the Tharsis volcanism (Reese et al., 2004). Although tectonic and geologic patterns indicate that the Tharsis rise was largely emplaced by the early Noachian (~4 Ga) (Anderson et al., 2001; Jakosky and Phillips, 2001; Phillips et al., 2001a), localized volcanism on the rise occurred within at least the last 100 mg (Tanaka et al., 1986), and perhaps within the last 10s of my (Hartman et al., 1999).

While an early phase of plate tectonics cannot be ruled out for Mars or Venus, the small size of Mars and perhaps the low volatile content of Venus' lithosphere prohibited an ongoing

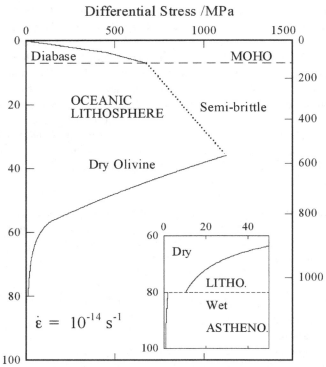

Figure 1. Strength envelope plot showing strength of rock as a function of depth for the oceanic lithosphere on Earth, assuming a crustal thickness of 6 km, thermal profile from Turcotte and Schubert 1982), and a strain rate of 10^{-14} s^{-1}. Solid lines show the data for dry behavior, following Byerlee's law at upper crustal conditions, and flow laws for olivine-rich rocks deforming in the power-law breakdown (Evans and Goetze 1979) and dislocation creep (Chopra and Paterson 1984) regimes. The inset shows the change in behavior in transitioning from the dry lithosphere to the wet asthenosphere (Hirth and Kohlstedt 1996).

plate tectonics regime. On Earth, the effect of water on the rheology of the lithosphere and mantle enables plate tectonics by reducing the strength of the plates and creating a low viscosity zone (Richards et al. 2001, and references therein). On Mars, the smaller diameter and more effective cooling likely resulted in the transition to a stagnant lid (Solomatov and Moresi, 2005, 2007), although as discussed below loss of volatiles might have contributed. For Venus, the leading hypothesis for triggering a transition from active to stagnant lid regimes is the loss of volatiles to increase the strength of the lithosphere and reduce the vigor of convection. For Venus, our understanding of its geologic history is hindered by the uniformly young age of its surface and uncertainties over rates of geologic activity. On Mars, the early tectonic history is in many places obscured by the scars of early bombardment and subsequent erosion and deposition. Even for Earth our knowledge of geologic history remains incomplete.

Whether plate tectonic operated in Earth's Archean remains a debated issue (Benn et al. 2006 and references therein). Uncertainties in terms of the complete geologic histories of Earth, Mars, and Venus will necessarily lead to speculations when it comes to comparisons of tectonic evolution. What is more certain is that, at present, the tectonic mode of these planets differs as does the volatile content of their atmospheres and, most likely, interiors. Our goal, in this review, is to explore the hypothesis that the effects of volatiles on lithospheric strength and convective mantle stress may be a fundamentally linked to the differences in the present day, and geologically recent, mode of tectonics on Earth and Venus. Although we will focus on changes in lithospheric stress, we acknowledge that other factors, e.g. lithospheric buoyancy [Davies 1992b], could also lead to transitions in tectonic style.

3. EFFECTS OF VOLATILES ON RHEOLOGY

3.1. Earth

It has been known for decades that the presence of water has a major effect on the mechanical behavior of rocks deforming in either the brittle or ductile regimes (see Kirby and Kronenberg 1987, Evans and Kohlstedt 1995, and Kohlstedt et al. 1995 for reviews). In particular the mechanical properties of silicate rocks are strongly affected by the presence of water, either as a free fluid phase, or dissolved as a trace impurity in individual minerals or grain boundaries. At the shallowest crustal levels on Earth, temperatures are too low for plastic deformation to occur, and most of the deformation is localized along fault structures. Within these zones, water acts to push open fractures and faults, so that slip is more easily activated and friction is reduced (e.g., Byerlee 1978, Evans and Kohlstedt 1995). The reaction of water moving along faults within basaltic or ultramafic rocks may also produce hydrous minerals locally, such as talc and serpentine that, because of their platy crystal structures, allow easier slip than the country rock (e.g., Escartin and Hirth 1997).

At greater depths, and higher temperatures, fully plastic processes will dissipate applied stresses at levels that are lower than required for brittle failure by fracture or frictional sliding. Within the fully plastic regime, water also acts to weaken the rocks, mostly as a dissolved phase along grain boundaries, where it facilitates grain-boundary diffusion and the motion of grain-boundary dislocations, or within individual mineral grains, where it promotes lattice diffusion, enhancing dislocation climb, which is controlled by atomic diffusion, and hence weakening the rock. Concentrations of water as low as a few tens of atomic parts per million dissolved in olivine, for example, have been shown to reduce the strength of the mineral by a factor of 2–3 in the laboratory, apparently due to facilitation of the dislocation climb process (Mackwell et al. 1985, Hirth and Kohlstedt 2003). While olivine is the best-quantified example of water-weakening of minerals deforming by dislocation creep, intracrystalline water weakening effects have been demonstrated in most silicate minerals of importance in the interior of Earth and other terrestrial planets. Between the fully brittle and fully plastic deformation regions, there is an extensive depth range where both plastic and brittle processes contribute to the deformation (the semi-brittle regime of Kohlstedt et al. (1995)). The detailed rheologies for this transitional regime are complex and difficult to model but will be affected by the presence of water.

A simplified illustration of the distribution of strength through the oceanic lithosphere can be provided by a strength envelope diagram (Goetz and Evans, 1979; Brace and Kohlstedt, 1980), where the maximum compressive stress that can be supported by the rock is plotted as a function of depth in Earth for a uniform strain rate and an appropriate temperature-depth profile (e.g., Goetze and Evans 1979, Brace and Kohlstedt 1980, Kohlstedt et al. 1995). An example of such a diagram is given in Figure 1, where a basaltic crust around 6 km thick overlays the olivine-dominated upper mantle. As can be seen, the strength of the crust remains defined by frictional sliding on pre-existing faults (Byerlee's Law) over its entire depth range; such slip is largely temperature and mineralogically independent (although water or platy hydrous minerals within the fault plane may reduce strength substantially). Deformation of the upper part of the mantle lithosphere is ill constrained, as it results from interplay of brittle and plastic deformation processes, mostly within semi-brittle or ductile fault zones. It is unlikely that stresses will grow sufficiently to result in appreciable seismicity.

Within the uppermost mantle, partial melting and melt extraction below the mid-ocean ridges results in depletion of lower melting temperature minerals, such as garnet and clinopyroxene. This partial melting not only extracts a mineral component, but also results in a depletion of water in the remaining solid minerals, as water partitions strongly into a melt phase (Hirth and Kohlstedt 1996). As the strength of olivine has been demonstrated to increase with decreasing water content and decreasing temperature, this peridotitic upper mantle residuum (the oceanic mantle lithosphere) is considerably stronger that the underlying undepleted mantle. Thus, only below around 40 km (600°C) for a strain rate of 10^{-14} s^{-1} do the peridotite rocks begin to deform by fully plastic deformation mechanisms. Below this depth, the rheology of peridotite will be controlled by deformation of olivine

in the power-law breakdown (low temperature creep) field (Evans and Goetze 1979); at higher temperatures, power-law creep will predominate. On the assumption that the transition from depleted, nominally dry to undepleted fertile mantle occurs at a depth of around 80 km, a significant step in strength occurs, as illustrated in the inset in Figure 1. The magnitude of this step is defined from experiments at relatively low water pressures, and may underestimate the contrast in strength under true mantle conditions.

Thus, the dehydrating effects of partial melting at mid-ocean ridges will tend to produce a stiff nominally dry lithospheric mantle overlying wet, weaker fertile mantle. Increasing temperature with depth below the lithosphere results in a steady decrease in rock strength. However, at greater depths, the effects of the pressure dependence of creep will begin to dominate over the thermal effects, producing an increase in viscosity with depth. Thus, there is a region below the lithosphere where there is a minimum in mantle viscosity for olivine-rich rocks. This zone, usually referred to as the asthenosphere, is the uppermost zone of mantle convective flow, which allows the lithospheric plates to essentially float on the convecting mantle and the motion of the plates to be largely mechanically decoupled from the convecting mantle. Potential decreases in oxygen fugacity and water fugacity (as the solubility of water in olivine increases but the concentration of water-derived species remains essentially constant) with depth may also act to strengthen olivine-rich lithologies with depth. Such steady increases in strength with depth make it difficult to define a clear lower boundary for the asthenosphere. Seismic data constrain the associated low viscosity zone to depths of 100–300 km (e.g., Hales et al., 1968; Grand and Helmberger, 1984).

Strength envelopes, such as Figure 1, are illustrative rather than quantitative tools for modeling the distribution of strength through the lithosphere. The assumptions of a uniform strain rate through the lithospheric column, distributed rather than localized deformation on fault systems (brittle or plastic), and no role for hydrous platy minerals on rheology all limit the applicability of such models to Earth. Nonetheless, these models remain instructive in illustrating the distribution of strength and mechanical behavior, and the role of water in mantle strength.

3.2. Venus

As discussed above, the D/H ratio suggests that Venus has lost very significant quantities of water. No direct information is available on water content of the interior, which must be inferred from models of the deformation of the lithosphere, laboratory experiments, and comparison to Earth and Mars. Strengthening of the crust and upper mantle on Venus due to a loss of water will partially compensate for the weakening effects of the higher surface temperatures.

The relaxation of topography provides a crude means to estimate the rheology of other planets. Such estimates are not highly constrained due to uncertainty about parameters such as crustal thickness, thermal gradient, composition, volatile content, and mantle viscosity. On Venus, gravity and topography data are used to estimate the local crustal thickness, and data from the Venera landers indicate basalt at the landing sites. The formation of the large plateaus and mountain belts on Venus has been studied to constrain the combination of crust and mantle viscosity structure, for either an upwelling or a downwelling model of formation. If the crust is dry and thus very strong, the lithospheric thickness and the mantle viscosity have to be Earth-like in order to allow plateaus to form over mantle downwellings in a reasonable time-scale of 10^8 yr (Lenardic et al., 1995; Kidder and Phillips, 1996). It is difficult to form plateaus over a mantle downwelling in a stagnant lid regime, regardless of the crustal rheology (Kidder and Phillips, 1996). However, if they are formed above a hot mantle plume, the plateau forms via crustal thickening so that crustal deformation is not a factor (Grimm and Phillips, 1991; Phillips et al., 1991; Phillips and Hansen, 1998). The rims observed on most plateaus have been proposed to form via topographic relaxation, but this is a strong function of both the degree of plateau compensation and rheology (Nunes et al., 2004).

Support of long wavelength topography can provide a means to probe the viscosity of the upper mantle. Both the global admittance spectrum (Kiefer et al., 1986; Kaula, 1990) and the very large gravity anomalies found at the large volcanic rises indicate that there is no low-viscosity zone under the lithosphere (Phillips, 1986; Kiefer and Hager, 1991; Smrekar and Phillips, 1991). The low viscosity zone beneath Earth's oceanic lithosphere has the effect of reducing topographic uplift due to mantle plumes, which in turn reduces the observed gravity signature above plumes (Parsons and Daly, 1983).

Constraints on the rheology of the crust on Venus can be derived from experimental studies of the mechanical behavior of appropriate lithologies. Crustal rheologies for Venus are generally based on experimentally deformed samples of diabase conducted using moderately coarse-grained rocks of basaltic composition (Mackwell et al., 1998). Rocks of basaltic composition are composed predominantly of plagioclase feldspar and pyroxene, both of which deform plastically by dislocation creep under the conditions that exist in the lower crust on Venus, and show weakening effects in the presence of water. In their study, Mackwell et al. (1998) demonstrated that dry diabase is much stronger than wet diabase and is,

in fact, comparable in strength to dry olivine rocks at deep crustal conditions on Earth or Venus. From their experiments, they obtained flow laws that describe the strength of the rocks in terms of strain rate, temperature, and oxygen fugacity. No pressure dependence was quantified, as pressure effects are expected to be very small over the depth range appropriate to crustal rocks on Venus. The activation energy from creep of 485 kJmol^{-1} and a stress exponent of 4.7 were determined for both rocks, even though one contained much more plagioclase than the other, and the chemistry of the pyroxenes was quite different in the two rocks. Because of a smaller pre-exponential constant, the diabase with more plagioclase was overall weaker than the pyroxene-rich diabase, consistent with measurements of the creep behavior of plagioclase (Shelton and Tullis 1981) and pyroxenite (Kirby and Kronenberg 1984, Boland and Tullis 1985, Bystricky and Mackwell 2001, Chen et al. 2006), which demonstrate strengths bracketing the results for the diabase.

Using data from experiments and our knowledge of the likely thermal and compositional conditions in the interior of Venus, we can develop strength envelope models that provide insight into the Venusian dynamics and tectonics (Phillips et al., 1997). Figure 2 illustrates one possible model for the lithosphere on Venus, based on the assumption of a surface temperature of 470 °C, a thermal gradient of 10 K/km, a crustal thickness of 25 km, and a strain rate of 10^{-15} s^{-1}. The least constrained of these parameters are the thermal gradient and the strain rate, while the crustal thickness varies significantly across the planet, ranging from 10–90 km (e.g. Simons et al., 1994; Anderson and Smrekar, 2006; Grimm and Hess, 1997). The illustration is useful, however, in providing some general perspectives on the distribution of rock strength in the Venusian interior. Superimposed on the diagram are "wet" rheologies for crustal and mantle rocks, providing a counterpoint to the dry data.

As for Earth's oceanic lithosphere, deformation in the uppermost Venusian crust is dominated by frictional sliding on pre-existing faults, and there is a significant depth range where rheologically poorly constrained semi-brittle mechanisms are operative. However, unlike for the oceanic crust, the high surface temperature and thick crust mean that there is a deep crustal region where fully plastic processes are operative, dominated by dislocation creep processes in the plagioclase component of the rocks. These dry rheologies (e.g., Chopra and Paterson 1984, Mackwell et al. 1998, Mei and Kohlstedt 2000) indicate that significant strength resides in both the crust and mantle, consistent with gravity measurements and apparently long-lasting high topography. There is no major zone of weakness in the lower crust, so that the jelly sandwich model (Banerdt and Golombek, 1988; Zuber and Parmentier, 1990) is not necessarily appropriate for a dry

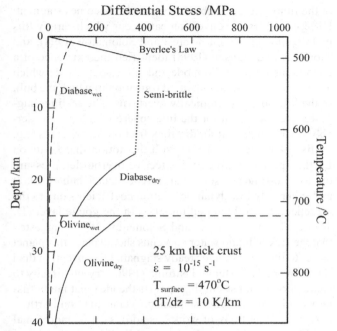

Figure 2. Strength envelope plot showing strength of rock as a function of depth for the Venusian lithosphere, assuming a crustal thickness of 25 km, a surface temperature of 470 °C, a thermal gradient of 10 K/km, and a strain rate of 10^{-15} s^{-1}. Solid lines show the data for dry behavior, following Byerlee's law at upper crustal conditions, and flow laws for Maryland Diabase (Mackwell et al. 1998), and olivine-rich rocks deforming in the power-law breakdown (Evans and Goetze 1979) and dislocation creep (Chopra and Paterson 1984) regimes. The dotted lines show deformation behavior under wet conditions for diabase (Shelton and Tullis 1981) and olivine-rich rocks (Chopra and Paterson 1984).

crust. It should be noted, however, that a weak lower crustal zone does become more apparent with increasing crustal thickness and increasing thermal gradient. Also, the contrast in strength between the crust and mantle is modest (around a factor of 3 for the model in Figure 2), so that mechanical decoupling between crust and mantle is limited.

The wet rheologies for basaltic composition rocks (Caristan 1982, Shelton and Tullis 1981), also plotted in Figure 2, suggest that the rocks would behave plastically almost to the surface, and that the mid- to lower crust would be very weak. Such predictions are clearly inconsistent with our observations of Venus, but point to the key role of water in defining tectonic processes on planetary surfaces.

3.3. Mars

Most of the data that exists on volatiles on Mars pertains to atmospheric composition and frozen surface deposits.

Estimating the volatile and incompatible element contents of the interior can only be done indirectly, and much of the scanty data available is extracted from Martian meteorites. Since they are hundreds of millions or billions of years old, they provide only snap shots of Martian history. Estimates of the Martian bulk silicate composition based on meteorite compositions have been made by several researchers. Both Dreibus and Wänke (1985) and Lodders and Fegley (1997) predict a high initial volatile content for Mars, higher than that of Earth. Karner et al. (2004) examined the trace and volatile contents of plagioclase grains from basalts on Earth, Mars, the Moon, and Vesta. Terrestrial and Martian plagioclases were enriched in K and Na relative to lunar plagioclases or those from Vesta, indicating that Earth and Mars have a similarly volatile-rich interior. The Eu/Sm ratio in these grains, which responds to oxygen fugacity, similarly indicates that the Earth and Mars have far higher oxygen fugacities than do the Moon or Vesta.

The meteorites, however, are highly heterogeneous and indicate that dissimilar source regions exist in the Martian mantle. Some meteorites indicate that their sources have been depleted for the majority of the planet's history (e.g., Longhi, 1991) while others are relatively enriched (Blichert-Toft et al., 1999; Herd et al., 2002; Borg *et al.*, 2002). Convection has apparently not been vigorous or long-lasting enough to homogenize source regions that were separated within the first tens to hundreds of millions of years of Martian evolution. Convection which is apparently less efficient than that of Earth's may imply that Martian mantle viscosity is higher than Earth's, perhaps caused by a lack of incompatible elements and volatiles, but other explanations are also possible. The isotopic range in the Martian mantle is far larger than Earth's, implying that Earth's mantle is either better mixed or differentatiated less thoroughly or more recently.

Some information on the Martian atmosphere is obtained from noble gases trapped in melt in the Martian meteorites: Bogard et al. (2001) has reported that the Martian atmospheric ^{40}Ar content is only ~2% of the ^{40}Ar expected to be produced by decay of ^{40}K over the last 4 Ga. The low ^{40}Ar may mean that the Martian crust and mantle are far lower in potassium than supposed, or degassing has been inefficient, or that atmospheric erosion has been highly efficient. There is no reason to expect Mars to be deficient in potassium, and evidence from meteorites and surface measurements implies that it is not. In the second case, a corollary is that the Martian mantle and crust are enriched in volatiles.

Melt inclusions from the meteorites that crystallized at some depth contain amphiboles, implying a significant initial magmatic water content. Johnson et al. (1991) and McSween and Harvey (1993) estimate 1 to 2% water in the primary magma. Experiments that produce pyroxene compositions similar to Shergotty's imply a preeruptive water content of 1.8% (Dann et al., 2001). These estimates of magmatic water can constrain the water content of the melting source, but only if melting extent and later fractionation can be estimated. Estimates of the water weight fraction for the Chassigny source by Jones (2004) and by McSween et al. (2001) are both near 10^{-4}.

As on Venus, relaxation of present-day topography also provides a gross estimate of the crustal and upper mantle rheology. A number of studies have focused on relaxation of the dichotomy boundary for this purpose. Since the dichotomy boundary formed so early in the history of Mars, there is considerable uncertainty about the appropriate thermal gradient in the first half By. Recent studies have all concluded that preservation of the dichotomy topography requires that the crust has somehow increased in viscosity since its initial formation, via rapid interior cooling (Nimmo, 2005), crustal hydrothermal cooling (Parmentier and Zuber, 2007), or volatile loss (Guest and Smrekar, 2007).

4. VOLATILES AND INTERIOR CONVECTION

In this section we review what is known about the relationship between the large-scale tectonic style of terrestrial planets and the mantle, lithosphere and atmospheric volatile content. There have been many studies that show that the strong temperature dependence of silicate viscosity favors a style of planetary convection in which the cold thermal boundary layer is thick and stagnant (e.g. Fowler, 1985; Morris and Canright, 1984; Davaille and Jaupart, 1993; Solomatov 1995). One characteristic of this style of convection is that it produces very little horizontal tectonic stress and, therefore, very different geological structures are created than those familiar on Earth. Solomatov and Moresi (1997, 2000) showed that the stagnant lid mode of convection should be stable in the long term for most of the terrestrial planets and sub-continental convection on Earth, even taking into account the expected stress-dependent creep mechanisms.

In comparative planetology of the terrestrial planets, the oceanic lithosphere of Earth is unusual in that it is demonstrably not stagnant, but instead actively recycles into the interior. The lithosphere of Venus may also be relatively young having been extensively resurfaced by volcanic activity in the last billion years; little is known about the tectonic style of the planet prior to the onset of this resurfacing, but Turcotte (1993) suggested that significant episodes of tectonic recycling of the lithosphere could have occurred prior to this time. Assuming that such recycling would occur when the convective stresses imposed on the lithosphere reach the yield strength, Moresi and Solomatov (1998) derived scalings relating Rayleigh number and the maximum yield strength

of the lithosphere possible before the lid becomes stagnant. Different styles of convection can occur which satisfy this mobilization condition: Moresi and Solomatov (1998), and Trompert and Hansen (1998) identified the existence of a continuously mobile lid mode and an episodic mode which appears close to the stagnant lid transition in which the bulk of the lithosphere periodically overturns. The scalings suggest a yield strength for the lithosphere which is considerably smaller than that for dry rock friction experiments (e.g. Tullis, 1986) but consistent with the notion that faults on Earth experience a reduced effective overburden pressure due to pore-fluid pressures (e.g. Turcotte and Schubert, 1982) and inherently low dynamic friction coefficients due to the presence of soft, hydrated minerals in the gouge layer (Di Toro et al, 2004). Moresi and Solomatov (1998) suggested that the lack of liquid water on Venus should place this planet closer to the transition to steady stagnant lid convection than Earth. Work by Foster and Nimmo (1996), trying directly to characterize fault strength, supports the idea that faults sustain significantly higher stresses on Venus than on Earth.

A significant body of work has addressed the additional conditions required to obtain a mobile lid style similar to that observed for Earth's oceanic plates. Zhong and Gurnis (1995,1996) emphasized the importance of narrow fault zones with low friction coefficient and non-Newtonian upper-mantle rheology for accurate descriptions of convergent and transform boundaries. Tackley (2000a,b) and Richards et al (2001) observed that the presence of a low-viscosity asthenosphere underneath the bulk of the moving plate is necessary to obtain both localized plate boundary deformation and a low-strain rate in the rest of the lithosphere. If water is the cause, then this suggest a link between a planet's internal volatile content and its ability to maintain an Earth like mode of plate tectonics (Kaula, 1990). This potential link between volatiles, the asthenosphere, and plate tectonics is made all the more intriguing by Venus' lack of recent plate tectonics and the lack of evidence for a Venusian asthenosphere (Phillips, 1986; Kiefer et al, 1986). Tackley also pointed out that strong localization depends upon a strain or strain-rate softening mechanism being available in the deforming lithosphere (for example, the strain-rate softening plasticity model of Bercovici, 1993). Bercovici (1998) demonstrated that this constitutive law can be derived from a physical model in which the interaction of voids with interstitial fluids creates a self-lubricating feedback between the mechanical stresses and the induced deformation.

Throughout the work that has been and continues to be done on elucidating the conditions required for plate tectonics, the initial assumption of Moresi and Solomatov (1998) is a reliable starting point: For a planet to have an active lid mode of mantle convection that allows for lithospheric recycling akin to subduction on Earth, the convective stresses imposed on the lithosphere must be capable of reaching the yield strength of the lithosphere itself. The yield strength can be strongly influenced by the presence of pore water and hydrated minerals in fault zones. This allows for the possibility that near-surface dehydration over time could alter lithospheric strength. Further, changes in convective vigor due to planetary cooling could cause significant changes in convective stress. Both these factors allow for the potential of planetary evolution that transitions from active to stagnant lid convection (Phillips and Hansen, 1998).

Stagnant lid convection appears to have shaped the geologic evolution of Venus profoundly. One hypothesis is that stagnant lid convection itself leads to a reduction in the amount of tectonic activity (Schubert et al., 1997). Alternatively, stagnant lid convection may have triggered resurfacing. The transition from plate tectonics to stagnant lid convection causes the mantle temperature to increase because the stagnant lid reduces the amount of heat lost relative to convection with plate tectonics. This increase in the temperature could result in large volumes of pressure release melting, thus resurfacing the planet (Reese et al., 1999). Another hypothesis is that a phase transition within the mantle could have triggered a global mantle avalanche from a two-layer mantle regime to whole mantle convection, allowing for complete resurfacing (Steinbach et al., 1993; Steinbach and Yuen, 1994).

The thermal evolution of Mars (Nimmo and Stevenson, 2000) was explored using parameterized thermal history models that also considered a transition from active to stagnant lid convection early in Mars' geologic history. Although the study of Nimmo and Stevenson (2000) did not provide a specific explanation as to why an early episode of plate tectonics ceased on Mars, a subsequent study suggested it could have been caused by increased mantle temperatures due to the formation of the hemispheric dichotomy (Lenardic et al., 2004). Increased mantle temperature would lower the mantle viscosity and, in turn, lower convective stress in the mantle causing it to drop below the lithospheric yield. More recently it has been suggested that increased mantle temperatures in the ancient Earth, due to a greater component of radioactive heating, could have caused convective stresses to approach the lithospheric yield stress resulting in an episodic mode of Precambrian plate tectonics (O'Neill et al., 2007).

Although the last two studies noted above remain hypothetical, they do highlight that a change in the yield stress of a planet's lithosphere is not the only factor that can cause a transition from a plate tectonic to a stagnant lid mode of convection; changes in convective stress can also initiate a transition. Convective stress is a strong function of average mantle viscosity, which depends exponentially on tempera-

ture. It is also a function of water content and this suggests another potential connection between tectonic styles and a planet's internal volatile content. The mantle budget of volatiles for a planet is determined by the balance between loss, due to volcanic degassing, and gain due, on Earth, to subduction of hydrated oceanic crust and lithosphere. The connection between volatile cycling through the mantle and the thermal evolution of a planet has not received significant modeling attention to date. The pioneering work of McGovern and Schubert (1989) addressed this coupling using a parameterized convection approach applied to Earth's thermal evolution. They assumed that plate tectonics operated throughout Earth's geologic history and did not address how volatile cycling might affect convective stresses. This issue remains a potentially rich one for future investigation.

This body of work reviewed above highlights the importance of the volatile distribution of a planet on its global tectonic style. Volatiles within the crust and shallow lithosphere can affect lithospheric yield stress. The ability of the lithosphere to deform through microcrack formation, and the ability of those cracks to coalesce into long-lived weak faults all depend on whether there is near surface water. Volatiles in the interior mantle also play a role by effecting bulk mantle viscosity and convective stress levels. Volatile loss from the mantle due to volcanic degassing can effect climate which can feedback and effect a planet's magmatic and tectonic history. Collectively, this short review suggests that future studies aimed at exploring the connection between volatiles and a planet's tectonic style will need to address not only the role of volatile induced weakening of the lithosphere but also the cycling of volatiles though the mantle over a planet's history and the effects this has on mantle melting, on mantle degassing, and on climate-solid planet feedbacks.

5. GEOCHEMICAL CONSEQUENCES OF STAGNANT LID CONVECTION

The mantle and lithospheric interactions hypothesized to occur on Venus are likely to be similar to terrestrial processes in stable continental lithospheres. Although stable continental lithosphere or a lithosphere that is part of a stagnant lid may not experience subduction or high strain, the lower lithosphere may be affected by mantle melting, and, if the right density and viscosity conditions are present, lithospheric instability. Recycling into the interior can be accomplished only through foundering of the lithosphere, either through brittle or ductile gravitational instability. Mantle melting is likely to occur mainly through adiabatic decompression melting in warm upwellings during mantle convection. If melting occurs, the melt will rise buoyantly until it encounters the lithosphere, where the thermal gradient rapidly cools the magma below its solidus unless sufficient magma pools to form a dike.

In the case of adiabatic melting, magma will percolate upward at rates specified by Darcy velocities (e.g. Wark et al., 2003) and freeze in pore spaces (Rubin, 1995), creating a lower lithospheric boundary that is both denser and less permeable. Because of the immediate departure of the geotherm from convective to conductive at the bottom of the thermal lithosphere, any small-degree partial melt will be exposed to lithospheric temperatures substantially below the melt's solidus within a few kilometers of rise (Figure 3). If melt volumes become significant, cracking from hydrostatic pressure may occur and dikes may form. They may similarly stall and freeze, or they may erupt, perhaps feeding a magma chamber as an intermediate step.

If frozen at sufficient depth in the lithosphere, the magma may crystallize as an eclogite, consisting of garnet and pyroxene, rather than the lower-pressure phases: pyroxene, olivine, and plagioclase. This process for creating a dense lithosphere has been suggested for Earth (Kay and Kay, 1993; Elkins-Tanton and Hager, 2000; Jull and Kelemen, 2001) and for Venus (Namiki and Solomon, 1993; Jull and Arkani-Hamed, 1995). The density anomaly between depleted peridotite and eclogite can be significant. The composition of a 14% melt from a fertile peridotite equilibrated at 3.0 GPa and 1515°C from Walter (1998) may approximate the high-pressure melt which would be produced at depth by adiabatic rise of a hypothetical Venusian mantle composition. If recast into an eclogite mode containing primarily clinopyroxene and garnet with olivine, its density at one atmosphere and 300K would be 3,545 kg/m^3. By comparison, a fertile peridotite may have a density of about 3,350 kg/m^3 and a depleted peridotite 3,290 kg/m^3. The eclogite is about 10% denser than the peridotite.

The processes of mantle melting and magma rise into the lithosphere produce compositional, thermal, and density anomalies that were not previously present. Magma produced from peridotite contains more iron as a fraction of iron plus magnesium than does its coexisting peridotite source. The magma will carry this iron enrichment into the lithosphere, creating a heterogeneous lithosphere and leaving behind a region of depleted mantle. The newly enriched lithosphere will be slightly more ductile because of the addition of the heat of fusion of the solidified magma. Such density variations are likely to lead to lithospheric instabilities on Venus (Elkins-Tanton et al., 2007).

5.1. Movement of Volatiles in a Stagnant lid Regime

Both mantle upwellings and associated melting and lithospheric delamination provide opportunities for volatile movement in a stagnant lid environment. The minerals that compose

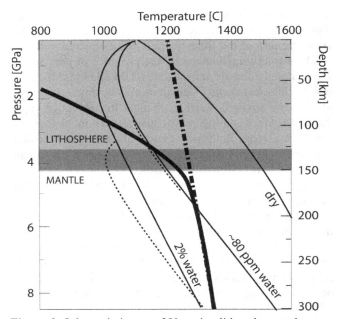

Figure 3. Schematic image of Venusian lithosphere and upper mantle. Solid bold line – temperature profile, with surface at 460°C and mantle assumed to be adiabatic with a potential temperature of 1200°C. Dash-dot bold line: adiabat at mantle potential temperature. Dotted lines = carbonated eclogite solidii for two different compositions; left from Dasgupta et al. (2004) and right from Yaxley and Brey (2004). Solid thin lines are peridotite solidii. Dry peridotite solidus: Hirschmann (2000). 80 ppm water solidus: 2% water solidus: Litasov and Ohtani (2003). Venus may be more likely to have a carbonated mantle than a water-bearing mantle, but the available solidii shown here from the literature aptly demonstrate the possible processes. Lithosphere is 150 km thick. Given the mantle potential temperature assumed and solidii as shown, dry mantle peridotite would not melt at all under the 150 km lithosphere, while peridotite with 2 wt% water would have begun to melt at 300 km depth, as would at least some highly carbonated eclogite. Assuming that the mantle composition is equivalent to peridotite with 80 ppm water, then rising mantle material will begin to melt at about 180 km depth and would percolate upward. The thick dark grey region at bottom of lithosphere is where melt rising along an adiabat would be in lithospheric country rock at temperatures above its solidus; at the top of this region any rising melt would be expected to be wholly frozen (though this is a simplification, since the melt would be fractionating as it rose through cooler and cooler lithosphere, and the evolving liquid would have a lower and lower temperature solidus as well as higher and higher viscosity). This region of frozen melt is the candidate region for eclogite formation.

the terrestrial mantle, and are widely assumed to compose the mantles of other terrestrial planets, include olivine, clinopyroxene, orthopyroxene, and garnet. These minerals are referred to as nominally anhydrous because they do not incorporate water or other volatiles into their crystal structures, but can accommodate a small fraction in crystal defects. The volatiles and incompatible elements critical to mantle viscosity, mantle melting temperatures, and resultant magma compositions include water (in the form of OH when incorporated into nominally anhydrous minerals), carbon dioxide, potassium, and sodium. Carbon dioxide and water behave differently in melts. Both can dissolve into silicate melts at quantities greater than 10 wt% at pressures above their degassing pressures (*e.g.* Papale, 1999). However, the majority of a silicate fluid's water content comes out of solution and forms bubbles at approximately 0.5 GPa, and carbon dioxide does so at ~3 GPa (*e.g.*, Newton and Sharp, 1975; Wyllie and Huang, 1976), leaving behind only a fraction in solution. Prior to degassing, both OH and CO partition between crystallizing solids and the evolving melts. Nominally anhydrous minerals can contain a dynamically and petrologically significant amount of OH (Koga et al., 2003; Aubaud et al., 2004; Forneris, 2003; Bell et al., 2004; Bolfan-Casanova, 2000). These minerals can contain as much as 1000 ppm water, and with contributions from minimal interstitial liquids, resulting mantle solids can retain 0.25 to 0.5 wt% water (Elkins-Tanton and Parmentier, 2005). Carbon dioxide, in contrast, cannot partition into mantle minerals in as significant quantities (Keppler, 2003) (see Table 1).

Despite Venus' low atmospheric water content, the interior likely retains some small fraction of water, carbon dioxide and other incompatible elements. A portion of the initial volatile content degassed during early melting following accretion to form the dense atmosphere. Part of the initial volatile content will be retained in the solid mantle, ranging from 100% in regions that have not been melted, to 1 to 2% in regions that have experienced melting depending upon the partition coefficient (Bolfan-Casanova and Keppler, 2000; Koga et al., 2003; Aubaud et al., 2004; Bell et al., 2004). The

interior of Venus is therefore likely to contain some fraction of volatiles to the present day.

The lithosphere of Venus consists of magmas, more or less processed, that originated from melts of its interior. Portions of Venus' lithosphere will therefore be volatile-bearing if the planetary interior is volatile-bearing. Sinking lithospheric material will carry incompatible elements (volatiles and alkalis in particular) back into the mantle. In magmas that freeze at shallow depths or even erupt onto the Venusian surface, high atmospheric pressure acts to hold water in the crust and mantle. Degassing of lavas is inefficient first because high heat transfer to the dense atmosphere quickly quenches lava, and second because Venus' atmospheric pressure allows basalt to retain a higher water content than Earth equivalent. Atmospheric pressure on Venus allows basaltic lava to retain on the order of 1 wt% water, while the same basalt at Venus surface pressure could contain only about 0.2 wt% of carbon dioxide (Papale, 1999).

Incompatible elements in the lithosphere are available to encourage melting in material sinking from the lithosphere in gravitational instabilities. The process of gravitational instability itself provides both opportunities for melting and the possible production of additional, compositionally distinct source regions: as the denser lithospheric anomalies sink into the underlying mantle, it is efficiently warmed conductively by the underlying upper mantle. If the unstable material contains quantities of volatiles or incompatible elements even in concentrations of less than a percent, these components may be released from the sinking material as its pressure and temperature increase, just as subducting slabs on Earth devolatilize as they sink into the mantle. These released incompatible elements can either cause the sinking material to melt or they may be released into the surrounding

Table 1. Approximate partitioning and saturation limits for water and carbon dioxide in nominally anhydrous mantle minerals. **WATER (OH)**

Mineral	Solid-melt coefficients	Saturation limit
olivine	0.002 [1, 2]	1000 ppm [1,4]
clinopyroxene	0.02 [1,2]	1000 ppm [3, 5]
garnet	0.0008 [4]	700 ppm [5]
ringwoodite (γ -spinel)	0.02 [from data in 5]	2.5 wt% [5,6]
majorite	0.003 [from data in 5]	675 ppm [5]

[1] Koga (2003) [2] Aubaud (2004) [3] Forneris and Holloway (2003) [4] Bell (2004) [5] Bolfan-Casanova (2000) [6] Ohtani et al. (2004)

CARBON

Mineral	Saturation limit
olivine	1 ppm [1]
clinopyroxene	0.5 ppm [1]
garnet	2 ppm [1]

[1] Keppler et al. (2003)

upper mantle and there trigger melting, in analogy to melting in the mantle wedge of a terrestrial subduction zone.

Dynamics and volatile and incompatible element transfer in a stagnant lid regime can thus create a variety of magmatic source regions. In the simplest case, compositions of source regions can include end-members of fertile and depleted peridotite and eclogite, their hydrated, carbonated, or alkali-enriched analogs, as well as intermediate sources consisting of mixtures of these. These sources can additionally be melted to different extents and at different temperatures and pressures, yielding a significant array of magma compositions that might closely approximate the range of compositions found on Earth.

In particular, eclogite melts may have significantly higher silica and lower magnesia contents than similarly derived peridotite melts. Pancake domes (Pavri et al., 1992) and the festoon flows (Moore et al., 1992) have the morphology of very viscous lavas, and could be formed from melts generated when lithosphere delaminates (Elkins-Tanton et al., 2007). In addition, the basalt-eclogite transition in the crust may limit crustal thickness and create density anomalies leading to lithospheric instability (Dupeyrat and Sotin, 1995). Anderson and Smrekar (2006) observe numerous small scale (500–1000 km) gravity variations in lithospheric properties that lack clear association with geologic surface features, which they interpret as possibly due to lower lithospheric delamination driven by the basalt-eclogite transition or other density anomalies.

Carbonated peridotite melts appear to have the lowest silica and highest magnesia, and therefore among the lowest viscosities, of any of this suite of potential magma types. Melts of carbonated peridotites and eclogites at low melt fractions or over certain pressure intervals produce coexisting immiscible silicate and carbonate melts (e.g. Dasgupta et al., 2006). Such magmas have been suggested to produce the extremely long (hundreds of km), narrow channels seen on Venus (Kargel et al., 1994; Treiman, 1995), though they require very large melt volumes to produce such long channels. Three Venusian compositions were measured by Soviet landers (Surkov, 1990). In most of their major element compositions these melts fall between eclogitic and peridotitic melt compositions, with the exception of one composition, which contains approximately 4 wt% K_2O and therefore represents a composition significantly enriched in incompatible elements.

6. STAGNANT LID CONVECTION AND MANTLE PLUMES

Stagnant lid convection has a strong effect on interior mantle temperatures and thus on boundary layer thickness. A hotter mantle creates a thinner lithosphere, which

leads to enhanced melting and volcanism (e.g. Smrekar and Parmentier, 1996; Nimmo and McKenzie, 1997), consistent with the widespread volcanism and possible volcanic resurfacing. The large volcanic rises share many of the characteristics of the terrestrial hotspots, and are likely to form above plumes originating at the CMB (e.g. Kiefer and Hager, 1991). As with the lithosphere, the relative affects of temperature versus volatiles is unknown. A dry interior (relative to Earth's) would have a higher viscosity and a thicker lid.

Thermal convection in Venus' mantle is likely to be driven by the internal heating generated by the decay of the long-lived radiogenic elements, the cooling from above and the presence of a hot core. Although the presence of a liquid core is not proven due to the lack of knowledge of Venus' moment of inertia, several factors imply that a liquid iron-rich core is probably still present. The similarity in size and density between Earth and Venus that suggests that an iron-rich core segregated very quickly after accretion. The fact that a smaller planet like Mars still has a liquid iron core (Yoder et al., 2003) suggests that Venus' core is unlikely to have solidified. Finally, a hot thermal boundary layer is needed to explain the plumes and volcanoes.

If there were no hot thermal boundary at the CMB, the convection pattern would be described by only cold plumes that would form at the cold boundary layer below the stagnant lid (Parmentier et al., 1994). The two boundary layer system was studied by Sotin and Labrosse (1999), who showed that although hot plumes can form at the CMB, they have a hard time reaching the upper thermal boundary layer because their buoyancy vanishes before they reach this layer. The heat balance is thus mostly controlled by the by the cold plumes. However, it must be noted that numerical simulations are difficult to realize since one must take into account the curvature of the planet and the temperature-dependence and other complexities of viscosity, and the effect of the mineral changes at the upper/lower mantle interface.

The number of plumes can be assessed following the study by Whitehead and Luther (1975). The wavelength (λ) of the plumes forming at the hot thermal boundary layer depends on the thickness (δ) of and the viscosity ratio ($\Delta\mu$) across the boundary layer:

$$\lambda = \frac{4\pi}{3}\delta\left(\Delta\mu\right)^{1/3}. \tag{1}$$

If variations in viscosity are only linked to temperature variations across the thermal boundary layer ($\Delta T_{CMB} = T_{CMB} - T_M$), the viscosity ratio can be expressed by:

$$\Delta\mu = \frac{\mu_M}{\mu_{CMB}} = \exp\left(\frac{Q\Delta T_{CMB}}{RT_{CMB}T_M}\right), \tag{2}$$

where μ_M and μ_{CMB} are the viscosity of the mantle above the hot thermal boundary layer and at the core-mantle boundary, respectively. The core surface area divided by the square of the wavelength gives an estimate for the number of plumes.

In a stationary convective regime, the thickness of the thermal boundary layer can be calculated from, and is relatively insensitive to, the value of the thermal boundary layer Rayleigh number (Deschamps and Sotin, 2000). It can be argued that Venus' mantle viscosity is smaller than Earth's because the mantle temperature in the stagnant lid regime is higher than in the plate tectonics regime. Then, the higher value of the thermal boundary layer Rayleigh number yields a smaller value of δ. Thus, the number of plumes would be larger and their diameter, which scales to the thickness of the thermal boundary layer, would be smaller, as compared to Earth.

On the other hand, one can argue that the viscosity is larger because Venus' mantle has lost all its volatiles. Although no strong constraints on mantle temperature and core temperature are available for Venus, reasonable values for these parameters can be assessed by comparison with Earth. In a recent study (Stofan and Smrekar, 2005), it was suggested that the 10 topographic rises identified on Venus are the surface expression of plumes starting at the core-mantle boundary. This number is similar to the number of terrestrial plumes (Courtillot et al. 2003), which suggests that the conditions at the core-mantle boundary are similar on Earth and Venus. The assumption that the cooling rate of Venus is limited by the presence of a stagnant lid suggests that the mantle temperature could be larger (up to 250 K) than that of Earth (Figure 4). It may be that the lower viscosity implied by this larger temperature could be balanced by the lack of volatiles in the mantle in order to produce roughly the same number of plumes on both planets.

Another variable is the presence of coronae and their contribution to mantle heat loss. Coronae are volcano-tectonic features with an average diameter of 200–300 km (Glaze et al., 2002). Analysis of their gravity and topography signatures suggests that approximately 20% of the over 500 coronae on Venus may be active (Johnson and Richards, 2003; Hoogenboom, Smrekar). Detailed modeling of corona formation suggests that up to 25% of Venus' heat loss budget (Smrekar and Stofan, 1997). The small average diameter of coronae as compared to large volcanic rises implies that they do not arise from the CMB, but instead come from a shallower interface, perhaps at the lower mantle – upper mantle boundary (Smrekar and Stofan, 1997; Stofan and Smrekar, 2005). If there is a significant boundary between the CMB and the surface, this may have a significant impact on boundary layer thicknesses.

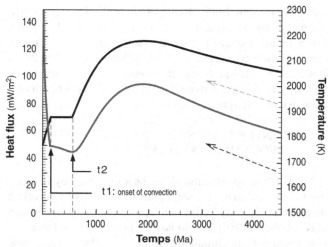

Figure 4. Temperature (upper curve) and heat flux (lower curve) evolution using scaling laws derived from studies of thermal convection with a stagnant lid. The presence of a stagnant lid reduces the cooling rate of the planet. For comparison, values for the Earth, where plate tectonics operates, are indicated at present time with dashed arrows indicating a cooling rate of 100 K/Gyr. The predicted difference in mantle temperature between Venus and the Earth is quite large (250 K). The thermal history during the first billion years is indicative and cannot be constrained. It is quite possible that the initial temperature was much larger. After t2, scaling laws are applicable.

Some workers have suggested that the combination of large volcanic rises and coronae on Venus implies a recent or on-going transition between active and stagnant lid regimes. Robin et al. (2007) observe a transition from the active to stagnant lid regime that occurs over a relatively long duration (several hundred m.y.). During this transition, which follows subduction of the cold lithosphere, a mixed mode of convection occurs in which both isoviscous and low viscosity thermal plumes occur. They suggest that these two scales of plumes may represent Venusian coronae and large volcanic rises. Johnson and Richards (2003) interpret coronae as small-scale upwellings arising at the core mantle boundary that are focused towards larger scale upwelling in high Rayleigh number, temperature-dependent viscosity simulations. The primary evidence of this type of pattern on Venus is the concentration of coronae in the Beta-Atla-Thenis (BAT) region.

On Mars, degree-1 or hemispherical convection has been proposed to explain both the dichotomy and the Tharsis Rise. Two primary mechanisms have been proposed to create such long wavelength plumes. Models of the spinel-perovskite endothermic phase transition in the mantle have shown that the phase change may act as a barrier to all but the longest wavelength convection (Tackley et al., 1993; Harder and Christensen, 1996; Breuer et al., 1997; Harder, 2000). Alternatively, a viscosity contrast between the upper and lower mantle, where the upper mantle viscosity is lower, can also produce a degree-1 plume (Zhong and Zuber, 2001; Roberts and Zhong, 2006). It is conceivable that the lower viscosity in the upper mantle of Mars could be a result of a decrease in volatile content via outgassing.

7. INTERACTIONS BETWEEN CLIMATE, VOLATILES, AND THE SOLID BODY

Volatile loss from the mantle to the atmosphere by volcanic degassing can influence solid planet dynamics by its effect on the bulk viscosity of the mantle, as explored by McGovern and Schubert (1989). It also has the potential to influence a planet's long-term climate by introducing added greenhouse gasses to the atmosphere. The coupling between climate and solid planet dynamics has received recent attention that has been strongly motivated by the exploration of Venus. Surface temperature changes, due to changes in atmospheric composition on Venus, could penetrate into the planets' near surface and cause thermal contraction/expansion capable of fracturing the upper portions of the solid planet (Anderson and Smrekar, 1999; Solomon et al., 1999). Taking the connection one step deeper, Phillips et al. (2001b) showed that if atmospheric induced temperature changes (e.g., Bullock and Grinspoon, 1999) were long lived they could penetrate into the mantle and thus effect mantle melting and volcanism. Not only would internal mantle temperature changes affect melting, they would also affect viscosity and thus convective stress. This opens the possibility that the climatic conditions of a planet could play a significant role in determining its global tectonic style, i.e., climate changes could initiate a transition from a plate tectonic like mode of mantle convection to a stagnant lid mode (Lenardic et al., 2007).

8. FUTURE MEASUREMENTS

A wealth of Soviet and American missions explored Venus from the late 1960s to early 1990s. After a long hiatus, the European mission Venus Express arrived in 2006 to study the composition and the circulation of the atmosphere. Surface missions are needed to address fundamental questions about the geologic evolution of the surface and interior of Venus. Other chapters also discuss valuable surface measurements that would provide insight on the geochemical evolution, including core differentiation, oxidation state of the crust, formation of geologic features above plumes, and the degree of melting (Treiman chapter) and the geologic evolution (Basilevsky and McGill chapter). Here we discuss those measurements that are aimed at

understanding the thermal and tectonic evolution as well as geochemical measurements aimed at the melting processes of the crust and lithosphere and the associated implications for volatile content and rheology. Seismic measurements would provide both interior structure and the present day levels of activity, both of which are essential for understanding the overall evolution. Heat flow from the interior is also general index of geologic activity, and provides an essential constraint on rheology.

8.1. Venus Seismology

A seismic network consisting of at least four stations on the surface of Venus would provide information on the present day level of seismicity, interior structure, and thus on the overall evolution of Venus. Although we do not know the present day rates of volcanism and seismicity on Venus, there are suggestions from atmospheric chemistry for relatively recent volcanism (Esposito, 1984). Thus venusquake swarms associated with magma intrusion in the crust may be expected. There is abundant evidence of faulting on Venus, although the faults will generally be drier than on Earth so that frictional sliding is likely to be more inhibited (e.g., Foster and Nimmo, 1996; Grindrod et al. 2005). Although the effects of water are hard to isolate on Earth, some studies have suggested the seismogenic zone is temperature limited (Hyndman and Wang, 1993; Hyndman et al., 1997). Thus lower water content implies a thicker seismogenic zone while the higher temperatures may imply lower magnitudes than on Earth. It is unclear which factor is dominant.

Estimates of crustal, elastic, and thermal lithospheric thickness based on gravity and topography data have inherent uncertainties due to both methodology and assumptions about density contrast and compensation mechanism. Numerous studies have provided estimates of these values, which have a large range, especially for thermal lithospheric thickness. Seismic network data would independently constrain crustal thickness in several locations, thus making interpolation to the rest of the planet much more accurate. This information would have far reaching implications for local tectonic studies, lithospheric rheology, and global crustal differentiation. The depth of seismicity, coupled with heat flow data, would help determine the local thermal gradient and elastic thickness. This provides a present day boundary condition for models of thermal evolution and interior convection, as well as a key constraint on models of tectonic deformation.

Analysis of the Love number from tracking data suggests that the core of Venus is liquid, but doesn't provide a firm constraint on core size. Seismic data could yield tight constraints on core size and thus composition. If coupled with improved tracking data, it should be possible to determine if the core is liquid, solid, or both. Along with the absence of plate tectonics, the absence of an interior dynamo stands out as a fundamental way in which Venus and Earth differ. Nimmo and Stevenson (2000) and Stevenson (2001) have proposed that the two processes are related. Improved information on the core would give new clues as to how and why Earth and Venus diverged.

8.2. Heat Flow

Energy loss from the interior is one of the most fundamental planetary processes. It provides significant constraints on the abundance of radiogenic isotopes, the thermal evolution and differentiation history, and the mechanical properties and thickness of the lithosphere. It is also an index of the present-day level of volcanic and tectonic activity. For Venus, knowledge of the present-day level of geologic activity and the thickness of the lithosphere are amongst the highest scientific priorities.

Knowledge of the present day heat flow would distinguish between various hypotheses of planetary evolution. A variety of heat flow predictions for Venus have been made over time, and reflect the evolution of our understanding of both tectonics on Venus and convection in general. Prior to mapping the surface with the Magellan radar, which showed no evidence of plate tectonics, a reasonable approach was to simply scale heat flow from Earth to Venus, giving a value of 78 mW m^{-2} (Solomon and Head, 1982). Parameterized and finite-amplitude convection models have also been used to predict heat flow values ranging from 41 to 67 mW m^{-2} (Turcotte et al., 1979; Phillips and Malin, 1983; Arkarni-Hamed and Toksoz, 1984).

Based on our current understanding of Venus and stagnant lid convection, a very low heat flow (<20 mW/m^2) would indicate that either stagnant lid convection or a very late stage of episodic global overturn is operating (Parmentier and Hess 1993; Turcotte 1993; Solomotov and Moresi 1996). An intermediate value of heat flow (20–40 mW/m^2) would be consistent with a stagnant lid modulated by convective heat flux (Phillips et al. 1997), or a scenario in which the lithosphere had overturned relatively recently. High values (> 40 mW/m^2) would indicate a thinner lithosphere and a robust heat loss mechanism, such as hotspots in combination with delamination (Smrekar and Stofan 1997). Leitner and Firneis (2006) have also proposed a new estimate of the heat flux on Venus of ~40 mW m^{-2}. This model assumes that Venus loses heat only via conduction and hot-spot volcanism and uses a series of arguments about the contribution from different volcanic features and the composition of the crust to constrain their estimate. Actual estimates of heat flow from the surface of Venus would provide extremely valuable insight into both the nature of convection and the thickness of the lithosphere.

Measuring heat flow on Venus would also provide invaluable information on lithospheric rheology and tectonic processes. After volatile content, thermal gradient is the parameter that most strongly influences lithospheric rheology. The thermal gradient is a critical variable in many models of lithospheric deformation because of the strong dependence of rheology on temperature that controls the depth of the brittle-ductile transition. Thus determining the thermal gradient is key to distinguishing between the effects of temperature versus volatile content. An estimate of the actual heat flow and the associated elastic lithospheric thickness would also be a valuable constraint on models of specific tectonic features since thermal gradient is such a large uncertainty. Improved estimates of elastic thickness would help determine if specific features formed recently or in a prior geologic period when Venus may have had a different (likely thinner) elastic lithosphere.

Ideally heat flow would be measured in multiple locations, perhaps in parallel with a seismic network. Regional variations due a variety of factors such as crustal thickness, composition, and geologic activity can influence the measured heat flux. Significant climatic changes are not predicted to have occurred within the last 5 my, and thus would not affect heat flow appreciably (Bullock and Grinspoon 1996, 1999).

8.3. Geochemistry

Geochemical measurements from a number of locations on the surface are critical to constraining Venusian evolution. To test the hypotheses that lavas on Venus can be created from peridotites and eclogites with a range of volatile and incompatible element compositions, the concentrations of specific oxides should be measured that could define the end-members of these ranges.

Measured Venusian surface compositions are compared in Table 2 with a range of melt compositions found on Earth. The terrestrial data in Table 2 is compiled from experiments with known and controlled initial conditions, and is therefore limited in range compared to the massive data available for natural samples. In natural samples the melting source is inferred but not known with absolute certainty, and so is less useful for this application. The compositions given in Table 2 do not include complications such as hydrous or carbonated fluids that carry significant alkali elements: On Earth, high alkali content in lavas implies an enriched fluid participant, since high alkalis are not compatible in peridotites or present in many eclogites.

Table 2. Ranges of terrestrial melt compositions and measured Venusian surface compositions, as a guide to future surface measurements. All data rounded and given in wt%. All data is from experiments except Venusian compositions.

	Dry terrestrial eclogite melts [1, 2]	Dry terrestrial peridotite melts [3, 4, 5, 6]	Carbonated terrestrial eclogite melts [7, 8, 9]*	Carbonated terrestrial peridotite melts [10, 11]	Venusian compositions [12]
SiO_2	50 – 58	46 – 52	33 – 63	7 – 43	45 – 49
Al_2O_3	14 – 19	9 – 18	2 – 13	0 – 11	16 – 18
TiO_2	1 – 7	0 – 1	3 – 19	~1	0 – 2
FeO	5 – 11	6 – 12	4 – 26	9 – 10	8 – 9
CaO	7 – 10	7 – 13	5 – 15	9 – 32	7 – 10
MgO	1 – 6	7 – 24	8 – 14	19 – 30	8 – 12
K_2O	0 – 1	na	0 - 1	trace	0 – 4
Na_2O	3 – 4	0 - 3	1 - 5	1 – 2	na

[1] Pertermann and Hirschmann (2003); [2] Pertermann et al. (2004); [3] Falloon et al. (1999); [4] Baker and Stolper (1994); [5] Walter (1998); [6] Kinzler and Grove (1992); [7] Dasgupta et al. (2006); [8] Hammouda (2003); [9] Yaxley and Brey (2004); [10] Gudfinsson and Presnall (2005); [11] Hirose (1997); [12] Surkov (1990).

* Note that immiscible carbonate melts are also produced in these experiments, but their compositions are not included here. na = not available

Important first-order measurements on Venusian lavas therefore include the following: 1) Alkali elements to test for volatile and incompatible element presence; 2) CaO, to test for the presence of carbonatitic magmas, 3) SiO2, to compare the primitive or evolved nature of magmas and to look for the presence of andesites, rhyolites or other non-primary magmas, and 4) FeO and MgO to test for primitiveness of magmas and to differentiate between eclogite and peridotite melts. Important potential types of crust to analyze include tessera terrain, hotspot rises and volcanic plains.

9. SUMMARY

In this chapter we have explored the consequences of volatile history on the evolution of terrestrial planets (excluding Mercury). The effect of volatiles on rheology can be as large or larger than that of interior temperature difference, thus has implications for tectonic deformation, melting processes, and interior convection. Clearly volatile history has shaped the geology of Mars, Venus, and Earth, but there are numerous open questions about both volatile evolution and how it couples to planetary evolution. Here we summarize what is known about volcanism, tectonism, and convection for the three planets and how these characteristics may relate to volatile history. We also point out apparently contradictory observations, and future measurements that could lead to a more complete understanding of the links between volatiles and interior evolution.

Our first order understanding is of the geologic evolution of the terrestrial planets is that Earth is in an active mode of convection while Venus and Mars are in a stagnant lid mode. The transition between active and stagnant lid convection occurs when convective stresses can no longer strongly deform the lithosphere. We infer that the early high

temperatures on Mars and Venus and the associated vigorous convection and thin lithospheres allowed for active lid, plate-tectonics like deformation. But no direct evidence of early plate tectonics is observed on either planet. On Mars, the dichotomy boundary has been proposed to be a plate tectonic boundary (Lenardic et al., 2004), but the interpretation is not unique. The preservation of huge impact basins on Mars indicates that any possible plate tectonic episode was relatively short lived (Frey, 2006). On Venus, active lid tectonics could have occurred within the last b.y. but resurfacing has erased clues to past history.

Assuming that Venus and Mars have undergone a transition between from active to stagnant lid convection, what caused the transition? What were the relative roles of volatiles and temperature? Very early in a planet's history the extremely high temperatures are likely to dominant the rheologic behavior. Cooling of the planet will increase the strength of the lithosphere and decrease the convective vigor of the interior. Mars clearly cooled more rapidly than Venus, consistent with an early transition.

After an initial period of cooling, volatiles can have as large, or even larger effect on tectonic deformation. For Mars this occurs within the first 1 b.y. (Guest and Smrekar, 2007). Although the early, temperature dominated stage of the three planets may have been similar, the geologic and volatile evolution have likely been different for the majority of the planets' history. On Earth, volatiles are abundant at the surface, in the atmosphere, and are continually being recycled back into the interior via subduction. On Mars, meteorite analyses largely support water in the interior. On Venus, we have only indirect evidence of interior volatile content, like the apparent lack of a low viscosity zone. Yet the atmospheres of both Mars and Venus indicate very substantial decreases in water content. On Mars, surface mineralogy also indicates an abrupt transition from minerals formed in or modified by water to an arid environment very early in Martian history (Bibring et al., 2005; Murchie et al., 2007).

What is responsible for the loss of water in the atmospheres of Venus and Mars? On Mars, loss of the magnetic field may have exposed the atmosphere to erosion by solar wind (e.g. Perezdetejada, 1992). Alternatively the water may simply have condensed from the atmosphere and be frozen in the crust. On Venus, possible causes of atmospheric water loss are less clear. The chemistry of the atmosphere is complex, and the volatile content may have evolved over time in response to outgassing events (Bullock and Grinspoon, 1996, 1999). One possibility is that outgassing to the atmosphere created greenhouse warming. A persistent increase in surface temperature over 100s of m.y. could affect mantle temperatures and thus viscosity and melting (Phillips et al., 2001b), possibly even initiating a change from active to stagnant lid convection (Robin et al., 2007). Fractures observed on the surface may have formed in response to such climate-induced temperature variations (Anderson and Smrekar, 1999; Solomon et al., 1999). However the alternative has also been proposed, namely that stagnant lid convection caused the interior to heat up leading to massive melting and volcanic resurfacing of Venus (Reese et al., 1999). The possibility that stagnant lid convection could lead to outgassing or vice versa demonstrates the complexity of the feedback between volatiles and interior evolution.

Another implication of a rise in interior temperature due stagnant lid convection may be the number of mantle plumes and thus the rate of interior outgassing. The differences in mantle plumes between the three planets are one manifestation of variations in interior convection. The number of large active plumes originating at the CMB on Venus and Earth is a topic of debate, but is probably similar (Stofan and Smrekar, 2005). To first order the match in the number of plumes implies a similarity in interior viscosity, possibly indicating a tradeoff between interior temperature and volatile content. In addition, Venus has over 500 coronae, which may be additional smaller scale (average diameter of 250–300 km) mantle upwellings originating at a shallower interface. The fact that coronae are not observed on Earth has been interpreted in different ways. One possibility is that the low viscosity zone and thick continental lithosphere on Earth prevent the surface manifestation of corona-sized plumes at Earth's surface (Smrekar and Stofan, 1997). Another is that coronae represent a manifestation of a transition from a active regime, during which the large scale volcanic rises were formed, to a stagnant stage, which should generate more plumes (Jellnik et al., 2002). Overall the implication of an Earth-like number of large scale plumes plus over 500 coronae on Venus is not at all

clear. The fact that some coronae and some large volcanic rises appear to be active is problematic for the stagnant lid transistion hypothesis. The comparable number of large volcanic rises or hotspots on Venus and Earth suggest a similar interior viscosity, yet there is evidence that Venus lacks a low viscosity asthenosphere. Nor is it clear why coronae are found only on Venus.

The Martian Tharsis rise is the largest plume in the solar system, as well as likely the longest lived (Phillips et al., 2001a; Hartmann et al., 1999). Both Tharsis and the dichotomy boundary have been proposed to form as a result of degree-one mantle plumes. The visocity in the upper mantle may need to be less than that of the lower mantle may be required to produce a degree-one plume (Roberts and Zhong, 2006). The long duration volcanism at Tharsis may have locally degassed the lithosphere and mantle lowering the viscosity (Guest and Smrekar, 2007).

Contrasts in tectonic style are a conspicuous difference between the three planets. The largest deformation zones on terrestrial continents are the mountain belts created near plate boundaries. The ocean basins and their fractures are also a direct result of active lid convection. On Venus the only clear mountain belts encircle a high plateau. There is no apparent evidence for an associated plate boundary. In addition there are numerous regions of intense deformation in the form of tessera plateaus as well as localized regions of tessera in the plains. The multiple directions of fracturing observed in tesserae are unlike anything observed on other terrestrial planets. Ridge belts and rift zones represent lower strain deformation zones. The multiple branches and local changes in orientations of even these more linear features suggest that they are not associated with a plate boundary.

A fundamental difference between active and stagnant lid modes is the opportunity for recycling of volatiles into the interior. On Earth, volatiles are carried back into the mantle by subduction. For stagnant lid planets, the only path is via lithospheric delamination, driven by either local heating of the lithosphere or by density differences due to phase changes or other compositional density differences. Volcanic features such as the extremely long channels and the festoons and pancake domes on Venus may offer evidence of the unusual chemical conditions that can be produced in the delamination melting environment (Elkins-Tanton et al., 2007). Additionally variations in crustal and elastic thickness on scales of ~500–1000 km without apparent correlation to surface geology may be a manifestation of lower lithospheric delamination on Venus (Anderson and Smrekar, 2006). On Mars, there are many significant gravity anomalies on Mars that have no obvious link to surface geology that may result from either volcanic intrusion or delamination.

We will learn more about atmospheric water and chemistry in general from the Venus Express mission. Future surface missions have the potential to revolutionize our understanding of the relationship between volatiles and interior processes. Geochemical measurements of elemental composition as well as mineralogy are now possible at much higher precision than at the time of the Venera landers. These measurements are key to understanding the history of the mantle, crust, and volatiles. Similarly, estimating the thickness of the crust, the thermal lithosphere, intierior heat flow, and the size of core would give essential constraints on thermal evolution.

It may never be possible to determine with certainty whether a decrease in volatiles or temperature or a combination caused the initial balance of forces to shift from an active to stagnant lid on Mars and Venus. For Venus, it may be possible for the mantle to reach high enough temperatures to shift back to an active lid. In any casewhat remains clear is that the transition to a stagnant lid has profound consequences for the interior convection, tectonic deformation, and volatile cycling. The greater our knowledge of how these processes function on Venus and Mars, the more insight we will have into not only how the volatiles and interior evolution are linked on Earth, but also on planets discovered around other suns.

Acknowledgements. The authors thank an anonymous reviewer, and in particular Roger Phillips, for their efforts in reviewing this manuscript. Their comments greatly improved the chapter. The authors also gratefully acknowledge NASA support in the form of grants from the Planetary Geology and Geophysics Program. The research by the first author of this chapter was carried out at the Jet Propulsion Laboratory, California Institute of Technology, under a contract with the National Aeronautics and Space Administration.

REFERENCES

Anderson, F. S., and S. E. Smrekar (1999), Tectonic effects of climate change on Venus, *J. Geophys. Res. Planets, 104*, 30,743–30,756.

Anderson, F. S., and S. E. Smrekar (2006), Global mapping of crustal and lithospheric thickness on Venus, *J. Geophys. Res. Planets, 111*(E8), E08006, doi:10.1029/2004JE002395.

Anderson, R. C., J. M. Dohm, M. P. Golombek, A. F. C. Haldermann, B. F. Franklin, K. L. Tanaka, J. Lias, and B. Peer (2001), Primary centers and secondary concentrations of tectonic activity through time in the western hemisphere of Mars, *J. Geophys. Res., 106*, 20,563–20,586.

Arkani-Hamed, J. (1994), On the thermal evolution of Venus, *J. Geophys. Res., 99*(E1), 2019–2034, doi:10.1029/93JE03172.

Arkani-Hamed, J., and M. N. Toksöz (1984), Thermal evolution of Venus, *Phys. Earth Planet In., 34*(4), 232–250 doi:10.1016/0031-9201(84)90065-7.

Arkani-Hamed, J., G. G. Schaber, and R. G. Strom (1993), Constraints on the thermal evolution of Venus inferred from Magellan data, *J. Geophys. Res., 98*(E3), 5309–5315, doi:10.1029/93JE00052.

Arndt, N., C. Chauvel, G. Czamanske, and V. Fedorenko (1998), Two mantle sources, two plumbing systems: tholeiitic and alkaline magmatism of the Maymecha River basin, Siberian flood volcanic province, *Contrib. Mineral Petr., 133*, 297–313.

Aubaud, C., E. H. Hauri, and M. M. Hirschmann (2004), Hydrogen partition coefficients between nominally anhydrous minerals and basaltic melts, *Geophys. Res. Lett., 31*(20), L20611, doi:10.1029/2004GL021341.

Baker, M. B., T. L. Grove, and R. Price (1994), Primitive basalts and andesites from the Mt. Shasta region, N. California: products of varying melt fraction and water content, *Contrib. Mineral Petr., 118*, 111–129.

Baker, M. B., and E. M. Stolper (1994), Determining the composition of high-pressure mantle melts using diamond aggregates, *Geochim. Cosmochim. Acta 58*, 2,811–2,827.

Banerdt, W. B., G. E. McGill, and M. T. Zuber (1997), Plains tectonics on Venus, in *Venus II: Geology, Geophysics, Atmosphere, and Solar Wind Environment*, edited by S. W. Bougher et al., pp. 901–930, Univ. of Arizona Press, Tucson, AZ.

Basilevsky, A.T. and G.E. McGill (2007), Surface evolution of Venus, this book

Basilevsky, A. T., J. W. Head, G. G. Schaber, and R. G. Strom (1997), The Resurfacing History of Venus, in *Venus II: Geology, Geophysics, Atmosphere, and Solar Wind Environment*, edited by S. W. Bougher et al., pp. 1047–1084, Univ. of Arizona Press, Tucson, AZ.

Bell, D. R., G. R. Rossman, and R. O. Moore (2004), Abundance and partitioning of OH in a high-pressure magmatic system: Megacrysts from the Monastery kimberlite, *J. Petrol., 45*, 1539–1564.

Benn, K., J.-C. Mareschal, and K. C. Condie (Eds) (2005), *Archean Geodynamic Processes, Geophys. Monogr. Ser.*, vol. 164, 320 pp., AGU, Washington, D. C.

Bercovici, D. (1993), A simple-model of plate generation from mantle flow, *Geophys. J. Int., 114*, 635–650.

Bercovici, D. (1998), Generation of plate tectonics from lithosphere-mantle flow and void-volatile self-lubrication, *Earth Planet. Sci. Lett., 154*, 139–151.

Bibring, J.-P., Y Langevin, A. Gendrin., B. Gondet, F. Poulet, M. Berthé, A. Soufflot, R. Arvidson, N. Mangold, J. Mustar, P. Drossart, , and the OMEGA team (2005), Mars Surface Diversity as Revealed by the OMEGA/Mars Express Observations, *Science, 307*(5715), 1576–1581, doi:10.1126/science.1108806.

Bindschadler, D. L., and J. W. Head (1991), Tessera terrain, Venus: Characterization and models for origin and evolution, *J. Geophys. Res., 96*(B4), 5889–5907, doi:10.1029/90JB02742.

Bindschadler, D.L., and E. M. Parmentier (1990), Mantle flow tectonics: the influence of a ductile lower crust and implications for the formation of topographic uplands on Venus, *J. Geophys. Res., 95*(B13), 21,329–21,344, 10.1029/90JB01602.

Bindschadler, D.L., G. Schubert, and W. M. Kaula (1992), Cold-spots and hotspots: Global tectonics and mantle dynamics of Venus. *J. Geophys. Res., 97*(E8), 13,495–13,532, doi:10.1029/92JE01165

Blichert-Toft, J., J. D. Gleason, P. Télouk, and F. Albarède (1999), The Lu-Hf geochemistry of shergottites and the evolution of the Martian mantle-crust system, *Earth Planet. Sci. Lett., 173*(1–2), 25–39, doi:10.1016/S0012-821X(99)00222-8.

Bogard, D. D., R. N. Clayton, K. Marti, T. Owen, and G. Turner (2001), Martian volatiles : Isotopic composition, origin, and evolution, *Space Sci. Rev., 96*(1–4), 425–458, doi:10.1023/A:1011974028370.

Boland, J. N., and T. E. Tullis (1986), Deformation behavior of wet and dry clinopyroxenite in the brittle to ductile transition region techniques, in *Mineral and Rock Deformation: Laboratory Studies, The Paterson Volume, Geophys. Monogr. Ser.,* vol. 36, edited by B. E. Hobbs and H. C.Heard, pp. 297–324, AGU, Washington, D. C.

Bolfan-Casanova, N., H. Keppler, and D. C. Rubie (2000), Water partitioning between nominally anhydrous minerals in the MgO-SiO_2-H_2O system up to 24 GPa: implications for the distribution of water in the Earth's mantle, *Earth Planet. Sci. Lett. 182*(3–4), 209-221, doi:10.1016/S0012-821X(00)00244-2.

Bond, T. M., and M. R. Warner (2006), Dating Venus: Statistical Models of Magmatic Activity and Impact Cratering, 37th Annual Lunar and Planetary Science Conference, Lunar and Planetary Institute, League City, Texas, 13–17 March, abstract no.1957.

Borg, L. E., L. E. Nyquist, H. Wiesmann, and Y. Reese (2002), Constraints on the petrogenesis of Martian meteorites from the Rb-Sr and Sm-Nd isotopic systematics of the lherzolitic shergottites ALH77005 and LEW88516, *Geochim. Cosmochim. Acta 66*(11), 2037–2053, doi:10.1016/S0016-7037(02)00835-9.

Brace, W. F., and D. L. Kohlstedt (1980), Limits on lithostatic stress imposed by laboratory experiments, *J. Geophys. Res., 85*, 6248–6252.

Breuer, D., D. A. Yuen, and T. Spohn (1997), Phase transitions in the Martian mantle: Implications for partially layered convection, *Earth Planet. Sci. Lett. 148*(3–4), 457–469, doi:10.1016/S0012-821X(97)00049-6.

Breuer, D., D. A. Yuen, T. Spohn, and S. Zhang (1998), Three-dimensional models of Martian mantle convection with phase transitions, *Geophys. Res. Lett., 25*, 229–232.

Bullock, M. A., and D. H. Grinspoon (1999), Global climate change on Venus, *Sci. Am., 280*, 50–57.

Bullock, M.A., and D.H. Grinspoon (1996), The stability of climate on Venus, *J. Geophys. Res., 101*(E3), 7521-7530, doi:10.1029/95JE03862.

Byerlee, J. D. (1978), Friction of rocks, *Pure Appl. Geophys., 116*, 615–626.

Bystricky, M., and S. Mackwell (2001), Creep of dry clinopyroxene aggregates. *J. Geophys. Res., 106*(B7), 13,443–13,454, doi:10.1029/2001JB000333, 2001.

Campbell, B.A. (1999), Surface formation rates and impact crater densities on Venus. *J. Geophys. Res., 104*(E9), 21,951–21,956, doi:10.1029/1998JE000607.

Campbell, D. B., J. W. Head, J. K. Harmon, and A. A. Hine (1984), Venus volcanism and rift formation in Beta Regio, *Science, 226*, 167–170.

Caristan, Y. (1982), The transition from high-temperature creep to fracture in Maryland Diabase. *J. Geophys. Res., 87*, 6781–6790.

Chen, S., T. Hiraga, and D. L. Kohlstedt (2006), Water weakening of clinopyroxene in the dislocation creep regime. *J. Geophys. Res. 111*(B8), B08203, doi:10.1029/2005JB003885.

Chopra, P. N., and M. S. Patterson (1984), The role of water in the deformation of dunite, *J. Geophys., Res., 89*, 7861–7876.

Connerney, J., M. H. Acuna, P. J. Wasilewski, N. F. Ness, H. Reme, C. Mazelle, D. Vignes, R. P. Lin, D. L. Mitchell, and P. A. Cloutier (1999), Magnetic lineations in the ancient crust of Mars, *Science, 284*(5415), 794–798, doi:10.1126/science.284.5415.794.

Courtillot, V., A. Davaille, J. Besse, and J. Stock (2003), Three distinct types of hotspots in the Earth's mantle, *Earth Planet. Sci. Lett., 205*(3-4), 295–308, doi:10.1016/S0012-821X(02)01048-8.

Dalton, J. A., and D. C. Presnall (1998), Carbonatitic melts along the solidus of model lherzolite in the system CaO-MgO-Al_2O_3-SiO_2CO_2 from 3 to 7 GPa, *Contrib. Mineral Petr., 131*(2–3), 123–135, doi: 10.1007/s004100050383.

Dann, J. C., A. H. Holzheid, T. L. Grove, and H. Y. McSween, Jr. (2001), Phase equilibria of the Shergotty meteorite: Constraints on the pre-eruptive water contents of martian magmas and fractional crystallization under hydrous conditions, *Meteorit. Planet. Sci., 36*, 793–806.

Dasgupta, R., M. M. Hirschmann, and A. C. Withers (2004), Deep global cycling of carbon constrained by the solidus of anhydrous, carbonated eclogite under upper mantle conditions, *Earth Planet. Sci. Lett., 227*(1–2), 73–85, doi:10.1016/j.epsl.2004.08.004.

Dasgupta, R., M. M. Hirschmann, and K. Stalker (2006), Immiscible transition from carbonate-rich to silicate-rich melts in the 3 GPa melting interval of eclogite + CO_2 and genesis of silica-undersaturated ocean island basalts, *J. Petrol., 47*, 647–671, doi:10.1093/petrology/egi088.

Davaille, A., and C. Jaupart (1993), Transient high-Rayleigh-number thermal convection with large viscosity variations, *J. Fluid. Mech., 253*, 141–166, doi:10.1017/S0022112093001740.

Davies, G. F. (1992a), Temporal variation of the Hawaiian plume flux, *Earth Planet. Sci. Lett., 113*(1–2), 277286, doi:10.1016/0012-821X(92)90225-K.

Davies, G. F. (1992b), On the emergence of plate tectonics, *Geology, 20*(11), 963–966, doi:10.1130/0091-7613(1992)020<0963:OTEOPT>2.3.CO;2.

Davies, G. F. (1998), Topography: a robust constraint on mantle fluxes, *Chem. Geol., 145*, 479–489.

Deschamps, F., and C. Sotin (2000), Inversion of two-dimensional numerical convection experiments for a fluid with a strongly temperature-dependent viscosity, *Geophys. J. Int., 143*(1), 204–218, doi:10.1046/j.1365-246x.2000.00228.x.

Di Toro, G., D. L. Goldsby, and T. E. Tullis (2004), Friction falls towards zero in quartz rock as slip velocity approaches seismic rates, *Nature, 427*(6973), 436–438, 10.1038/nature02249.

Donahue, T. M., D. H. Grinspoon, R. E. Hartle, and R. R. Hodges, Jr. (1997), Ion/neutral escape of hydrogen and deuterium: Evolution of water, *Venus II: Geology, Geophysics, Atmosphere, and Solar Wind Environment*, edited by S. W. Bougher et al., pp. 385–414, Univ. of Arizona Press, Tucson, AZ.

Dreibus, G., and H. Wänke (1985), Mars: A volatile-rich planet, *Meteoritics, 20*, 367–381.

Ducea, M., and J. Saleeby (1998), A case for delamination of the deep batholithic crust beneath the Sierra Nevada, California, *Int. Geol. Rev., 40*, 78–93.

Dupeyrat, L., and C. Sotin (1995), The effect of the transformation of basalt to eclogite on the internal dynamics of Venus, *Planet. Space Sci., 43*, 909–921, doi:10.1016/0032-0633(94)00227-I.

Elkins-Tanton, L. T. and B.H. Hager (2000), Melt intrusion as a trigger for lithospheric foundering and the eruption of the Siberian flood basalt, *Geophys. Res. Lett., 27*(23), 3937–3940, doi:10.1029/2000GL011751.

Elkins-Tanton, L. T., and E. M. Parmentier (2005), The fate of water in the Martian magma ocean and the formation of an early atmosphere, 36th Annual Lunar and Planetary Science Conference, Lunar and Planetary Institute, League City, Texas, 14–18 March, abstract no. 1906.

Elkins-Tanton, L. T., and T. L. Grove (2003), Evidence for deep melting of hydrous, metasomatized mantle: Pliocene high potassium magmas from the Sierra Nevadas, *J. Geophys. Res. 108*(B7), 2350, doi:10.1029/2002JB002168.

Elkins-Tanton, L. T., E. M. Parmentier, and P. C. Hess (2005), The formation of ancient crust on Mars through magma ocean processes, *J. Geophys. Res. 110*, E12S01, doi:10.1029/2005JE002480.

Elkins-Tanton, L. T., S. Zaranek, and E. M. Parmentier (2005), Early magnetic field and magmatic activity on Mars from magma ocean overturn, *Earth Planet. Sci. Lett., 236*(1-2), 1–12, 10.1016/j.epsl.2005.04.044.

Elkins-Tanton, L. T., S. E. Smrekar, P. C. Hess, and E. M. Parmentier (2007), Volcanism and volatile recycling on a one-plate planet: Applications to Venus, *J. Geophys. Res., 112*(E4), E04S06, doi:10.1029/2006JE002793.

Escartin, J., and G. Hirth (1997), Effects of serpentinization on the lithospheric strength and style of normal faulting at slow-spreading ridges, *Earth Planet. Sci. Lett. 151*, 181–189.

Esposito, L. W. (1984), Sulfur dioxide -Episodic injection show evidence for active Venus volcanism, *Science, 223*, 1072–1074.

Evans, B., and C. Goetze (1979), Temperature variation of hardness of olivine and its implication for polycrystalline yield stress, *J. Geophys. Res., 84*, 5505–5524.

Evans, B., and D. L. Kohlstedt (1995), Rheology of rocks, in *Rock Physics and Phase Relations: A Handbook of Physical Constants*, edited by T.J. Ahrens, pp. 148–164, AGU, Washington, D.C..

Falloon, T. J., D. H. Green, L. V. Danyushevsky, and U. H. Faul (1999), Peridotite melting at 1.0 and 1.5 GPa: An experimental evaluation of techniques using diamond aggregates and mineral mixed for determination of near-solidus melts, *J. Petrol., 40*, 1,343–1,375.

Farmer, G. L., A. F. Glazner, and C. R. Manley (2002), Did lithospheric delamination trigger late Cenozoic potassic volcanism in the southern Sierra Nevada, California? *Geol. Soc. Am. Bull., 114*, 754–768.

Fedorenko, V. A., and G. K. Czamanske (1997), Results of new field and geochemical studies of the volcanic and intrusive rocks of the Maymecha-Kotuy area, Siberian flood-basalt province, Russia, *Int. Geol. Rev., 39*, 479–531.

Foley, C. N., M. Wadhwa, L. E. Borg, P. E. Janney, R. Hines, and T. L. Grove (2005), The early differentiation history of Mars from 182W-142Nd isotope systematics in the SNC Meteorites, *Geochim. Cosmochim. Acta., 69*, 4557–4571.

Foley, C. N., M. Wadhaw, L. E. Borg, P. E. Janney, R. Hines, and T.L. Grove (2005), The early differentiation history of Mars from W-182-Nd-142 isotope systematics in the SNC meteorites, *Geochim. Cosmochim. Acta., 69*, 4557–4571.

Forneris, J. F., and J. R. Holloway (2003), Phase equilibria in subducting basaltic crust: implications for H_2O release from the slab, *Earth Planet. Sci. Lett., 214*(1–2), 187–201, doi:10.1016/S0012-821X(03)00305-4.

Foster, A., and F. Nimmo (1996), Comparisons between the rift systems of East Africa, Earth and Beta Regio, Venus, *Earth Planet. Sci. Lett., 143*, 183–195.

Fowler, A. C. (1985), Fast thermoviscous convection, *Stud. Appl. Math., 72*, 189–219.

Frey, H. V., S. E. Sakimoto, and J. Roark (1998), The MOLA topographic signature at the crustal dichotomy boundary zone on Mars, *Geophys. Res. Lett., 25*(24), 4409–4412, doi:10.1029/1998GL900095.

Frey, H. V. (2006), Impact constraints on, and a chronology for, major events in early Mars history, *J. Geophys. Res., 111*(E8), E08S91, doi:10.1029/2005JE002449.

Frey, H. V., J. H. Roark, K. M. Shockey, E. L. Frey, and S. E. H. Sakimoto (2002), Ancient lowlands on Mars, *Geophys. Res. Lett., 29*(10), 1384, doi:10.1029/2001GL013832.

Frey, H., and R. A. Schultz (1988), Large Impact Basins and the Mega-Impact Origin for the Crustal Dichotomy on Mars, *Geophys. Res. Lett., 15*, 229–232.

Gaetani, G. A, and T. L. Grove (1998), The influence of water on melting of mantle peridotites, *Contrib. Mineral. Petr., 131*, 323–346.

Ghent, R. R., R. J. Phillips, V. L. Hansen, and D. C. Nunes (2005), Finite element modeling of short-wavelength folding on Venus: Implications for the plume hypothesis for crustal plateau formation, *J. Geophys. Res., 110*(E8), E11006, doi:10.1029/2005JE002522.

Goetze, C. and B. Evans (1979), Stress and Temperature in the bending lithosphere as constrained by experimental rock mechanics, *Geophys. J. Int., 59*, 463–478.

Grimm, R. E. (1994), Recent deformation rates on Venus, *J. Geophys. Res., 99*(E11), 23,163–23,171, doi:10.1029/94JE02196.

Grimm, R. E., R. J. and Phillips (1991), Gravity anomalies, compensation mechanisms, and the geodynamics of western Ishtar Terra, Venus, *J. Geophys. Res., 96*, 8305–8324.

Grindrod, P. M., F. Nimmo, E. R. Stofan, and J. E. Guest (2005), *J. Geophys. Res., 110*(E12), E12002, doi:10.1029/2005JE002416.

Grinspoon, D. H. (1993), Implications of the high D/H ratio for the sources of water in Venus' atmosphere, *Nature, 363*, 428–431.

Gudfinnsson, G. H., and D. C. Presnall (2002), The minimum potential temperature of the Hawaiian mantle is about 1420°C, *Eos Trans. AGU, 83*(47), F1415, Fall Meeting Suppl Abstract V52D-06.

Gudfinnsson, G. H., and D. C. Presnall (2005), Continuous gradations among primary carbonatitic, kimberlitic, melilititic, basaltic, picritic, and komatiitic melts in equilibrium with garnet lherzolite at 3-8 GPa, *J. Petrol., 46*(8), 1645–165, doi:101093/petrology/egi029

Guest, A., and S. E. Smrekar (2007), Constraints on the thermal and volatile evolution of Mars from relaxation models of the Martian dichotomy and elastic thickness predictions, submitted.

Hammouda, T. (2003), High-pressure melting of carbonated eclogite and experimental constraints on carbon recycling and storage in the mantle, *Earth Planet. Sci. Lett., 214*, 357–368.

Hansen, V. L., and R. J. Phillips (1993), Tectonics and volcanism of Eastern Aphrodite Terra: No subduction, no spreading, *Science, 260*, 526–530.

Hansen, V. L., B. K. Banks, and R. R. Ghent (1999), Tessera terrain and crustal plateaus, *Geology, 27*, 1071–1074.

Hartmann, W. K., M. Malin, A. McEwen, M. Carr, L. Soderblom, P. Thomas, E. Danielson, P. James, and J. Veverka (1999), Evidence for recent volcanism on Mars from crater counts, *Nature, 397*, 586–589.

Hauck, S. A., R. J. Phillips, and M. H. Price (1998), Venus: Crater distribution and plains resurfacing models, *J. Geophys. Res., 103*, 13,635–13,642.

Hauri, E. H., G. A. Gaetani, and T. H. Green (2006), Partitioning of water during melting of the Earth's upper mantle at H_2O undersaturated conditions, *Earth Planet. Sci. Lett. 248*, 715–734.

Head, J. W., J. F. Mustard, M. A. Kreslevsky, R. Milliken, and D. R. Marchant (2003), Recent ice ages on Mars, *Nature, 426*, 797–802.

Head, J. W., L. S. Crumpler, J. C. Aubele, J. E. Guest, and R. S. Saunders (1992), Venus Volcanism—Classification of volcanic features and structure, associations, and global distribution from Magellan data, *J. Geophys. Res. Planets, 97*, 13,153–13,197.

Herd, C. D. K., L. E. Borg, J. H. Jones, and J. J. Papike (2002), Oxygen fugacity and geochemical variations in the Martian basalts: Implications for Martian basalt petrogenesis and the oxidation state of the upper mantle of Mars, *Geochim. Cosmochim. Acta., 66*, 2025–2036.

Herrick, R. R. (2006), 37th Annual Lunar and Planetary Science Conference, Lunar and Planetary Institute, League City, Texas, 13–17 March, abstract no. 1588.

Herrick, R. R. (1994), Resurfacing history of Venus, *Geology, 22*, 703–706

Herrick, R. R., and N. K. Forsberg-Taylor (2003), The shape and appearance of craters formed by oblique impact on the Moon and Venus, *Meteoritics & Planetary Science*, 38, 1551–1578.

Hirose, K. (1997), Partial melt compositions of carbonated peridotite at 3 GPa and the role of CO_2 in alkali-basalt magma generation, *Geophys. Res. Lett., 24*(22), 2837–2840, doi:10.1029/97GL02956.

Hirschmann, M. P. (2000), Mantle solidus: Experimental constraints and the effects of peridotite composition, *Geochem. Geophys. Geosyst., 1*(10), doi:10.1029/2000GC000070.

Hirth, G., and D. L. Kohlstedt (1996), Water in the oceanic upper mantle: Implications for rheology, melt extraction and the evolution of the lithosphere, *Earth Planet. Sci. Lett. 144*, 93–108.

Hirth, G., and D. L. Kohlstedt (2003), Rheology of the Upper Mantle and the Mantle Wedge: A View From the Experimentalists, in *Inside the Subduction Factory, Geophys. Monogr. Ser.*, vol. 138, 83–105, AGU, edited by J. Eiler, AGU, Washington, D.C.

Hyndman, R. D., and K. Wang (1993), Thermal constraints on the zone of major thrust earthquake failure—the cascadia subduction zone, *J. Geophys. Res., 98*, 2039–2060.

Hyndman, R. D., M. Yamano, and D. A. Oleskevich (1997), The seismogenic zone of subduction thrust faults, *Island Arc, 6*, 244–260.

Jakosky, B. M., and R. J. Phillips (2001), Mars' volatile and climate history, *Nature, 412*, 237–244.

Johnson, M. C., M. J. Rutherford, and P. C. Hess (1991), Chassigny petrogenesis: Melt compositions, intensive parameters, and water contents of martian magmas, *Geochim. Cosmochim. Acta., 55*, 349–366.

Johnson, C. L., and M. A. Richards (2003), A conceptual model for the relationship between coronae and large-scale mantledynamics on Venus, *J. Geophys. Res., 108*(E6), 5058, doi:10.1029/2002JE001962.

Jones, J. H. (2004), The Edge of Wetness: The Case for Dry Magmatism on Mars, 35th Annual Lunar and Planetary Science Conference, Lunar and Planetary Institute, League City, Texas, 15–19 March, abstract no. 1798.

Jull, M., and P. B. Kelemen (2001), On conditions for lower crustal convective instability, *J. Geophys. Res., 106*, 6,423–6,446.

Jull, M. G., and J. Arkani-Hamed (1995), The implications of basalt in the formation and evolution of the mountains on Venus, *Phys. Earth Planet. In., 89*, 163–175.

Karato, S., and H. Jung (1998), Water, partial melting and the origin of the seismic low velocity and high attenuation zone in the upper mantle, *Earth Planet. Sci. Lett., 157*, 193–207.

Kargel, J. S., R. L. Kirk, B. Fegley, Jr., and A. H. Treiman (1994), Carbonate-sulfate volcanism on Venus?, *Icarus, 112*(1), 219–252, doi:10.1006/icar.1994.1179.

Karner, J., J. J. Papike, and C. K. Shearer (2004), Plagioclase from planetary basalts: Chemical signatures that reflect planetary volatile budgets, oxygen fugacity, and styles of igneous differentiation, *Am. Mineral., 89*, 1101–1109.

Kaula, W. M. (1975), The seven ages of a planet, *Icarus, 26*(1), 1–15 doi:10.1016/0019-1035(75)90138-4.

Kaula, W.M. (1990), Venus: A contrast in evolution to Earth, *Science, 247*, 1191–1196.

Kaula, W.M. (1995), Venus reconsidered, *Science, 270*, 1460–1464.

Kay, R. W., and S. M. Kay (1993), Delamination and delamination magmatism, *Tectonophysics, 219,* 177–189.

Ke, Y., and V. S. Solomatov (2006), Early transient superplumes and the origin of the Martian crustal dichotomy, *J. Geophys. Res., 111*(E10), E10001, doi:10.1029/2005JE002631.

Keppler, H., M. Wiedenbeck, and S. S Shcheka (2003), Carbon solubility in olivine and the mode of carbon storage in the Earth's mantle, *Nature, 424,* 414–416.

Kidder, J. G., and R. J. Phillips (1996), Convection-driven sub-solidus crustal thickening on Venus, *J. Geophys. Res., 101,* 23,181–23,194.

Kiefer, W. S., and B. H. Hager (1991), A mantle plume model for the equatorial highlands of Venus. *J. Geophys. Res., 96,* 20,947–20,966.

Kiefer, W. S., and B. H. Hager (1991), Mantle downwelling and crustal convergence—A model for Ishtar Terra, Venus, *J. Geophys. Res., 96,* 20,967–20,980.

Kiefer, W., M. A. Richards, B. H. Hager, and B. G. Bills (1986), A dynamic model of Venus' gravity field, *Geophys. Res. Lett., 13,* 14–17.

Kinzler, R. J., and T. L. Grove (1992), Primary magmas of mid-ocean ridge basalts 2. Applications, *J. Geophys. Res., 97,* 6,907–6,926.

Kirby, S. H., and A. K. Kronenberg (1984). Deformation of clino-pyroxenite: Evidence for a transition in flow mechanisms and semibrittle behavior, *J. Geophys. Res., 89,* 3177–3192.

Kleine, T., K. Mezger, H. Palme, and C. Munker (2004), The W iso-tope evolution of the bulk silicate Earth: constraints on the timing and mechanisms of core formation and accretion, *Earth Planet. Sci. Lett., 228*(1–2), 109–123, 10.1016/j.epsl.2004.09.023.

Klemme, S., J. D. Blundy, and B. J. Wood (2002), Experimen-tal constraints on major and trace element partitioning during partial melting of eclogite, *Geochim. Cosmochim. Acta., 66,* 3,109–3,123.

Koga, K., E. Hauri, M. Hirschmann, and D. Bell (2003), Hydrogen concentration analysis using SIMS and FTIR: Comparison and calibration for nominally anhydrous minerals, *Geochem. Geo-phys. Geosyst., 4*(2), 1019, doi:10.1029/2002GC000378.

Kohlstedt, D. L., B. Evans, and S. J. Mackwell (1995), Strength of the lithosphere: Constraints imposed by laboratory experiments, *J. Geophys. Res., 100,* 17,587–17,602.

Leitner, J. J., and M. G. Firneis (2006), A Review of Venusian Surface Heat Flow Estimates, AGU Chapman Conference on Exploring Venus as a Terrestrial Planet, AGU, Key Largo, Flor-ida, 13–17 February.

Lenardic, A., A. M. Jellinek, and L.-N. Moresi (2007), The poten-tial of a climate change induced transition in the global tectonic style of a terrestrial planet, *Earth Planet. Sci. Lett.,* in revision.

Lenardic, A., F. Nimmo, and L. Moresi, Growth of the hemispheric dichotomy and the cessation of plate tectonics on Mars, *J. Geo-phys. Res., 109,* E02003, doi:10.1029/2003JE002172, 2004.

Lenardic, W., W. M. Kaula, and D. L. Bindschadler (1995), Some effects of a dry crustal flow law on numerical simulations of coupled crustal deformation and mantle convection on Venus, *J. Geophys. Res., 100,* 16,949–16,957.

Litasov, K., and E. Ohtani (2003), Stability of various hydrous phases in CMAS pyrolite-H_2O system up to 25 GPa, *Phys. Chem. Mineral., 30,* 147–156.

Lodders, K., and B. Fegley Jr. (1997), An oxygen isotope model for the composition of Mars, *Icarus, 126,* 373–394.

Longhi, J. (1991), Complex magmatic processes on Mars: Inferences from the SNC meteorites, in *Lunar and Planetary Science Confer-ence, 21st, Houston, TX, Mar. 12–16, 1990, Proceedings (A91-42332 17-91),* 695–709, Lunar and Planetary Institute, Houston, Texas.

Mackwell, S. J., D. L. Kohlstedt, and M. S. Paterson (1985), The role of water in the deformation of olivine single crystals, *J. Geophys. Res., 90*(B13), 11,319–11,333.

Mackwell, S. J., M. E. Zimmerman, and D. L. Kohlstedt (1998), High-temperature deformation of dry diabase with applications to tectonics on Venus, *J. Geophys. Res., 103*(B1), 975–984.

Malin, M. C., and K. Edgett (2000), Evidence for recent ground-water seepage and surface runoff on Mars, *Science, 288*(5475), 2330–2335, doi:10.1126/science.288.5475.2330.

Malin, M. C., K. S. Edgett, L. V. Posiolova, S. M. McColley, and E. Z. Noe Dobrea (2006), Present-day impact cratering rate and contemporary gully activity on Mars, *Science, 314*(5805), 1573–1577, doi:10.1126/science.1135156.

Masursky, H., E. Eliason, P. G. Ford, G. E. McGill, G. H. Petten-gill, G. G. Schaber, and G. Schubert (1980), Pioneer Venus radar results: Geology from images and altimetry, *J. Geophys. Res., 85,* 8232–8260.

McGill, G. (1994), Hotspot evolution and Venusian tectonic style, *J. Geophys. Res., 99*(E11), 23,149–23,161.

McGill, G. E., S. J. Steenstrup, C. Barton, and P. G. Ford (1981), Continental rifting and the origin of Beta-Regio Venus, *Geophys. Res. Lett., 8,* 737–740.

McGovern, P. J., and G. Schubert (1989), Thermal evolution of the Earth: Effects of volatile exchange between atmosphere and inte-rior, *Earth Planet. Sci. Lett., 96*(1–2), 27–37, doi:10.1016/0012-821X(89)90121-0.

McKinnon, W. B., K. J. Zahnle, B. A. Ivanov., and H. J. Melosh (1997), Cratering on Venus: Models and observations, in *Venus II: Geology, Geophysics, Atmosphere, and Solar Wind Environ-ment,* edited by S. W. Bougher, D. M. Hunten, and R. J. Phillips, 969–1014, Univ. of Arizona Press, Tucson.

McSween, Jr., H., Y., and R. P. Harvey (1998), An evaporation model for formation of carbonates in the ALH84001 martian meteorite, *Int. Geol. Rev., 40,* 774–783.

McSween, Jr., H., Y., T. L. Grove, R. C. F. Lentz, J. C. Dann, A. H. Holzheid, L. R. Riciputi, and J. G. Ryan (2001), Geo-chemical evidence for magmatic water within Mars from pyroxenes in the Shergotty meteorite, *Nature, 409,* 487–490, doi:10.1038/35054011.

Médard, E., and T. L. Grove (2006), Early hydrous melting and degassing of the Martian interior, *J. Geophys. Res., 111,* E11003, doi:10.1029/2006JE002742.

Mei, S., and D. L. Kohlstedt (2000), Influence of water on plastic deformation of olivine aggregates 1. Diffusion creep regime, *J. Geophys. Res., 105*(B9), 21,457–21,470. doi:10.1029/2000JB900179.

Moore, H. J., J. J. Plaut, P. M. Schenk, and J. W. Head (1992), An unusual volcano on Venus, *J. Geophys. Res.*, *97*(E8), 13,479–134,493.

Moresi, L.-N., and V. S. Solomatov (1998), Mantle convection with a brittle lithosphere: Thoughts on the global tectonic styles of the Earth and Venus, *Geophys. J. Int., 133*, 669–682.

Morris, S., and D. Canright (1984), A boundary layer analysis of Bernard convection in a fluid of strongly temperature-dependent viscosity, *Phys. Earth Planet. Int., 36*, 355–373.

Namiki, N., and S. C. Solomon (1993), The gabbro eclogite phase transition and the elevations of mountain belts on Venus, *J. Geophys. Res., 98*(8), 15,025–15,031.

Newton, R. C., and W. E. Sharp (1975), Stability of forsterite+CO_2 and its bearing on the role of CO_2 in the mantle, *Earth Planet. Sci. Lett., 26*, 239–244.

Nimmo, F., and D. McKenzie (1997), Convective thermal evolution of the upper mantles of Earth and Venus, *Geophys. Res. Lett., 24*(12), 1539–1542.

Nimmo, F. (2005), Tectonic consequences of Martian dichotomy modification by lower crustal flow and erosion, *Geology, 33*(7), 533–536, doi:10.1130/G21342.1.

Nimmo, F., and D. J. Stevenson (2000), Influence of early plate tectonics on the thermal evolution and magnetic field of Mars, *J. Geophys. Res., 105*(E5), 11,969–11,979, doi:10.1029/1999JE001216.

Author(s) TBD (1990), title TBD, publisher TBD, Norwood, Chichester, England.

Nunes, D. C., R. J. Phillips, C. D. Brown, and A. J. Dombard (2004), Relaxation of compensated topography and the evolution of crustal plateaus on Venus, *J. Geophys. Res., 109*, E01006, doi:10.1029/2003JE002119.

O'Neill, C., A. Lenardic, L.-N. Moresi, T. Torsvik, and C. A. Lee (2007), Episodic precambrian subduction, *Nature,* in press.

Ohtani, E., K. Litasov, T. Hosoya, T. Kubo, and T. Kondo (2004), Water transport into the deep mantle and formation of a hydrous zone, *Phys. Earth Planet. Inter., 143–144*, 255–269, doi:10.1016/j.pepi.2003.09.015.

Owens, T. (1992), The composition and early history of the atmosphere of Mars, in *Mars*, edited by. H. H. Kieffer, B. M. Jakosky, C. W. Snyder, and M. S. Matthews, 818–834, Univ. Arizona Press, Tucson.

Papale, P. (1999), Modeling of the solubility of a two-component $H_2O + CO_2$ fluid in silicate liquids, *Am. Mineral., 84*, 477–492.

Parmentier, E. M., and P. C. Hess (1992), Chemical differentiation of a convecting planetary interior: Consequences for a one plate planet such as Venus, *Geophys. Res. Lett. 19*(20), 2015–2018.

Parmentier, E. M., C. Sotin, and B. Travis (1994), Turbulent 3D thermal convection in an infinite Prandtl number, volumetrically heated fluid: implications for mantle dynamics, *Geophys. J. Int, 116*, 241–251.

Parmentier, E. M., and M. T. Zuber (2007), Early evolution of Mars with mantle compositional stratification or hydrothermal crustal cooling, *J. Geophys. Res., 112*, E02007, doi:10.1029/2005JE002626.

Parsons, B., and S. Daly (1983), The relationship between surface topography, gravity anomalies, and temperature structure of convection, *J. Geophys. Res., 88*, 1129–1144.

Pavri, B., J. W. Head, III, K. B. Klose, and L. Wilson (1992), Steep-sided domes on Venus: Characteristics, geologic setting, and eruption conditions from Magellan data, *J. Geophys. Res., 97*(E8), 13,445–13,478.

Pertermann, M., and M. M. Hirschmann (2003), Anhydrous partial melting experiments on MORB-like eclogite: Phase relations, phase compositions and mineral-melt partitioning of major elements at 2–3 GPa, *J. Petrol., 44*, 2,173–2,201, doi:10.1093/petrology/egg074.

Pertermann, M., M. M. Hirschmann, K. Hametner, D. Günther, and M. W. Schmidt (2004), Experimental determination of trace element partitioning between garnet and silica-rich liquid during anhydrous partial melting of MORB-like eclogite, *Geochem. Geophys. Geosyst. 5*, Q05A01, doi:10.1029/2003GC000638.

Phillips, R. J. (1986), A mechanism for tectonic deformation on Venus, *Geophys. Res. Lett., 13*, 1141–1144.

Phillips, R., J., C. L. Johnson, S. J. Mackwell, P. Morgan, D. T. Sandwell, and M. T. Zuber (1997), Lithospheric mechanics and dynamics of Venus, in *Venus II: Geology, Geophysics, Atmosphere, and Solar Wind Environment*, edited by S. W. Bougher, D. M. Hunten, and R. J. Phillips, 1163–1204, Univ. of Arizona Press, Tucson.

Phillips, R. J., and M. C. Malin (1983), The interior of Venus and tectonic implications, in *Venus*, edited by D. M. Hunten, L. Colin, T. M. Donahue and V. I. Moroz, 159–214, Univ. Arizona Press, Tucson.

Phillips, R. J., and V. L. Hansen (1998), Geological evolution of Venus: Rises, plains, plumes, and plateaus, *Science, 279*(5356), 1492–1497, doi:10.1126/science.279.5356.1492.

Phillips, R. J., and V. L. Hansen (1994), Tectonic and magmatic evolution of Venus, *Ann. Rev. Planet. Sci., 22*, 597–654.

Phillips, R. J., and N. R. Izenberg (1995), Ejecta correlations with spatial crater density and Venus resurfacing history, *Geophys. Res. Lett., 22*, 1517–1520.

Phillips, R. J., R. E. Grimm, and M. C. Malin (1991), Hotspot evolution and the global tectonics of Venus, *Science, 252*, 651–658.

Phillips, R. J., M. A. Bullock, and S. A. Hauck, II (2001), Climate and interior coupled evolution on Venus, *Geophys. Res. Lett., 28*, 1779–1782.

Phillips, R. J., R. F. Raubertas, R. E. Arvidson, I. C. Sarkar, R. R. Herrick, N. R. Izenberg, and R. E. Grimm (1992), Impact craters and Venus resurfacing history, *J. Geophys. Res., 97*(E10), 15,923–15,948.

Reese, C. C., V. S. Solomatov, and L.-N. Moresi (1999), Non-Newtonian stagnant lid convection and magmatic resurfacing of Venus, *Icarus, 139*, 67–80, doi:10.1006/icar.1999.6088.

Reese, C. C., V. S. Solomatov, J. R. Baumgardner, D. R. Stegman, and A. V. Vezolainen (2004), Magmatic evolution of impact-induced Martian mantle plumes and the origin of Tharsis, *J. Geophys. Res., 109*, E08009, doi:10.1029/2003JE002222.

Richards, M. A., W. S. Yang, J. R. Baumgardner, and H. P. Bunge (2001), Role of a low-viscosity zone in stabilizing plate tectonics:

Implications for comparative terrestrial planetology, *Geochem. Geophys. Geosyst., 2*(8), U1–U16, doi:10.1029/2000GC000115.

Righter, K., and I. S. E. Carmichael (1996), Phase equilibria of phlogopite lamprophyres from western Mexico: biotite-liquid equilibria and P-T estimates for biotite-bearing igneous rocks, *Contrib. Min. Pet., 123*(1), 1–21, doi:10.1007/s004100050140.

Robin, C. M. I., A. M. Jellinek, V. Thayalan, and A. Lenardic (2007), Transient mantle convection on Venus: The paradoxical coexistence of highlands and coronae in the BAT region, *Earth Planet. Sci. Lett., 256*, 100–119, doi:10.1016/j.epsl.2007.01.016.

Rubin, A. M. (1995), Propagation of magma-filled cracks, *Ann. Rev. Earth Planet. Sci., 23*, 287–336, doi:10.1146/annurev.ea.23.050195.001443.

Sato, K. (1997), Melting experiments on a synthetic olivine lamproite composition up to 8 GPa: Implications to its petrogenesis, *J. Geophys. Res., 102*(B7), 14,751–14,764.

Schaber, G. G., R. G. Strom, H. J. Moore, L. A. Soderblom, R. L. Kirk, D. J. Chadwick, D. D. Dawson, L. R. Gaddis, J. M. Boyce, and J. Russell (1992), Geology and distribution of impact craters on Venus: What are they telling us?, *J. Geophys. Res., 97(E8), 13,257–13,301*.

Schubert, G., D. L. Turcotte, and P. Olsen (2001), *Mantle Convection in the Earth and Planets*, 900 pp., Cambridge Univ. Press, Cambridge, England.

Schubert, G., V. S. Solomatov, P. J. Tackley, and D. L. Turcotte (1997), Mantle convection and the thermal evolution of Venus, in *Venus II: Geology, Geophysics, Atmosphere, and Solar Wind Environment*, edited by S. W. Bougher, D. M. Hunten, and R. J. Phillips, 879–900, Univer. of Arizona Press, Tucson, Arizona.

Sclater, J. G., C. Jaupart, D. Galson (1980), The heat-flow through oceanic and continental-crust and the heat-loss of the Earth, *Rev. Geophys. Space Phys., 18*, 269–312.

Senske, D. A., G. G. Schaber, and E. R. Stofan (1992), Regional Topographic rises on Venus: Geology of western Eistla Regio and comparisons to Beta Regio and Atla Regio, *J. Geophys. Res., 97*(8), 13,395–13,420.

Shelton, G., and J. A. Tullis (1981), Experimental flow laws for crustal rocks, *Trans. Amer. Geophys. Union, 62*, 396.

Simons, M., S. C. Solomon, and B. H. Hager (1994), Global variations in the geoid/topography admittance of Venus, *Science, 264*(5160), 798–803, doi:10.1126/science.264.5160.798.

Sleep, N. H. (1994), Martian plate tectonics, *J. Geophys. Res., 99*(E3), 5639–5656.

Smrekar, S. E., and R. J. Phillips (1991), Venusian highlands: Geoid to topography ratios and their implications, *Earth Planet. Sci. Lett., 107*, 582–597, doi:10.1016/0012-821X(91)90103-O.

Smrekar, S. E., and E. M. Parmentier (1996), The interaction of mantle plumes with thermal and chemical boundary layers: Applications to hotspots on Venus, *J. Geophys. Res., 101*(B3), 5397–5410.

Smrekar, S. E., and E. R. Stofan (1997), Coupled upwelling and delamination: A new mechanism for coronae formation and heat loss on Venus, *Science, 277*, 1289–1294.

Smrekar, S. E., and S. C. Solomon (1992), Gravitational spreading of high terrain in Ishtar Terra, Venus, *J. Geophys. Res., 97*(E10), 16,121–16,148.

Solomatov, V. S., and L.-N. Moresi (2000), Scaling of time-dependent stagnant lid convection: Application to small-scale convection on Earth and other terrestrial planets, *J. Geophys. Res., 105*(B9), 21,795–21,818.

Solomatov, V. S., and L.-N. Moresi (1997), Three regimes of mantle convection with non-Newtonian viscosity and stagnant lid convection on the terrestrial planets, *Geophys. Res. Lett., 24*(15), 1907–1910, doi:10.1029/97GL01682.

Solomatov, V. S. and L.-N. Moresi (1996), Stagnant lid convection on Venus, *J. Geophys. Res., 101*(E2), 4737–4754, doi:10.1029/95JE03361.

Solomatov, V. S. (1995), Scaling of temperature- and stress-dependent viscosity convection, *Phys. Fluids, 7*, 266–274.

Solomon, S. C. (1993), The geophysics of Venus, *Phys. Today, 46*(7), 48–55.

Solomon, S. C., O. Aharonson, J. M. Aurnou, W. B. Banerdt, M. H. Carr, A. J. Dombard, H. V. Frey, M. P. Golombek, S. A. Hauck II, J. W. Head, III, B. M. Jakosky, C. L. Johnson, P. J. McGovern, G. A. Neumann, R. J. Phillips, D. E. Smith, and M. T. Zuber (2005), New perspectives on ancient Mars, *Science, 307*(5713), 1214–1220, doi:10.1126/science.1101812.

Solomon, S. C., and J. W. Head (1982), Mechanisms of lithospheric heat transport on Venus: Implications for tectonic style and volcanism, *J. Geophys. Res., 87*, 9236–9246.

Solomon, S. C., M. A. Bullock, and D. H. Grinspoon (1999), Climate change as a regulator of tectonics on Venus, *Science, 286*(5437), 87–90, doi:10.1126/science.286.5437.87.

Solomon, S. C., S. E. Smrekar, D. L. Bindschadler, R. E. Grimm, W. M. Kaula, G. E. McGill, R. J. Phillips, R. S. Saunders, G. Schubert, S. W. Squyres, and E. R. Stofan, (1992), Venus tectonics: An overview of Magellan observations: *J. Geophys. Res., 97*(E8), 13,199–13,256.

Sotin, C., and S. Labrosse (1999), Thermal convection in an isoviscous, infinite Prandtl number fluid heated from within and from below: applications to the transfer of heat through planetary mantles, *Phys. Earth Planet. Int., 112*, 171–190.

Steinbach, V., D. A. Yuen, and W.-L. Zhao (1993), Instabilities from phase-transitions and the timescales of mantle thermal evolution, *Geophys. Res. Lett., 20*(12), 1119–1122.

Steinbach, V., and D. A. Yuen (1994), Effects of depth-dependent properties on the thermal anomalies produced in flush instabilities from phase-transitions, *Phys. Earth Planet. Int., 86*(1–3), 165–183.

Stevenson, D. J. (2001), Mars' core and magnetism, *Nature, 412*, 214–219, doi:10.1038/35084155.

Stixrude, L., and C. Lithgow-Bertelloni (2005), Mineralogy and elasticity of the oceanic upper mantle: Origin of the low-velocity zone, *J. Geophys. Res., 110*(B3), B03204, doi:10.1029/2004JB002965.

Stofan, E. R., A. W. Brian, and J. E. Guest (2005), Resurfacing styles and rates on Venus: assessment of 18 venusian quadrangles, *Icarus, 173*(2), 312–321, doi:10.1016/j.icarus.2004.08.004.

Stofan, E. R., and S. E. Smrekar (2005), Large topographic rises, coronae, large flow fields and large volcanoes on Venus: Evidence for mantle plumes? in *Plates, Plumes, and Paradigms*, edited by G. R. Foulger, J. H. Natland, D. C. Presnall, and D. L. Anderson, Geological Society of America Special Paper 388, 841.

Strom, R. G., G. G. Schaber, and D. D. Dawson, (1994), The global resurfacing of Venus, *J. Geophys. Res., 99*(E5), 10,899–10,926.

Surkov, Y. A. (1990), *Exploration of Terrestrial Planets from Spacecraft: Instrumentation, Investigation, Interpretation*, 2nd edition, 446 pp. John Wiley, New York.

Tackley, P. J. (2000), Self-consistent generation of tectonic plates in time-dependent, three-dimensional mantle convection simulations, 1. Pseudoplastic yielding, *Geochem. Geophys. Geosyst., 1*(8), doi:10.1029/2000GC000036.

Tackley, P. J. (2000), Self-consistent generation of tectonic plates in time-dependent, three-dimensional mantle convection simulations, 2, Strain weakening and asthenosphere, *Geochem. Geophys. Geosyst., 1*(8), doi:10.1029/2000GC000043.

Tanaka, K. L. (1986), The stratigraphy of Mars, *J. Geophys. Res., 91*(B13), E139-E158.

Treiman, A. H. (1995), Ca-rich carbonate melts: A regular-solution model, with applications to carbonatite magma + vapor equilibria and carbonate lavas on Venus, *Am. Mineral., 80*, 115–130.
Trieman chapter

Trompert, R., and U. Hansen (1998), Mantle convection simulations with rheologies that generate plate-like behaviour, *Nature, 395*, 686–689, doi:10.1038/27185.

Tullis, T. E. (1986), Friction and faulting, *Pure Appl. Geophys.*, 1241–1249.

Turcotte, D. L. (1993), An episodic hypothesis for venusian tectonics, *J. Geophys. Res., 98*(E9), 17,061–17,068, doi:10.1029/93JE01775.

Turcotte, D. L., and G. Schubert (1982), *Geodynamics*, 450 pp., John Wiley and Sons, New York. Second edition published by Cambridge University Press (2001).

Turcotte, D. L., F. A. Cooke, and R. J. Willeman (1979), Parameterized convection within the moon and the terrestrial planets, 10th Lunar and Planetary Science Conference, Houston, Tex., 19–23 March, 2375–2392, Pergamon Press, Inc., New York.

Turcotte, D. L., G. Morein, D. Roberts, and B. D. Malamud (1999), Catastrophic resurfacing and episodic subduction on Venus, *Icarus, 139*, 49–54, doi:10.1006/icar.1999.6084.

Walter, M. J. (1998), Melting of garnet peridotite and the origins of komatiite and depleted lithosphere, *J. Petrol., 39*, 29–60, doi:10.1093/petrology/39.1.29.

Wark, D. A., C. A. Williams., E. B. Watson, and J. D. Price (2003), Reassessment of pore shapes in microstructurally equilibrated rocks, with implications for permeability of the upper mantle, *J. Geophys. Res., 108*(B1), 2050, doi:10.1029/2001JB001575.

Watters, T. R., C. J. Leuschen, J. J. Plaut, G, Picardi, A. Safaeinili, S. M. Clifford, W. M. Farrell, A. B. Ivanov, R. J. Phillips, and E. R. Stofan (2006), MARSIS radar sounder evidence of buried basins in the northern lowlands of Mars, *Nature, 444*, 905–908, doi:10.1038/nature05356.

Whitehead, J. A., and D. S. Luther (1975), Dynamics of laboratory diapir and plume models, *J. Geophys. Res., 80*(B5), 705–717.

Wilhelms, D. E., and S. W. Squyres (1984), The martian hemispheric dichotomy may be due to a giant impact, *Nature, 309*, 138–140, doi:10.1038/309138a0.

Williams, J.-P., and F. Nimmo (2004), Thermal evolution of the Martian core: Implications for an early dynamo, *Geology, 32*(2), 97–100, doi:10.1130/G19975.1.

Wise, D. U., M.P. Golombek, and G. E. McGill (1979), Tectonic evolution of Mars, *J. Geophys. Res., 84*, 7934–7939.

Wyllie, P. J., and W.-L. Huang (1975), Peridotite, kimberlite, and carbonatite explained in the system CaO-MgO-SiO_2-CO_2, *Geology, 3*(11), 621–624, doi:10.1130/0091-7613(1975)3<621: PKACEI>2.0.CO;2.

Yaxley, G. M. and G. P. Brey (2004), Phase relations of carbonate-bearing eclogite assemblages from 2.5 to 5.5 GPa: implications for petrogenesis of carbonatites, *Contrib. Min. Pet., 146*(5), 606–619, doi:10.1007/s00410-003-0517-3.

Yoder, C. F., A. S. Konopliv, D. N. Yuan, E. M. Standish, W. M. Folkner (2003), Fluid core size of mars from detection of the solar tide, *Science, 300*, 299–303, doi:10.1126/science.1079645.

Zhong, S. J., and M. Gurnis (1996), Interaction of weak faults and non-newtonian rheology produces plate tectonics in a 3D model of mantle flow, *Nature, 383*, 245–247, doi:10.1038/383245a0.

Zhong, S. J., and M. Gurnis (1995), Towards a realistic simulation of plate margins in mantle convection, *Geophys. Res. Lett., 22*(8), 981–984.

Zhong, S.-J., and M. T. Zuber (2001), Degree-1 mantle convection and the crustal dichotomy on Mars, *Earth Planet. Sci. Lett., 189*(1), 75–84.

Atmospheric Composition, Chemistry, and Clouds

Franklin P. Mills

*Research School of Physical Sciences and Engineering and The Fenner School of Environment
and Society, Australian National University, Canberra, Australia*

Larry W. Esposito

Laboratory for Atmospheric and Space Physics, University of Colorado, Boulder, Colorado, USA

Yuk L. Yung

*Division of Geological and Planetary Sciences, California Institute of Technology,
Pasadena, California, USA*

Venus' atmosphere has a rich chemistry involving interactions among sulfur, chlorine, nitrogen, hydrogen, and oxygen radicals. The chemical regimes in the atmosphere range from ion-neutral reactions in the ionosphere to photochemistry in the middle atmosphere to thermal equilibrium chemistry and surface-atmosphere reactions in the lower atmosphere. This variety makes Venus an important planet to understand within the context of terrestrial-like planets, both in our own solar system and outside it. The primary chemical cycles are believed known but surprisingly few details about these cycles have been fully verified by concurrence among observations, experiments, and modeling. Good models have been developed that account for many properties of the cloud layers, but the size distribution, shape, and composition of the majority of the aerosol mass are still open issues. This chapter reviews the state of knowledge prior to the Venus Express mission for the composition, chemistry, and clouds of the neutral atmosphere on Venus. Observations by instruments on Venus Express, in combination with ground-based observations, laboratory experiments, and numerical modeling, should answer some of the major open questions regarding the composition, chemistry, and clouds of Venus' atmosphere.

1. INTRODUCTION

Interest in the composition of Venus' atmosphere began as early as Lomonosov's report from observations of the 1761 transit that Venus had an atmosphere and the pro-posal that there were clouds in the atmosphere [*Huggins and Miller*, 1864]. In the latter half of the 20th century, numerous space missions and extensive ground-based telescopic observations provided significant insight into the composition and chemistry of Venus' atmosphere. The most intense study of Venus' photochemistry was in the decade from the early 1970's to the mid-1980's when the modern understanding of Earth's stratospheric chemistry also was being derived. Indeed, much of the chlorine and

Exploring Venus as a Terrestrial Planet
Geophysical Monograph Series 176
Copyright 2007 by the American Geophysical Union.
10.1029/176GM06

sulfur chemistry identified in Earth's stratosphere was first proposed in early models of Venus' photochemistry [*Prinn*, 1985]. Photochemical modeling has had a renewed burst of interest since 2000. Throughout the 1980's and 1990's, and continuing to the present, ground-based observations showed that the abundances of trace species, intensity of airglow emission, temperatures, and wind velocities could vary markedly on timescales from hours to years. This variability has introduced a new dimension, time, into our understanding of chemistry in Venus' atmosphere, but few modeling studies have tackled this aspect and major uncertainties still surround the basic chemical state of the atmosphere. Observational advances in the last 20 years enabled retrieval of cloud characteristics and trace species abundances, many for the first time, from the base of the upper cloud layer to the lowest scale height above the surface. The resurgence of interest in the chemistry of Venus' atmosphere since 2000 has brought together advances from experimental measurements, ground-based observations, and numerical modeling. This chapter reviews the current state of knowledge of the chemistry of Venus' atmosphere (prior to the Venus Express mission), outlines current uncertainties, provides recommendations of future directions for research, and compares Venus' atmospheric chemistry to that of other terrestrial-like planets.

Venus' atmosphere can be divided into regions based on its composition, chemistry, and clouds. The upper atmosphere, above ~ 110 km, has low densities and overlaps with the ionosphere so that photodissociation, ion-neutral, and ion-ion reactions are increasingly dominant as one goes to higher altitudes. The middle atmosphere, ~ 60–110 km, receives sufficiently intense ultraviolet (UV) radiation from the sun that its chemistry is dominated by photon-driven processes, termed "photochemistry." The lower atmosphere, below ~ 60 km, receives little UV radiation from the sun, and due to the high atmospheric temperatures the chemistry is controlled increasingly by thermal processes, termed thermodynamic equilibrium chemistry or "thermochemistry," as one goes lower in the atmosphere. Bridging across the boundary between the lower and middle atmospheres are the cloud and haze layers which extend from ~ 30–90 km with the main cloud deck lying at ~ 45–70 km. The clouds define a transition region that reflects the competition between the middle atmosphere, dominated by photochemistry, and the lower atmosphere, dominated by thermochemistry [*Esposito et al.*, 1997]. This region is also where lightning could occur and produce NO [*Krasnopolsky*, 2006a] and where heterogeneous chemistry on aerosol and cloud particle surfaces may be important [*Mills et al.*, 2006]. Finally, the lowest scale height of the atmosphere is the region where surface-atmosphere interactions may dominate.

Three dominant chemical processes have been identified in the Venus atmosphere, which we will term the CO_2 cycle, the sulfur oxidation cycle, and the polysulfur cycle. The CO_2 cycle includes the photodissociation of CO_2 on the day side, transport of a significant fraction of the CO and O to the night side, production of O_2, emission of highly variable oxygen airglow on both the day and night sides, and conversion of CO and O_2 into CO_2 via catalytic processes. The sulfur oxidation cycle involves the upward transport of SO_2, oxidation of a significant fraction of the SO_2 to form H_2SO_4, condensation of H_2SO_4 and H_2O to form a majority of the mass comprising the cloud and haze layers, downward transport of sulfuric acid in the form of cloud droplets, evaporation of the cloud droplets, and decomposition of H_2SO_4 to produce SO_2. There is solid observational evidence for both the CO_2 and the sulfur oxidation cycles. The polysulfur cycle is more speculative but plausible based on existing laboratory data and limited observations. It involves the upward transport of sulfur as either SO_2 or OCS, photodissociation to produce S, formation of polysulfur (S_x) via a series of association reactions, downward transport of S_x, thermal decomposition of S_x, and reactions with oxygen and CO to produce SO_2 and OCS, respectively. Each of the cycles involves a number of trace species, such as ClO_x, HO_x, NO_x, and SO_x. The three cycles may (and probably do) interact through these trace species; see Plate 1. The strength of these links between the cycles in existing models depends on parameters that have significant uncertainties and few constraints from direct observational evidence.

The first part of this chapter summarizes recently published observations of the composition of Venus' atmosphere. The second and third parts discuss sulfur chemistry and the cloud layers, respectively, including the sulfur oxidation and polysulfur cycles. The fourth part discusses the chlorine chemistry that is believed to control the CO_2 cycle. The final sections discuss linkages among the chemical cycles, comparisons to other terrestrial-type planets, and recommendations for future research.

2. SUMMARY OF OBSERVATIONS AND RECENT DEVELOPMENTS

2.1 Venera 15 Infrared Spectra: Observational Results and Interpretation

VENERA 15 observed mainly the northern hemisphere, except for one orbital session when the spacecraft's orientation was changed to obtain a few tens of spectra in the equatorial region and southern midlatitudes. Absorption bands of three atmospheric gases (CO_2, H_2O and SO_2) and also

H_2SO_4 aerosols are clearly visible. The very strong 15 µm CO_2 band has different morphology at different locations, mostly owing to variability of the temperature profiles. Band center emission arises from atmospheric layers at about 90 km altitude. Differences appearing in the continuum likely reflect variability in the vertical structure of the upper clouds and haze.

Interpretation of Fourier Spectrometer (FTS) data [*Moroz et al.*, 1985, 1986, 1990; *Spankuch et al.*, 1985, 1990; *Zasova et al.*, 1985, 1989, 1993; *Schaefer et al.*, 1987, 1990; *Linkin et al.*, 1985; *Ignatiev et al.*, 1999; *Koukouli et al.*, 2005] gives the following results:

(1) Upper cloud and upper haze particles consist of sulfuric acid water solutions, confirmed by the good agreement of synthetic and observed spectra.

(2) The particle size distribution corresponds to the measured "mode 2" from the Pioneer Venus cloud particle size spectrometer (LCPS), with parameters as proposed by *Pollack et al.* [1980]. That is, a log-normal distribution with mean cross-section weighted size r = 1.05 µm and variance σ = 1.21. Possibly some other size distributions could also successfully match the data.

(3) CO_2 pure gas transmission functions for a set of channels within the 15 µm band and computed aerosol transmission functions allow an iterative procedure (e.g., using relaxation as by *Zasova et al.* [1989]) to simultaneously retrieve the temperature and aerosol profiles.

(4) The retrieved temperature and aerosol profiles from step (3) yield synthetic spectra for the SO_2 and H_2O bands, which allow derivation of their abundances. These abundances are in good agreement with SO_2 measured in the UV [*Zasova et al.*, 1993].

(5) The mean H_2O abundance was 12 ± 5 ppm with only a slight enhancement at equatorial latitudes and no clear day-night distinction [*Koukouli et al.*, 2005].

2.2. Water Vapor Measurements

Observed water vapor abundances in the middle atmosphere of Venus have shown strong variability—both spatially and temporally. The variability does not result from differences in retrieval technique or spectroscopic data [*Koukouli et al.*, 2005], and temporal variability in the absolute abundance and the spatial variations of the abundance has been reported by multiple instruments [*Barker*, 1975; *Koukouli et al.*, 2005; *Sandor and Clancy*, 2005] so both seem likely to be real.

In the lower atmosphere, spectra obtained by the entry probes on Venera 11, 13, and 14 were reanalyzed [*Ignatiev et al.*, 1997] using high temperature spectroscopic databases and a line-by-line radiative transfer code to assess

the previously reported inconsistencies between the results obtained from initial analyses of the Venera entry probe spectra [*Moroz et al.*, 1979, 1980; *Moshkin et al.*, 1983; *Young et al.*, 1984] and from ground-based observations of near-infrared emission from Venus' nightside [*Pollack et al.*, 1993]. The new analyses by *Ignatiev et al.* [1997] found results that are very similar to those derived from ground-based observations by *Pollack et al.* [1993]. *Ignatiev et al.* [1997] concluded the H_2O mixing ratio at 10–48 km was either close to a constant value of 30 ± 10 ppm or has a weak minimum at 10–20 km. In the clouds, they found the H_2O mixing ratio was in the range 30–50 ppm and below 5 km altitude, the mixing ratio for H_2O probably increases to 50–70 ppm. *Ignatiev et al.* [1997] also concluded that the much larger mixing ratios determined by in situ experiments on the Venera entry probes are not likely to be reliable.

2.3. Sulfur Dioxide Measurements

SO_2 was first detected in the atmosphere of Venus by *Barker* [1979] from the ground, and it was subsequently confirmed by *Stewart et al.* [1979] and *Conway et al.* [1979]. These observations indicated that the abundances of SO_2 in the 1978–1979 period were larger than the previously established upper limits [*Owen and Sagan*, 1972] by orders of magnitude. Continuous observations by Pioneer Venus from 1978 to 1986 show a steady decline in the cloud top SO_2 abundance toward values consistent with previous upper limits [*Esposito et al.*, 1988]. This decline has been confirmed by International Ultraviolet Explorer (IUE) observations [*Na et al.*, 1990] and by Hubble observations [*Na and Esposito*, 1995], see Plate 2 and Figure 1. Analysis of UV spectra from the Hubble Space Telescope (HST) Goddard High Resolution Spectrometer (GHRS) gives an SO_2 abundance of less than 25 ppb at the cloud tops [*Na and Esposito*, 1996]. See Figure 1 for the time history of SO_2 cloud top measurements. Explanations that have been advanced for the likely rapid increase and observed slow decline of SO_2 include active volcanism [*Esposito*, 1984], changes in the effective eddy diffusion within the cloud layers [*Krasnopolsky*, 1986, p. 147], and changes in atmospheric dynamics [*Clancy and Muhleman*, 1991]. It is important to note that the volcano hypothesis uses the volcanic eruption as a source of buoyancy that allows the abundant SO_2 below the Venus clouds to break through the stable upper cloud layer. The entrained SO_2 is then observable remotely at the cloud top in the UV. Similarly, because the observed SO_2 mixing ratio may differ by as much as four orders of magnitude from the base to the top of the cloud layers [*Berteaux et al.*, 1996; *Esposito et al.*, 1997], a small change in the effective eddy diffusion within the cloud layers may significantly alter the cloud top

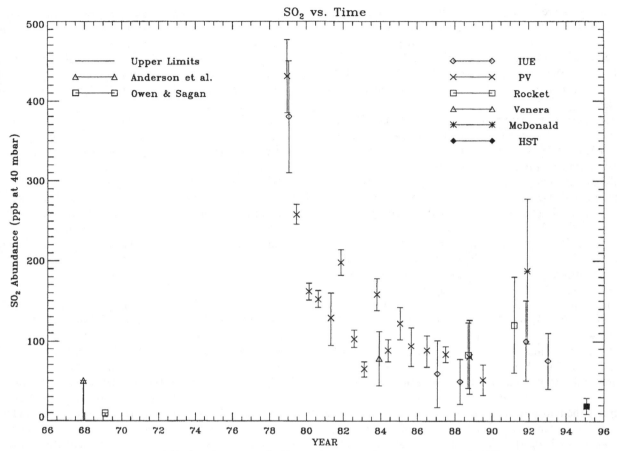

Figure 1. SO_2 abundance near the Venus cloud tops. (From *Esposito et al.*, 1997)

abundance of SO_2 [*Krasnopolsky*, 1986, p. 147]. The SO_2 abundance below the clouds varies much more slowly, related to the amount of volcanic activity over geologic time scales of millions of years. See *Fegley et al.* [1997a].

The observed rate of change in the cloud top abundance of SO_2 is significantly slower than would be expected if the change were due entirely to photochemical processes. The photochemical lifetime of SO_2 at 68–70 km altitude is about 3×10^3 s, but SO_2 is rapidly produced again from SO, so the time scale for net loss of SO_2 via oxidation to SO_3 is much longer, 10^6–10^7 s [*Winick and Stewart*, 1980; *Krasnopolsky and Pollack*, 1994; *Pernice et al*, 2004], which is comparable to or longer than the 1–2×10^6 s time scale for vertical transport due to eddy diffusion in models [*Mills*, 1999a]. The photochemical lifetime against oxidation to SO_3 depends on the column abundance of SO_2 and the production rate for SO_3, both of which have uncertainties.

The changes in SO_2 above and within the clouds of Venus may have a significant effect on the photochemistry of the clouds of Venus. Pioneer Venus observations have shown that

the clouds of Venus are created by the photochemical processes that oxidize upwelling SO_2 [*Winick and Stewart*, 1980; *Krasnopolsky and Parshev*, 1981abc; 1983; *Yung and DeMore*, 1981; 1982]. Thus any significant changes in SO_2 may have an effect on the chemistry and dynamics of the clouds.

2.4. Clouds and Hazes

VENERA-15 FS spectra provided a strong confirmation of aqueous solution (75–85%) of sulfuric acid as the particulate material in the upper clouds. Otherwise, such good coincidence of synthetic and observed spectra would be impossible. The first *in situ* detection of sulfuric acid was made in experiments SIGMA-3, a gas chromatograph, and MALAHIT, a mass spectrometer. The analysis of SIGMA-3 results yields the average mass loading of sulfuric acid about 1 mg/m^3, estimated for heights 48–54 km and "much lower" at about 54 km [*Porshnev et al.*, 1987]. Conclusions from MALAHIT analysis are even less definite and not in a good agreement with SIGMA-3 results; they give an average mass loading between 2 and 10 mg/m^3 [*Surkov et al.*, 1987]. Both experiments can say nothing about the upper cloud.

Very interesting, but puzzling, results were obtained by elemental X-ray analysis of the thin layers collected by the IPF experiment [*Andreichikov et al.*, 1987], which was an aerosol analyzer on VEGA. Three elements were identified: sulfur, chlorine and phosphorus. Sulfur and chlorine in cloud particles had been detected previously by similar measurements, but not phosphorus. It is clear that some P-bearing substance can be important as a particulate in lower clouds. Phosphoric acid H_3PO_4 is a likely candidate for this substance and phosphorous anhydride P_4O_6 may be the gas responsible for its production [*Andreichikov*, 1987]. A critical review of these data was given by *Krasnopolsky* [1989]. The key conclusion in the paper is that the lower subcloud boundary at the level about 33 km registered on VENERA 8 and later on some other missions can be explained by phosphoric acid particles dominating in the subclouds. The refractive index 1.7 estimated for some of the particles registered in ISAV-A experiment can be understood if they are sulfur. Some evidence for the presence of sulfur aerosols in clouds is also found in the analysis of the SIGMA-3 and IPF results.

2.5. Ground-Based Observations of the Lower Atmosphere

Important new information from near IR measurements of the Venus nightside probes the deeper Venus atmosphere through windows in the CO_2 absorption spectrum. Discovery of nightside emissions at 1.7 and 2.3 μm on Venus by *Allen and Crawford* [1984] has been extremely important for further progress in study of the chemical composition. These emissions were readily identified [e.g., *Krasnopolsky*, 1986, p. 181] as spectral windows to the lower atmosphere. Later, windows at 1.31, 1.27, 1.18, and 1.01 μm were found, and lines of CO_2, H_2O, HDO, SO_2, CO, OCS, HCl, and HF have been identified. Radiation in each window is formed at a particular altitude range, and by comparing data for the same species at different windows, it becomes possible to obtain vertical profiles. Furthermore, lines of different strengths were observed in the 2.3 μm window, and this allowed *Pollack et al.* [1993] to derive both the mixing ratio and its gradient.

Pollack et al. [1993] found the mixing ratio for HF was 1–5 ppb at 33.5 km with no evidence for a vertical gradient, HCl had a mixing ratio of 0.48 ± 0.12 ppm at 23.5 km with no evidence for a vertical gradient, and H_2O had a constant mixing ratio of 30 ± 10 ppm at 10–40 km. The results for HF from *Pollack et al.* [1993], *Young* [1972], and *Connes et al.* [1967] suggest the mixing ratio for HF is approximately constant below the cloud tops. A more refined analysis of 1.7 μm spectra and earlier telescopic observations suggest the mixing ratio of HCl decreases from 1.0 ppm near 5 km to 0.4 ppm at the cloud tops [*Dalton et al.*, 2000; *Connes et al.*, 1967; *Young*, 1972].

The currently recommended mixing ratio for H_2O is 30 ± 10 at 15–45 km and 30 ± 15 at 0–15 km [*Taylor et al.*, 1997]. Other contemporaneous analyses [*Ignatiev et al.*, 1997; *Meadows and Crisp*, 1996] are roughly consistent with this recommendation (Plate 3), but H_2O close to the surface is harder to constrain [*de Bergh et al.*, 2006].

Pollack et al. [1993] also found the mixing ratios of CO and OCS were equal to 23 ± 5 and 4.4 ± 1 ppm at 36 and 33 km, respectively, with gradients of 1.2 ± 0.45 and –1.58 ± 0.3 ppm/km, respectively. Modeling by *Krasnopolsky and Pollack* [1994], Plate 3, agrees well with the retrieved mixing ratios and gradients for CO and OCS. In their modeling, OCS reacts with SO_3 to produce CO above ~ 33 km. Subsequent observations [*Marcq et al.*, 2005; 2006] found a global average CO mixing ratio of 24 ± 2 ppm at 36 km with a vertical gradient of 0.6 ± 0.3 ppm/km and a larger abundance of CO at 20–40° S latitude than at 0–20° S. *Marcq et al.* [2005; 2006] also found a global average OCS mixing ratio of 0.55 ± 0.15 ppm at 36 km with a vertical gradient of -0.28 ± 0.1 ppm/km and a smaller abundance of OCS at 20–40° S latitude than at 0–20° S. The newer observations of CO are quantitatively consistent with the earlier results where the observational altitudes agree, although the best fit vertical gradient for CO is a factor of two smaller in the newer results. The smaller gradient in CO mixing ratio from *Marcq et al.* [2005; 2006] is more consistent with measurements by instruments on Pioneer Venus and Venera 12 [*Bezard and de Bergh*, 2007] as shown in Plate 3. The best fit mixing ratio and the vertical gradient for OCS from the newer observations are a factor of eight and five, respectively, smaller than the earlier results. The gradients for CO and OCS in the newer results, however, are still approximately equal in magnitude and opposite in sign. Further, the spatial anticorrelation between CO and OCS mixing ratios observed by *Marcq et al.* [2005;6] is qualitatively consistent with the anticorrelation in the vertical gradients of CO and OCS seen by *Pollack et al.* [1993] and *Marcq et al.* [2005;6] and suggests conversion between CO and OCS, possibly due to surface buffering [*Fegley and Treiman*, 1992], thermochemical equilibrium chemistry [*Krasnopolsky and Parshev*, 1979], or kinetic conversion of OCS to CO_2 and CO [*Krasnopolsky and Pollack*, 1994].

2.6. Contemporaneous Reviews of Atmospheric Composition

Two contemporaneous review articles provide additional details on observations that have been made of the chemical

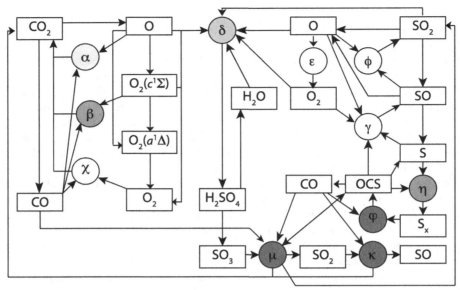

Plate 1. Schematic diagram for the atmospheric chemistry on Venus. Catalytic schemes are indicated by the Greek letters in circles. α = Schemes (E), (G1), (G2), (G3), and (N) in this chapter and the heterogeneous oxidation of CO on aerosols [*Mills et al.*, 2006]. β = CO + O₂(c) scheme [*Slanger et al.*, 2006; *Mills et al.*, 2006]. χ = Scheme (H) in this chapter and Schemes XIIab in *Yung and DeMore* [1982]. δ = Schemes (H) and (J) in this chapter and Schemes VIab and XIIab in *Yung and DeMore* [1982]. ε = Scheme (D). φ = Reactions (6), (7), and (41). γ = Reactions (3), (4), and (5). η = Reactions (8), (9), and (17) and the chlorosulfane schemes in *Mills and Allen* [2007]. φ = net Reaction (14). μ = net Reactions (12) and (13), and Reactions (15) and (16). κ = Reaction (18). The catalytic schemes with white background have been confirmed by laboratory chemical kinetic studies. Those in shades of red have not been fully confirmed. The degree of laboratory confirmation is indicated by the lightness of the shade of red. The darkest red have received no confirmation in laboratory studies; those in light red are largely but not completely confirmed.

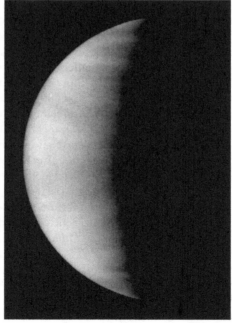

Plate 2. Venus image from HST at 218 nm. (Space Telescope Science Institute)

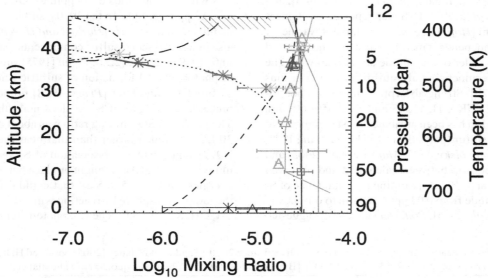

Plate 3. Predicted and observed composition of the lower atmosphere. Calculated abundances from a chemical kinetics model (model 1) [*Krasnopolsky and Pollack*, 1994] for H_2O (solid black line), CO (short dashed black line), OCS (dotted black line), H_2SO_4 (long dashed black line), and SO_3 (dash dotted black line). Calculated abundances from a thermochemical equilibrium model for the conditions believed to exist at the surface are shown for CO (black triangle) [*Fegley et al.*, 1997b] and OCS (black asterisk) [*Hong and Fegley*, 1997]. Retrieved H_2O abundances are the magenta solid line [*Meadows and Crisp*, 1996], the green solid line [*Ignatiev et al.*, 1997], the red solid line with red squares [*Pollack et al.*, 1993], and the blue square [*Marcq et al.*, 2006]. In site measurements of CO are the green triangles [*Gel'man et al.*, 1979; *Oyama et al.*, 1980]. Retrieved CO abundances are the blue triangles [*Marcq et al.*, 2006] and the red triangle [*Pollack et al.*, 1993]. Retrieved OCS abundances are the blue asterisks [*Marcq et al.*, 2006] and the red asterisk [*Pollack et al.*, 1993]. The range of H_2SO_4 vapor abundances retrieved from observations is indicated by the magenta hatched region [*Bezard and de Bergh*, 2007]. Recommended values [*Taylor et al.*, 1997] for CO are the cyan triangles and for OCS are the cyan asterisks.

composition of Venus' lower atmosphere [*Bezard and de Bergh*, 2007] and Venus' atmosphere below 100 km altitude [*de Bergh et al.*, 2006].

3. SULFUR CHEMISTRY

Sulfur chemistry is critical to the composition of the Venus atmosphere, and four sulfur species have been firmly identified: SO_2, SO, OCS, and H_2SO_4 (vapor and in aerosols). Recently published measurements of their abundances are summarized in Table 1, Table 3, *de Bergh et al.* [2006], and *Bezard and de Bergh* [2007]. H_2S was reported by Pioneer Venus below 20 km [*Hoffman et al.*, 1980], but it was never confirmed by an independent measurement and the reported value is at least an order of magnitude larger than would be expected from thermochemical equilibrium calculations [*Fegley et al.*, 1997b]. Strong absorption in spectrophotometer data from VENERA 11, 12, 13 and 14 at 450–600 nm between 10 and 30 km has been attributed most commonly to gaseous elemental sulfur (polysulfur, S_x) [*Moroz et al.*, 1979; *Sanko*, 1980; *Krasnopolsky*, 1987; *Maiorov et al.*, 2005] and there is good agreement between calculated and observed *Venera*-11 spectra at 3–19 km assuming the abundance of S_3 increases with altitude from 0.03 ppbv at 3 km to 0.1 ppbv at 19 km [*Maiorov et al.*, 2005]. *Pollack et al.* [1980] suggested

SO_2 could be at least partially responsible for the observed absorption at 400–500 nm below the cloud layers, but they did not publish the detailed results of their computations and no subsequent publication has either confirmed or refuted the importance of absorption by SO_2 below 40 km. The presence of thiozone (S_3) and polysulfur (S_x) in the clouds has been inferred [*Porshnev et al.*, 1987], but there has not been a definitive detection of S_x in the middle atmosphere.

The chemical scheme proposed by *Prinn* [1975, 1978, 1979] is based on a prediction by *Lewis* [1970] of sulfur species with mixing ratios of 60 ppm for OCS, 6 ppm for H_2S, and 0.3 ppm for SO_2. In the 1970s, sulfuric acid was clearly identified in the clouds [*Hansen and Hovenier*, 1974], while a search for gaseous sulfur components was not successful until 1979 (see Sec. (2.3)). *Prinn* [1975] suggested a scheme of photochemical formation of sulfuric acid from carbonyl sulfide OCS and later [*Prinn*, 1978] proposed the inverse processes leading to OCS and elemental sulfur from H_2SO_4. The predicted SO_2 mixing ratio was about a few ppm above 30 km and much larger than that near the surface. *Prinn* [1979] supposed that dissociation of S_3 and S_4 by the near ultraviolet ($\lambda \approx 350$ nm) might produce hot sulfur atoms with energies close to 1.5–2.5 eV that could drive chemistry in the lower atmosphere by reacting with H_2O and CO_2 to form SH, OH, SO, and CO. The UV photometer on the Venera 14

Table 1. Gas-phase species abundance observations published since 2000. †*Sandor and Clancy* [2005] observed HDO and assumed the above cloud D/H ratio was 157 ± 15 times SMOW [*Bjoraker et al.* 1992] to obtain the quoted H_2O abundances.

Species	Mixing Ratio (v/v)	Altitude	Year/Location	Notes
Cloud-top H_2O	(10 ± 5)–(90 ± 15) ppm	~ 110–160 mbar (~ 62–64 km)	1978–1979	*Koukouli et al.* [2005]
Cloud-top H_2O	~ 12 ± 5 ppm	~ 120–200 mbar (~ 60–64 km)	1983	*Koukouli et al.* [2005]
H_2O (inferred from HDO†)	(0.0 ± 0.06)–(3.5 ± 0.3) ppm	65–100 km	1998–2004	*Sandor and Clancy* [2005]
SO_2	430–10 ppb	40 mbar (~ 68 km)	1968–1995	See Figure 1
SO_2	≤ 50 ppm	35–55	1996, polar regions	*Jenkins et al.* [2002]
SO_2	≤ 100 ppm	35–55	1996, low latitudes	*Jenkins et al.* [2002]
SO_2	≤ 50 ppm	Below lower cloud	1996, 1999	*Butler et al.* [2001]
$H_2SO_4(g)$	0–9 ppm	35–55	1996	*Jenkins et al.* [2002]
$H_2SO_4(g)$	1-30 ppm	47–58 km	1983–1984, mid to high latitudes	*Gubenko et al.* [2001]
$H_2SO_4(g)$	1–2.5 ppm	At and below lower cloud	1996, 1999	*Butler et al.* [2001]
HCl	1–0.4 ppm	5–70 km		*Dalton et al.* [2000]
NO	5.5 ± 1.5 ppb	≤ 60 km	2004	*Krasnopolsky* [2006a]
NO flux	$(6 \pm 2) \times 10^7$ cm^{-2} s^{-1}		2004	*Krasnopolsky* [2006a]
S_3	0.03–0.1 ppb	3–19 km	1978	*Maiorov et al.* [2005]
CO	2×10^{-6}–2×10^{-4}	10–90 km	2003–2004	*Marcq et al.* [2005]
OCS	2×10^{-7}–10^{-5}	20–48 km	2003–2004	*Marcq et al.* [2005]

Table 2. Airglow observations published since 2000.

Airglow Band	Intensity	Altitude	Year/Location	Notes
O (^{1}S – ^{1}D) 557.7 nm	($\leq$ 10) - 167 R	~ 100 km	1999, 2001, 2002	*Slanger et al.* [2006]
O (^{1}D – ^{3}P) 630 nm	< 20 R		1999	*Slanger et al.* [2001]
O$_2$ (c$^1\Sigma$ – X$^3\Sigma$) Herzberg II	5.1 kR	mesosphere	1999	*Slanger et al.* [2001]
O$_2$ (c$^1\Sigma$ – X$^3\Sigma$) Herzberg II	3.1 kR	mesosphere	2001	*Slanger et al.* [2006]
O$_2$ (A$^{\prime 3}\Delta$ – a$^1\Delta$) Chamberlain	120 R	mesosphere	1999	*Slanger et al.* [2001]

Table 3. Cloud and haze observations published since 2000.

Observation	Value	Altitude	Year/Location	Notes
Cloud top pressure	5 – 30 mbar		1978 - 1991	*Braak et al.* [2002]
Cloud-top unit optical depth	110 – 160 mbar		1978 – 1979	*Koukouli et al.* [2005]
Cloud-top unit optical depth	120 – 200 mbar		1983	*Koukouli et al.* [2005]
Upper haze column abundance	0.25 – 1.5 μm^{-2}	> cloud top	1978-1991	*Braak et al.* [2002]
H$_2$SO$_4$ Droplet column abundance	15 cm^{-2}	> cloud top	1999 - 2004	*Mallama et al.* [2006]
H$_2$SO$_4$ Droplet diameter	2.1 μm	> cloud top	1999 - 2004	*Mallama et al.* [2006]

Table 4. Predicted gas-phase species which have not been observed in the middle atmosphere.

Predicted Species	Maximum Abundance	Altitude	Column Abundance	Notes
ClO	4 x 10^7 cm^{-3}	70 km	1 x 10^{14} cm^{-2}	*Pernice et al.* [2004]
Cl$_2$	6 x 10^{10} cm^{-3}	60 km	6 x 10^{16} cm^{-2}	*Pernice et al.* [2004]
ClC(O)OO	1 x 10^8 cm^{-3}	84 km	7 x 10^{13} cm^{-2}	*Pernice et al.* [2004]
COCl$_2$	4 x 10^8 cm^{-3}	66 km	7 x 10^{14} cm^{-2}	*Pernice et al.* [2004]
ClSO$_2$	3 x 10^{10} cm^{-3}	58 km	6 x 10^{15} cm^{-2}	*Pernice et al.* [2004]
S$_2$O	1 x 10^{10} cm^{-3}	64 km	2 x 10^{15} cm^{-2}	*Na and Esposito* [1997]

lander, however, found that the intensity of NUV radiation (320–390 nm) decreased by a factor of ~ 7 between 60 and 48 km [*Ekonomov et al.,* 1983]. This suggests the sulfur atoms produced from photodissociation of S$_3$ and S$_4$ in the lower atmosphere would have energies below 0.5 eV, which may not be sufficient to drive all of the lower atmosphere reactions proposed by Prinn [*Krasnopolsky and Pollack,* 1994], although there may still be sufficient energy to drive Prinn's proposed reaction S + CO$_2$ $\rightarrow$ SO + CO.

As noted above, there are two parts to the chemistry of sulfur species in the atmosphere of Venus. In the lower atmosphere and on the surface, the chemistry is dominated by thermodynamic equilibrium chemistry. Above the cloud tops, the chemistry is driven by photochemistry. Thus, the partitioning of sulfur among the different species represents a competition between thermodynamic equilibrium chemistry in the lower atmosphere and photochemistry above the cloud tops. We first discuss the photochemistry of sulfur in the middle atmosphere and then the thermal equilibrium chemistry of sulfur in the lower atmosphere.

3.1. Sulfur Photochemistry

The primary sulfur species in the middle atmosphere is SO$_2$, which undergoes either oxidation to SO$_3$ (which subsequently forms H$_2$SO$_4$ and condenses to form the cloud layers) if there is an additional source of oxygen atoms (i.e., photodissociation of CO_2) or disproportionation to form both SO_3 and polysulfur, S_x, if there is no additional source of oxygen atoms. The most reducing species of sulfur that has been observed in the *lower* atmosphere, however, is OCS. OCS may be converted to SO_2 either in the lower atmosphere, as is discussed in Sec. (3.2), or in the middle atmosphere where OCS readily undergoes photolysis:

$$OCS + h\nu \rightarrow CO + S(^1D) \tag{1}$$

where $S(^1D)$ is the first electronic excited state of the S atom. The most likely fate of $S(^1D)$ is quenching:

$$S(^1D) + CO_2 \rightarrow CO_2 + S(^3P) \tag{2}$$

The S atom gets oxidized to SO by reacting with O and O_2

$$S + O + M \rightarrow SO + M \tag{3}$$
$$S + O_2 \rightarrow SO + O \tag{4}$$

Direct oxidation of OCS is also possible via

$$OCS + O \rightarrow SO + CO \tag{5}$$

Further oxidation to SO_2 can proceed via the three-body reaction

$$SO + O + M \rightarrow SO_2 + M \tag{6}$$

Catalytic oxidation by ClO is also possible

$$SO + ClO \rightarrow SO_2 + Cl \tag{7}$$

The rate for Reaction (7) has been measured in laboratory studies [*Sander et al.*, 2002], and Reaction (7) accounts for 10–20 % of the loss of SO at 66–80 km altitude in recent photochemical models [*Pernice et al., 2004; Mills and Allen*, 2007]. Note that the net result is the oxidation of S to SO_2, and eventually to H_2SO_4. Plate 1 illustrates the central themes of the chemistry of Venus for the oxidation of CO and SO_2.

The ultimate fate for most SO_2 in the middle atmosphere is oxidation to H_2SO_4. The spectroscopic limit on OCS above 60 km is 10 ppb while the observed SO_2 mixing ratio at 40 mb (~70 km) is ~20–400 ppb, so oxidation of SO_2 that has been transported upward from the lower atmosphere is the primary source for H_2SO_4. Photochemical models by *Winick and Stewart* [1980], *Krasnopolsky and Parshev* [1981abc; 1983], and *Yung and DeMore* [1981; 1982] show that sulfuric acid is produced mostly in a thin layer of 2 km depth centered at 62 km. The H_2SO_4 eventually forms aerosols [*Winick and Stewart,* 1980; *Krasnopolsky and Parshev,* 1981abc; 1983; *Yung and DeMore,* 1981; 1982; *Esposito et al.,* 1988] that are transported by the meridional circulation to the polar region where they descend into the lower atmosphere [*Imamura and Hashimoto,* 1998]. Thermodynamic equilibrium reactions in the lower atmosphere decompose the sulfuric acid and turn the H_2SO_4 back to SO_2. This cycle of oxidation of SO_2 to H_2SO_4 followed by condensation, subsidence, evaporation, and decomposition has been termed the "fast atmospheric sulfur cycle" [*von Zahn et al.,* 1983].

In addition to oxidation to SO_2 and eventually to H_2SO_4, there is another possible fate for sulfur compounds in the atmosphere of Venus: formation of polysulfur. S atoms generated by the photolysis of SO and OCS can react via

$$S + OCS \rightarrow S_2 + CO \tag{8}$$

to produce S_2, or S_2 may be produced via coupled S-Cl chemistry involving chlorosulfanes [*Mills and Allen*, 2007]. Production of S_3 is possible through successive addition reactions such as

$$S + S_2 + M \rightarrow S_3 + M \tag{9}$$

S_3 is the chemical analog of ozone, known as thiozone. As the number of sulfur atoms increases, the polyatomic sulfur compounds tend to have lower saturation vapor pressures. It is convenient to name all sulfur species beyond S_3 "polysulfur" or S_x. The production of S_x is part of what has been termed the "slow atmospheric sulfur cycle" [*von Zahn et al.,* 1983], which is completed by decomposition reactions in the lower atmosphere (Sec. (3.2)). In the UV region S_x absorbs strongly, and it may be the principal constituent of the unidentified UV absorber in the upper atmosphere of Venus (Sec. (6)).

Whether the ultimate fate of sulfur in the middle atmosphere is H_2SO_4 or S_x depends crucially on the branching for the S atom, such as Reaction (4) versus Reaction (8). For this pair of reactions, the H_2SO_4 path would be preferred if

$$[O_2] > (k_8 / k_4) [OCS] \tag{10}$$

Thus when the O_2 abundance exceeds this value, production of oxidized sulfur species is favored, and ultimately H_2SO_4 is produced. When the O_2 abundance is below this value, production of polysulfur becomes possible. Recent work on chlorosulfane chemistry [*Mills and Allen*, 2007] suggests the critical factor determining oxidation of sulfur versus polysulfur formation may be the oxygen to chlorine ratio, $[O_2]/[Cl_2]$, but the concept described here remains valid.

Since the source of O_2 is photolysis of CO_2 in the upper atmosphere and the source of OCS is thermodynamic equilibrium chemistry in the lower atmosphere and the surface, we can imagine that the oxygen content of the subsiding and upwelling air parcels could be quite different. This may indeed be the explanation for the patchiness and the transience of the UV markers in the cloud tops of Venus.

3.2. Sulfur Chemistry Below the Clouds

Pioneer Venus, VENERA, and ground-based observations showed that the main sulfur species in the lower atmosphere (< 60 km) is SO_2 with a mixing ratio close to 150 ppm below the upper cloud [*Bertaux et al.*, 1996; *Bezard et al.*, 1993; *Oyama et al.*, 1980]. At these heights, photons with $\lambda > 450$–500 nm are the only ones available, and the energy of sulfur atoms released by photolysis of S_3 and S_4 may be too low to drive endothermic reactions, so the chemistry is increasingly controlled by thermodynamic equilibrium chemistry as one goes deeper in the atmosphere. The sulfur chemistry in the lower atmosphere has been summarized in terms of three cycles, the fast and slow atmospheric cycles and the geologic cycle [*von Zahn et al.*, 1983]. The middle atmosphere portions of the fast and slow atmospheric sulfur cycles were described in Sec. (3.1). The coupling between the photochemical and thermodynamic equilibrium parts of the atmospheric cycles is shown in Plate 1.

The details of the kinetic chemistry that completes the three sulfur cycles in the lower atmosphere still are largely unknown and only one kinetic model has been proposed to date. The lower atmosphere portion of the fast atmospheric cycle has been described by the net reactions [*von Zahn et al.*, 1983]

$$H_2SO_4 \rightarrow H_2O + SO_3 \tag{11}$$
$$\text{Net: } SO_3 + CO \rightarrow SO_2 + CO_2 \tag{12}$$
$$\text{Net: } H_2SO_4 + CO \rightarrow H_2O + SO_2 + CO_2 \tag{A}$$

and the lower atmosphere portion of the slow atmospheric sulfur cycle has been described by the net reactions [*von Zahn et al.*, 1983]

$$H_2SO_4 \rightarrow H_2O + SO_3 \tag{11}$$
$$\text{Net: } SO_3 + 4\,CO \rightarrow OCS + 3CO_2 \tag{13}$$
$$\text{Net: } H_2SO_4 + 4CO \rightarrow H_2O + OCS + 3CO_2 \tag{B}$$

with the polysulfur branch of the slow atmospheric sulfur cycle being balanced by the net reaction [*von Zahn et al.*, 1983]

$$\text{Net: } CO + (1/n)S_n \rightarrow OCS \tag{14}$$

In this formulation, the net Reactions (A), (B), and (14) balance the upper atmosphere portions of the fast and slow atmospheric sulfur cycles, so there is no net production or loss of any sulfur species in the atmosphere. In a recently accepted manuscript, *Krasnopolsky* [2007] correctly points out that the original formulation of the slow atmospheric cycle [*Prinn*, 1975], which involved conversion of OCS to CO via Reaction (1), is not consistent with the observed gradient of OCS below 40 km because OCS photolyzes at wavelengths less than 285 nm and these wavelengths are absorbed by CO_2, SO_2, and SO in or above the cloud layers. The more generalized formulation of the fast and slow atmospheric cycles [*von Zahn et al.*, 1983] quoted above, however, is consistent on a column integrated basis with the kinetic model in *Krasnopolsky* [2007].

Krasnopolsky and Pollack [1994] proposed a partial chemical kinetics model for the atmospheric sulfur chemistry at 20–40 km altitude that may be summarized on a column-integrated basis as:

$$2(H_2SO_4 \rightarrow SO_3 + H_2O) \tag{11}$$
$$SO_3 + CO \rightarrow SO_2 + CO_2 \tag{15}$$
$$SO_3 + OCS \rightarrow CO_2 + (SO)_2 \tag{16}$$
$$(SO)_2 + OCS \rightarrow CO + S_2 + SO_2 \tag{17}$$

The net process, $2H_2SO_4 + 2OCS \rightarrow 2H_2O + 2CO_2 + S_2 + 2SO_2$, is true only on a column-integrated basis. The relationships among the reactions at any specific altitude are more complicated, as shown in Plate 3 along with the observational data from *Pollack et al.* [1993], *Marcq et al.* [2005; 2006], and other sources. This scheme, which has been updated and extended in a recently accepted manuscript [*Krasnopolsky*, 2007], is the only kinetic model proposed for the lower atmosphere of Venus that is consistent with SO_2 as the dominant sulfur species. The partial model did not consider sources of OCS and other processes in the S_x–CO–OCS system, although *Krasnopolsky and Pollack* [1994] did consider

$$CO + SO_2 \rightarrow CO_2 + SO \tag{18}$$

which they said would proceed in the lowest 20 km of the atmosphere. Krasnopolsky and Pollack's scheme resulted in net destruction of OCS, which implied either the atmospheric sulfur chemistry is completed below 20 km or the atmosphere is evolving or both. Krasnopolsky and Pollack emphasized that this was only a partial solution to the problem of explaining the processes which govern profiles of OCS, CO, H_2SO_4, and SO_3, and their study did not cover some aspects of sulfur chemistry below 25 km. *Krasnopolsky* [2007] proposes a more complete model of the chemistry in Venus' lower atmosphere, which includes thermochemistry in the lowest 10 km, transport downward of photochemical products from the middle atmosphere, and photodissociation of S_3. The *Krasnopolsky* [2007] model closes the atmospheric sulfur chemistry and does not result in net production or loss of any sulfur species.

The Krasnopolsky and Pollack [1994] model is an excellent fit to the Pollack et al. [1993] and most other measurements at 30–45 km altitude. There is increasing disagreement between *Krasnopolsky and Pollack*'s calculations and observations for CO as one goes toward the surface, and extrapolation of the observed CO abundances to the surface suggests CO may be in thermochemical equilibrium near the surface, based on calculations by *Fegley et al.* [1997b]. The *Krasnopolsky* [2007] model appears to agree with observations of OCS and CO over a broader altitude range than the *Krasnopolsky and Pollack* [1994] model. The differing H_2O profiles inferred from observations by *Ignatiev et al.* [1997] and *Meadows and Crisp* [1996] for the lowest 10 km, however, indicate further measurements are needed in the lowest scale height to properly assess the state of the chemistry at those altitudes.

The fast and slow atmospheric sulfur cycles both imply there must be a significant flux of CO from the middle atmosphere to the lower atmosphere, which is balanced by an upward flux of CO_2. The observations summarized in Section 2.5 and Plate 3, however, are more supportive of conversion between CO and OCS, at least at 30–45 km altitude. The model in *Krasnopolsky* [2007] requires a smaller flux of CO down from the middle atmosphere and proposes a mechanism for converting between CO and OCS at 30–45 km altitude, but the CO flux is still larger than in photochemical models of the middle atmosphere [*Mills et al.*, 2006; *Mills and Allen*, 2007]. The exchange of species across the cloud layers is an area that requires more study via both modelling and observations.

Although the *Krasnopolsky and Pollack* [1994] and *Krasnopolsky* [2007] kinetic models give good fits to extant observations, that alone should not be construed as a substitute for rigorous laboratory demonstration of the reality of Reactions (15), (16), (17), and (18). Laboratory data for many of the key reactions in this kinetic model are nonexistent or highly uncertain. Another possibility that may duplicate the net results of these reactions is heterogeneous chemistry, which could occur on aerosol particles in the lower haze layer below the lower cloud layer.

The geological cycle involves reactions between atmospheric gases and surface minerals. These chemical weathering reactions proceed slowly compared to those comprising the atmospheric cycles, so significant disequilibria may exist between the surface and the atmosphere. There have been no measurements of the surface mineralogy, although we know the abundances of Mg and heavier elements from x-ray

fluorescence spectrometers on the VENERA 13 and 14 and Vega 2 landers [*Surkov et al.*, 1984; 1986]. These elemental abundances are typical of erupted basalts. The absence of definitive data has led to speculation about what surface minerals may exist, how they may buffer the atmosphere, and how the surface-atmosphere system may have interacted over the course of Venus' evolution. The general form of the proposed reactions is [*Johnson and Fegley*, 2002]

$$\text{atmospheric gas + surface rock} \rightarrow \text{product gas} + \text{chemically weathered rock} \qquad (19)$$

Surface rocks that have been proposed include carbonate ($CaCO_3$), wollastonite ($CaSiO_3$), anhydrite ($CaSO_4$), pyrite (FeS_2), pyrrhotite (Fe_7S_8), magnetite (Fe_3O_4), and tremolite ($Ca_2Mg_5Si_8O_{22}(OH)_2$). A number of good laboratory experiments have provided information on how rapidly chemical weathering would occur for minerals that may (or could have) existed on the surface of Venus. The experiments show that the rates for many of the posited reactions can be sensitive functions of temperature, oxygen fugacity, and/or the gaseous sulfur abundance [See Chapter 2 by *Treiman et al.*].

4. CLOUDS

The Venus clouds and hazes have enormous vertical extent, with a lower haze down to ~30 km and an upper thick haze up to 90 km altitude; the entire system covers a vertical depth of ~60 km, with an average visibility in the Venus clouds better than several km. The main cloud deck extends from ~70 km (the level of unit optical depth in the ultraviolet) down to altitudes between 45 and 50 km.

4.1. Cloud Structure, Properties, and Formation

Spacecraft *in situ* measurements allow us to divide the cloud system into upper, middle and lower clouds [see the review by *Esposito et al.*, 1983], Figure 2. Based on the Pioneer Venus and VENERA nephelometer results and the LCPS (Pioneer Venus Cloud Particle Size Spectrometer) measurements, it appears that the middle and upper cloud structure are planetwide features. In all cases the opacity is higher in the middle than upper cloud, typically by a factor of 2. The lower cloud is well defined and highly variable from location to location. Sharp layers are evident at the Pioneer Venus Large and Night probe sites.

The clouds within the main deck would all be thin stratoform in terrestrial classification. Instabilities are slight and latent convection potential is negligible [see *Knollenberg et al.*, 1980]. Only the middle cloud region appears to have any potential for convective overturning.

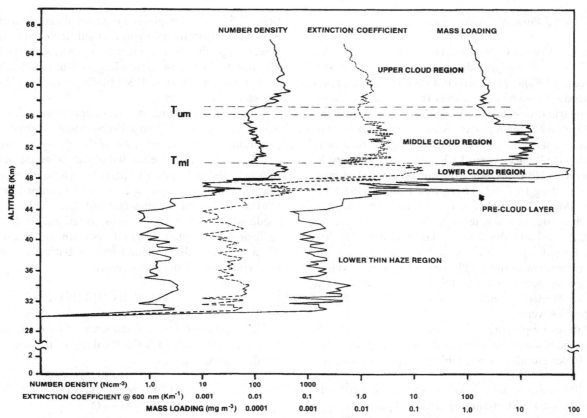

Figure 2. Cloud property vertical profiles. T_{ml} is the middle-lower cloud transition. T_{um} is the upper-middle cloud transition. [From *Knollenberg et al.*, 1980]

Considering the downward flux of sulfuric acid droplets, it is possible to understand why the clouds divide into three layers. Due to the photochemical formation of sulfuric acid (mostly in a thin layer of 2 km depth centered at 62 km based on models [*Winick and Stewart,* 1980; *Krasnopolsky and Parshev,* 1981abc, 1983; *Yung and DeMore,* 1981; 1982]), its flux increases steeply with decreasing altitude in the upper cloud layer which ends near 59 km (the measurements give 57 km). This increase correlates with the increasing H_2O mixing ratio while the concentration of sulfuric acid is relatively constant.

The flux of liquid sulfuric acid is constant in the middle cloud layer (57–50 km according to the measurements).

In the models by *Krasnopolsky and Pollack* [1994], the lower cloud layer forms at 48–47 km, where the predicted flux of liquid sulfuric acid exceeds that in the middle cloud layer by a factor of 4–7. The H_2O mixing ratio is rather constant, and thus the water fraction continues to decrease by a factor of 3 until the lower cloud boundary is reached. This corresponds to the acid concentration increasing to 97–98% at the lower cloud boundary. A strong gradient of gaseous sulfuric acid drives an upward flux which condenses and forms a strong downward flux of liquid sulfuric acid with the sum of both fluxes being constant in the lower and middle cloud layers. According to *Krasnopolsky and Pollack* [1994], this is the mechanism of the formation of the lower cloud layer. Thus, the upper cloud is created by photochemical production, the middle cloud by droplet growth and sedimentation, and the lower cloud by condensation enhancing the downward flux.

Krasnopolsky and Pollack [1994] developed a method to calculate the lower cloud boundary, with results between 48.4 km and 46.6 km. This is in good agreement with the PV radio occultation observations [*Cimino,* 1982], which show the boundary varying from 47–48 km at the low and middle latitudes and 47–43 km at high and subpolar latitudes.

The location of the lower boundary varies due to variations of the H_2SO_4 vapor mixing ratio, the water vapor mixing ratio below the clouds, and temperature and pressure. *Krasnopolsky and Pollack* [1994] found the boundary variations are produced mostly by variations of the sulfuric acid abundance and temperature. Variations of density in the lower cloud layer reflect variations of gaseous sulfuric acid, because water contributes only slightly to the sulfuric acid aerosol in the lower cloud layer.

4.2. The Mode 3 Particle Controversy

The Pioneer Venus LCPS measurement of larger, so called "Mode 3" particles has provided a controversy that is still unresolved; see *Esposito et al.* [1983]. The starting point for the Mode 3 controversy comes from direct evidence for asymmetric (possible crystalline) particles provided by *Knollenberg and Hunten* [1980]. *Knollenberg et al.* [1980] further state that only such crystals of high aspect ratio could satisfy the Pioneer Venus LCPS, LSFR, and LN results simultaneously. However, since the largest amount of mass (~80% according to *Knollenberg and Hunten* [1980]) is within the Mode 3 particles, it is extremely important to verify their existence and determine their composition.

The LCPS undoubtedly detected large particles, but the evidence for solid particles is indirect. There were internal inconsistencies in the LCPS measurements as well as inconsistencies between the LCPS measurements and the measurements made by other instruments. Some of these inconsistencies were:

1. Calculations employing LCPS size distributions do not give the backscatter observed by the PV nephelometer in the lower clouds if reasonable refractive indices are used.
2. The LCPS size distributions do not yield the optical depths derived by the LSFR, assuming spherical particles.
3. Overlapping size ranges of the LCPS give conflicting measurements in the lower clouds.

In addition, independent VENERA results show some oddities at the same altitudes:

1. VENERA nephelometer phase function measurements are inconsistent with spherical particles having reasonable refractive indices in the lower cloud.
2. X-ray fluorescence measurements [*Surkov*, 1979] show about ten times as much chlorine as sulfur in the Venus clouds.

The various inconsistencies can be explained by the simple hypotheses that Mode 3 is composed of solid, nonspherical particles. However this explanation requires an abundant gas-phase chemical in the clouds as the source for these particles. No such gas has yet been discovered.

Toon et al. [1982] reexamined the evidence that solid particles form a distinctive size mode. They find that Mode 3 is defined by a discontinuity located between two size ranges of the LCPS. Although this could be real, it could also be the result of a small calibration shift of the PV instrument. A shift in the calibration removes the discontinuity, along with the internal inconsistency of the LCPS. The revised size spectrum is consistent with the VENERA and Pioneer Venus optical data in the lower clouds; all the modes can be composed of sulfuric acid droplets without any solid par-

ticles. The only unexplained data are those showing large amounts of chlorine compared to sulfur in the clouds. We note, though, that the more recent Soviet measurements from VENERAs 13 and 14 show a large sulfur to chlorine ratio, the opposite of *Surkov*'s [1981] findings. The VEGA landers detected no large particles.

From the data in hand, it seems impossible to disprove the existence of Mode 3. Two self-consistent, alternative interpretations of the data exist. Accepting the spacecraft observation at face value, we are led to the existence of a mode of large solid non-spherical particles whose composition is unknown and whose source vapor has escaped detection. On the other hand, we may conclude that the large particle mode is merely the (mis-measured) tail end of the Mode 2 sulfuric acid droplets. This allows a simple understanding of the source of all the cloud particles, but at the cost of disbelieving some of the measurements.

5. CHLORINE CHEMISTRY

The most important consequence of photochemistry in the Venus atmosphere is the breaking of the strong O-CO bond, yielding O_2,

$$2(CO_2 + h\nu \rightarrow CO + O) \qquad (20)$$
$$O + O + M \rightarrow O_2 + M \qquad (21)$$

$$\text{Net: } 2CO_2 \rightarrow 2CO + O_2 \qquad (C)$$

The central problem of Venus photochemistry, thus, is the very low observational limit on the abundance of O_2. The observed large production rate of O_2 [*Connes et al.*, 1979; *Crisp et al.*, 1996] combined with the low upper limit on its abundance implies either a large reservoir in which oxygen is sequestered or fast oxidation of CO and/or SO_2, or all three. The abundance of O_2 is determined by the sulfur oxidation cycle, as described in Sec. (3.1), and the CO_2 cycle, which is the dominant chemical cycle above the cloud layers (>~ 70 km altitude). Hydrogen, chlorine, and sulfur chemistries are all linked to the CO_2 cycle, but chlorine chemistry is believed to provide the critical pathway for catalyzing oxidation of CO to CO_2 [*Krasnopolsky and Parshev*, 1981abc; 1983; *Yung and DeMore*, 1981; 1982; *Pernice et al.*, 2004].

Observations in 1966 [*Connes et al.*, 1967] detected a substantial (0.4–0.6 ppm at 67–70 km altitude [*Connes et al.*, 1967; *Young*, 1972]) abundance of HCl. HCl should be the dominant source of hydrogen and chlorine radicals at ~ 75–95 km because (1) the water vapor abundance should be suppressed by condensation onto the concentrated sulfuric acid particles in the upper haze layer [*Mills*, 1999a] and (2) photodissociation of H_2O is strongly attenuated below 100 km by CO_2 photoabsorption while photodissociation of HCl is still significant at 75 km.

Detection of HCl led to proposals that CO oxidation occurs via Reactions (22) and (23) [*Prinn*, 1971; *Sze and McElroy*, 1975].

$$ClOO + CO \rightarrow ClO + CO_2 \qquad (22)$$
$$ClO + CO \rightarrow Cl + CO_2 \qquad (23)$$

Subsequent laboratory studies, however, showed these reactions proceed too slowly to be significant. The next advances in understanding chlorine chemistry in the Venus atmosphere came in the early 1980s after VENERA 11 and 12 and near the beginning of the Pioneer Venus mission. Three modeling studies of increasing complexity and accuracy were published. *Winick and Stewart* [1980] introduced the first model with both sulfur and chlorine chemistry, including the now classic chlorine cycle for converting odd oxygen to molecular oxygen, Reactions (25) and (26).

$$O + O_2 + M \rightarrow O_3 + M \qquad (24)$$
$$Cl + O_3 \rightarrow ClO + O_2 \qquad (25)$$
$$ClO + O \rightarrow Cl + O_2 \qquad (26)$$

$$\text{Net: } 2O \rightarrow O_2 \qquad (D)$$

This cycle was so efficient it completely balanced the new sulfur cycles involving Reaction (4) that *Winick and Stewart* [1980] introduced to break the O-O bond. Consequently, the predicted O_2 column abundance was a factor of 50 larger than the then extant upper limit (and a factor of 150 larger than the current upper limit).

The next two modeling studies independently developed the chemistry of the chloroformyl radical (ClCO) and identified it as a potentially significant intermediary in the production of CO_2. These schemes involving ClCO are the only ones proposed to date that operate efficiently in the 75–90 km altitude region where CO_2 photodissociation is largest. The two schemes introduced by *Krasnopolsky and Parshev* [1981abc; 1983] for oxidation of CO via ClCO are

$$Cl + CO + M \rightarrow ClCO + M \qquad (27)$$
$$ClCO + O \rightarrow CO_2 + Cl \qquad (28)$$

$$\text{Net: } CO + O \rightarrow CO_2 \qquad (E)$$

$$Cl + CO + M \rightarrow ClCO + M \qquad (27)$$
$$ClCO + O_2 \rightarrow CO_2 + ClO \qquad (28)$$
$$ClO + O \rightarrow Cl + O_2 \qquad (26)$$

$$\text{Net: } CO + O \rightarrow CO_2 \qquad (F)$$

Neither of these breaks the O-O bond in O_2. There are two major problems with the Krasnopolsky and Parshev model. First, the equilibrium constant for ClCO in the compilation from which they drew their laboratory data [*Kondratiev*, 1971] was incorrect due to a typographical error and implied ClCO was two orders of magnitude more stable than laboratory studies had found. Second, independent laboratory studies [*Yung and DeMore*, 1982] found the mechanism postulated for Reaction (28) was incorrect. Production of CO_2 instead proceeds via the peroxychloroformyl radical (ClC(O)OO or $ClCO_3$), which is produced by Reaction (29).

$$ClCO + O_2 + M \rightarrow ClC(O)OO + M \qquad (29)$$

However, the Krasnopolsky and Parshev model calculated a much smaller O_2 abundance than any other model developed in the 1980's.

Yung and DeMore [1981; 1982] independently introduced three schemes for oxidation of CO via ClCO. Two are

$$Cl + CO + M \rightarrow ClCO + M \qquad (27)$$
$$ClCO + O_2 + M \rightarrow ClC(O)OO + M \qquad (29)$$
$$ClC(O)OO + Cl \rightarrow CO_2 + ClO + Cl \qquad (30)$$
$$ClO + O \rightarrow Cl + O_2 \qquad (26)$$

$$\text{Net: } CO + O \rightarrow CO_2 \qquad (G1)$$

$$Cl + CO + M \rightarrow ClCO + M \qquad (27)$$
$$ClCO + O_2 + M \rightarrow ClC(O)OO + M \qquad (29)$$
$$ClC(O)OO + O \rightarrow CO_2 + O_2 + Cl \qquad (31)$$

$$\text{Net: } CO + O \rightarrow CO_2 \qquad (G2)$$

Neither of these schemes breaks the O-O bond in O_2 but *Yung and DeMore*'s [1981; 1982] third scheme, which is discussed in Sec. (7.1), does. The primary failing of the *Yung and DeMore* [1981; 1982] models was that the calculated column abundance of ground-state O_2 was larger than the upper limit derived from observations made in 1983 [*Trauger and Lunine*, 1983].

Reaction (29) is the key step in what are believed to be the dominant gas-phase pathways for production of CO_2 in the Venus atmosphere, Figure 3. Production of CO_2 via pathways that include Reaction (29) accounted for 80% of the column total CO_2 production in the *Yung and DeMore* [1982] model C, and the four pathways in *Pernice et al.* [2004] that are most important for oxidation of CO are Schemes (E), (G1), (G2), and (G3).

$$Cl + CO + M \rightarrow ClCO + M \qquad (27)$$
$$ClCO + O_2 + M \rightarrow ClC(O)OO + M \qquad (29)$$

$$ClC(O)OO + h\upsilon \rightarrow CO_2 + ClO \qquad (32)$$
$$ClO + O \rightarrow Cl + O_2 \qquad (26)$$

$$\text{Net: } CO + O \rightarrow CO_2 \qquad (G3)$$

Scheme (G1) accounts for 54% of the column total CO_2 production in the *Pernice et al.* [2004] model while Schemes (E), (G2), and (G3) account for 19, 10, and 12%, respectively, of the column total CO_2 production in that model. The contributions of each of these pathways to oxidation of CO in the $+2\sigma$ model in Table 5 are comparable.

Recent calculations [*Mills*, 1998; *Pernice et al.*, 2004] have shown that a model using these gas-phase pathways can be brought into better agreement with the extant upper limit on O_2 [*Trauger and Lunine*, 1983; *Mills*, 1999b; Krasnopolsky, 2006b] by enhancing the stability of ClCO within its experimental uncertainty. However, none of the models that oxidize CO only via gas-phase processes agrees with the published 2σ upper limits on the column abundance of ground-state O_2 [*Trauger and Lunine*, 1983; *Krasnopolsky*, 2006b]. The smallest observational upper limit is equivalent to a column abundance of 8×10^{17} cm^{-2} [*Krasnopolsky*, 2006b], but there is disagreement over the interpretation of this observation [*Mills*, 1999b; *Krasnopolsky*, 2006b]. In addition, the improvement in the agreement with the O_2 observational upper limit achieved in later models [*Mills*, 1998; *Pernice et al.*, 2004] may create a disagreement with the CO vertical profile, but the observational constraints for CO from ground-based studies are not definitive, particularly on the day side and in the upper cloud [*Mills and Allen*, 2007].

There are three critical steps in the primary gas-phase pathway for production of CO_2, Figure 3: formation of ClCO; association of ClCO with O_2 to form $ClC(O)OO$ before ClCO thermally decomposes; and formation of CO_2 from $ClC(O)OO$. All three critical steps now have been observed in laboratory studies, so this reaction mechanism may be considered validated. The formation of ClCO and its thermal stability were studied most recently by *Nicovich et al.* [1990]. The assessed uncertainties in their results for heat of formation and reaction enthalpy are ~ 1.5 kcal/mol and in the reaction entropy is ~ 5 cal/mol/K [*Sander et al.*, 2002]. Formation of $ClC(O)OO$ via Reaction (29) was demonstrated in cryogenic

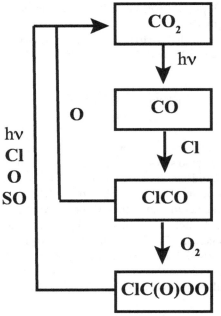

Figure 3. Schematic showing primary pathways for production of CO_2 via chlorine chemistry. The reaction $ClCO + O \rightarrow CO_2 + Cl$ accounts for 15 and 20% of the column total CO_2 production in the $+0.5\sigma$ and $+2.0\sigma$ models from Table 5, respectively.

matrix experiments [*Pernice et al.*, 2004] and a gas phase rate for production in laboratory studies has been reported [*Yung and DeMore*, 1982]. Other experiments reported in *Pernice et al.* [2004] demonstrated the formation of CO_2 from photodissociation of $ClC(O)OO$, derived the UV absorption spectrum for $ClC(O)OO$, and determined that $ClC(O)OO$ is thermally stable at temperatures below 350 C.

Substantial uncertainties remain, however, in the rates at which each of the three critical steps occur in Venus' atmosphere. The experimental uncertainty in the equilibrium constant for ClCO can be expressed in terms of Equation (33) for 200–300 K [*Sander et al.*, 2002]

$$f(T) = f(298 \text{ K}) \exp[\Delta B((1/T) - (1/298))], \qquad (33)$$

where f = uncertainty, T = temperature, ΔB = uncertainty in the equilibrium constant's exponential factor which

Table 5. Sensitivity of modeled O_2 column abundance to thermal stability of ClCO in baseline chlorine chemistry model [*Pernice et al.*, 2004].

ClCO stability	$+2\sigma$ [a]	$+1\sigma$	$+0.5\sigma$	Nominal	-0.5σ	-1σ	-2σ [b]
O_2 column abundance (10^{18} cm^{-2})	2.0	2.4	3.1	8.6	27	39	49

[a] Increased the thermal stability of ClCO by twice the assessed uncertainty in its equilibrium constant [*Sander et al.*, 2002].
[b] Decreased the thermal stability of ClCO by twice the assessed uncertainty in its equilibrium constant [*Sander et al.*, 2002].

is 500 K for ClCO, and f(298) = 5 for ClCO. Although small by experimental standards, the uncertainty in the equilibrium constant for ClCO translates into a large uncertainty in the predicted O_2 profile [*Mills*, 1998] and in the calculated O_2 column abundance, as illustrated in Table 5. No experimental uncertainty has been reported for Reaction (29) and only one laboratory study of this reaction rate is published [*Yung and DeMore*, 1982]. No experimental measurements exist for the rates at which ClC(O)OO reacts with any of the important radicals in the Venus atmosphere (e.g., O, SO, Cl), although production of CO_2 from photodissociation of ClC(O)OO has been demonstrated [*Pernice et al.*, 2004].

None of the chlorine radicals has yet been observed in Venus' atmosphere. Their detection would provide a major confirmation of this proposed model.

6. UNIDENTIFIED UV AND BLUE ABSORPTION

Currently, the only positively identified species in the visible atmosphere that absorb in the near UV are SO_2 and SO. These gases can account for the absorption observed at wavelengths short of 320 nm at all altitudes [*Esposito*, 1980; *Pollack et al.*, 1980] and for the absorption observed at 320–390 nm below 55 km [*Ekonomov et al.*, 1983]. However, they cannot account for the absorption observed in the upper cloud at wavelengths longward of 320 nm, nor can they account for the absorption observed at 400–500 nm [*Esposito*, 1980; *Pollack et al.*, 1980; *Ekonomov et al.*, 1983]. In addition, these other absorbers must explain the phase angle dependence of the UV dark markings [*Barker et al.*, 1975] as well as their short lifetime above the clouds (from hours to days, see *Esposito et al.* [1983]). They must also be consistent with the visible-wavelength solar flux observations of *Tomasko et al.* [1980] which show absorption at 58–62 km, and little absorption below. Near UV solar flux absorption results from VENERA 14 [*Ekonomov et al.*, 1983; 1984] provide an additional constraint. *Esposito and Travis* [1982] noted the correlation between dark markings seen longward of 320 nm and SO_2 enhancements seen at 207 nm. This means that in addition to the absorption spectrum, a good candidate must also match constraints on its vertical distribution, lifetime, and correlation with SO_2 enhancement. This last correlation could be either chemical or dynamical because the SO_2 observable in the middle UV is likely the result of local upwelling [*Esposito and Travis*, 1982]. We briefly review the suggested candidates below.

S_8. S_8 was proposed by *Hapke and Nelson* [1975] and *Young* [1977, 1983] as the second absorber since it absorbs strongly in the UV. However, *Pollack et al.* [1980] showed that the spectral characteristics of S_8 were inconsistent with those of the second absorber. Another shortcoming of S_8 as the second absorber is its vertical profile. S_8 is not expected to disappear rapidly below the upper cloud layer since it precipitates as a solid, and the idea of these particles hiding inside sulfuric acid aerosols has been discounted by *Young* [1983]. Thus, the vertical profile of S_8 does not match that of the second absorber. Furthermore, the chemical lifetime of S_8 above the clouds is much longer than the timescale of the dark markings, thus it is difficult to explain the rapid disappearance (lifetime <3 hours) of small scale dark markings [e.g., *Rossow et al.*, 1980].

S_3 and S_4. *Toon et al.* [1982] suggested metastable sulfur allotropes, S_3 and S_4, as the most likely candidate for the second absorber. The absorption cross sections of S_3 and S_4 peak around 400 and 520 nm, respectively. The combination of these two sulfur gases with SO_2 provides a very close match to the albedo of Venus. The peak in the absorption cross sections of S_3 around 400 nm lines up with a kink in the albedo spectrum of Venus. S_3 and S_4 are metastable, and once produced they quickly relax to S_8 which exists as particulates. In this model, the S_8 would therefore contribute less. It is not clear if this contradicts Young's assertion (above). S_8 particles could then become incorporated into the sulfuric acid aerosols and fall out of the cloud region. This scenario thus explains the short lifetime of the dark features and the absence of the second absorber below the upper clouds. Furthermore, these sulfur allotropes can account for the high real refractive index of the upper cloud material, and the bimodal size distribution observed in the Venus clouds. When there is little oxygen in the atmosphere, sulfur allotropes can be produced from SO_2 photolysis instead of sulfuric acid [*Prinn*, 1975, 1985], so S_3 and S_4 may be produced in areas where sulfur vapor is enriched. However, there has been no positive detection of S_3 or S_4 to date. Further, these allotropes are rapidly photo-dissociated, giving lifetimes close to 1 second in the upper cloud. One further problem with the above scenario is that sulfur particles would still absorb UV photons below the upper clouds.

Cl_2. *Pollack et al.* [1980] proposed Cl_2 as a candidate for the second absorber. Pollack et al. showed that about 1 ppm of Cl_2 in the upper cloud could produce the brightness contrast seen in the UV images of the clouds. Cl_2 may exist in the atmosphere of Venus since it is produced from photodissociation of HCl which was detected in the atmosphere of Venus by *Connes et al.* [1967]. Some of the difficulties of Cl_2 being the second absorber include the relative scarcity of HCl in the atmosphere. The photochemical calculations of *Winick and Stewart* [1980], *Krasnopolsky and Parshev* [1981abc; 1983], and *Yung and DeMore* [1981; 1982] showed that the amount of Cl_2 produced from HCl photolysis is two orders of magnitude smaller than 1 ppm. Furthermore, the vertical profile shows no maximum around 60 km, or rapid decline below 58 km. Finally, there is no obvious chemical or dynamical connection between Cl_2 and SO_2.

FeCl$_3$. Ferric Chloride is a possible candidate to explain absorption in the cloud layer. *Krasnopolsky* [1985, 1986] showed that many properties of the clouds can be explained if condensation of Fe_2Cl_6 vapor occurs at 47.5 km at the PV sounder probe site. This means that this species' mixing ratio is equal to 15 ppbv below 47.5 km. The calculated profile of the $FeCl_3$ condensate coincides with that of the mode 1 particles in the lower and middle cloud layer. The mode 1 $FeCl_3$ particles can then be transported by eddy diffusion to the upper cloud layer where they serve as condensation centers for the mode 2 H_2SO_4 particles. These particles are liquid below 62–63 km, and the ratio of the $FeCl_3$ flux to the H_2SO_4 production rate corresponds to a solution with concentration of $FeCl_3$ close to 1%. It is this solution which can explain the 320–500 nm absorption [*Zasova et al,*, 1981]. The reaction between $FeCl_3$ and concentrated H_2SO_4 is rather slow at temperatures 250–280 K at 62–58 km, and the lifetime of the solution is close to the precipitation time of one month. Colorless ferric sulfate, $Fe_2(SO_4)_3$, replaces $FeCl_3$ near 58 km. In thermochemical equilibrium:

$$Fe_2(SO_4)_3 + 6HCl + 3CO \leftrightarrows$$
$$Fe_2Cl_6 + 3H_2O + 3SO_2 + 3CO_2 \qquad (34)$$

it is favorable to form Fe_2Cl_6 vapor throughout the atmosphere. $FeCl_3$ in aerosol particles was observed by X-ray fluorescent spectroscopy from the VENERA 14 and VEGA landers [*Petryanov et al.*, 1981; *Andreichikov et al.*, 1987; *Krasnopolsky*, 1989]. The transition from $FeCl_3$ to colorless ferric sulfate predicted to occur at 58 km is close to the lower boundary of absorption observed by *Ekonomov et al.* [1983, 1984] on VENERA 14.

SCl$_2$. *Krasnopolsky* [1986] found a mixture of sulfur aerosol and SCl_2 fits well the absorption, but the required mixing ratio for SCl_2 of 10^{-7} exceeds his estimate of its mixing ratio by at least an order of magnitude. The SCl_2 mixing ratio in the models whose results are shown in Table 5 and in *Pernice et al.* [2004] is less than 10^{-9} with the maximum in the lowest layer at 58 km altitude.

Croconic acid. *Hartley et al.* [1989] proposed croconic acid ($C_5O_5H_2$) as a candidate for the second absorber based on their measurement of its absorption spectrum. They carried out radiative transfer calculations to show that sulfuric acid aerosols mixed with 2.5% of croconic acid can produce the observed UV contrast in the clouds of Venus. According to *Hartley et al.* [1989], croconic acid can be produced by reactions involving CO and HO_x, for example, in the equilibrium

$$C_5O_5H_2 + CO_2 \leftrightarrows 6CO + H_2O \qquad (35)$$

Its production is proportional to $[CO]^6$ and should be maximum near 40 km, because this acid decomposes at 40 km (T = 150°C). All these facts reduce the possibility of $C_5O_5H_2$. *Widemann et al.* [1993] reported a negative detection of croconic acid in the spectra of Venus' atmosphere taken with the VEGA probe.

Ammonium pyrosulfite. *Titov* [1983] studied the formation of aerosol in a mixture of SO_2, NH_3, and H_2O, and found that ammonium pyrosulfite, $(NH_4)_2S_2O_5$, resulted from the mixture. He studied the absorption spectrum of $(NH_4)_2S_2O_5$ and found that the mixture of NH_3 with H_2SO_4 has spectral characteristics similar to that of the second absorber. *Krasnopolsky* [1985] points out that formation of ammonium pyrosulfate is enhanced at low temperatures. This would predict an anti-correlation between the albedo at 365 nm and the radiance at 11.5 μm (this latter wavelength is sensing the upper cloud temperature), which is not seen in PV observations.

Nitrosylsulfuric acid. $NOHSO_4$ [*Watson et al.*, 1979; *Sill*, 1983]. Unfortunately, it only has weak blue absorbtion and requires the presence of NO and NO_2 with abundances that seem incompatible with the upper limit of 6 ppb for NO_2 and with the expected production of NO by lightning if it really occurs on Venus. These negative comments are still true, although *Krasnopolsky* [2006a] has observed 5.5 ppb of NO below 60 km.

S$_2$O. *Hapke and Graham* [1985, 1989] proposed that disulfur monoxide (S_2O) and polysulfur oxides (PSO) may be responsible for the UV markings in the clouds of Venus. They measured the relative reflectance of S_2O frost at 77K, and found that it has a low reflectivity in the wavelength region from 200 to 500 nm. *Na and Esposito* [1997] estimated the chemical lifetime and vertical distribution of S_2O, both of which match the second absorber, although calculations by *Mills* [1998] found S_2O abundances significantly smaller than those calculated by *Na and Esposito* [1997] for reasons that have not been resolved. Its obvious chemical connection with SO_2 could explain the correlations of the dark markings with SO_2 enhancements. Unfortunately, we do not have a good spectrum measured for the gas phase. In addition, recent *ab-initio* calculations [*Groves and Lewars*, 2000] suggest $(SO)_2$, the precursor to S_2O, may not be as stable as has been assumed but there are no experimental data.

Of the candidates discussed above, elemental sulfur and ferric chloride solution in sulfuric acid are the most promising candidate species for explaining the observed absorption at 320–500 nm, perhaps even in combination. S_2O may also be important but there are significant uncertainties in both its spectrum and its chemistry. Further laboratory studies are required to address existing uncertainties in the spectral characteristics of the candidate species and their chemistry. Venus Express images and spectra may allow progress on this topic. For example, a correlation of dark markings in VIRTIS-M data at multiple wavelengths where S_x is a strong absorber would be good evidence for the presence of S_x [*Carlson*, 2007]. However, in situ measurements within the clouds provide the only prospect for a secure identification. This must await a future Venus mission.

7. EMERGING DEVELOPMENTS: COUPLING AMONG CYCLES AND NITRIC OXIDE

7.1. Coupling Among Chemical Cycles

The three chemical cycles couple at a number of key points. First, as discussed in Sec. (2.6), the middle and lower atmosphere portions of the two sulfur cycles are believed to balance each other, on the assumption that the atmosphere is near a long-term equilibrium point. However, little modeling has been done across the boundary between photochemistry and thermochemistry and there appear to be inconsistencies between the best fit solutions derived for each altitude region. *Esposito et al.* [1997] provide an excellent discussion of these issues, so only a summary is provided here. Model 1 from *Krasnopolsky and Pollack* [1994] assumed a lower production rate for H_2SO_4 (2.2×10^{12} cm^{-2} s^{-1}), placed the lower cloud boundary at 48.4 km (agreeing with *in-situ* data), and predicts the H_2O mixing ratio is 30 ppm at 30 km (agreeing with *Pollack et al.* [1993]). Model 1 can be roughly correlated with *Yung and DeMore's* [1982] Model C, which has good kinetic data for the middle atmosphere. Model 2 from *Krasnopolsky and Pollack* [1994] assumed a higher production rate for H_2SO_4 (6×10^{12} cm^{-2} s^{-1}), placed the lower cloud boundary at 46.5 km (agreeing with radio occultations), and predicts the H_2O mixing ratio is 90 ppm at 30 km (close to *Moroz et al.* [1983]). Model 2 can be roughly correlated with *Krasnopolsky and Parshev's* [1981abc; 1983] photochemical model. A newer analysis of the *Venera* optical spectra using a high temperature spectroscopic database [*Ignatiev et al.*, 1997] found an abundance for H_2O at 30 km that agrees well with the interpretation of ground-based observations [*Pollack et al.*, 1993], so Model 1 is now strongly favored on that basis. In addition, *Yung and DeMore's* [1982] Model C has much firmer photochemical data. The predicted O_2 column abundance from Model C, however, exceeds the spectroscopic upper limit by an order of magnitude, although more recent models based on the *Yung and DeMore* [1981; 1982] Model C chemistry [*Pernice et al.*, 2004] show better agreement. *Krasnopolsky and Parshev's* [1981abc; 1983]

model overpredicts the O_2 column abundance by only a factor of two but key kinetic data underlying the calculations were not correct (Sec. (5)). Modeling and data over the depth of the cloud layers are needed to resolve these apparent conflicts.

The CO_2 and sulfur oxidation cycles are also linked near the cloud tops in a manner that can efficiently break the O-O bond in O_2 [*Yung and DeMore*, 1981; 1982]. The essence of the chemical scheme may be summarized as

$$Cl + CO + M \rightarrow ClCO + M \qquad (27)$$
$$ClCO + O_2 + M \rightarrow ClC(O)OO + M \qquad (29)$$
$$ClC(O)OO + Cl \rightarrow CO_2 + ClO + Cl \qquad (30)$$
$$SO_2 + h\nu \rightarrow SO + O \qquad (36)$$
$$SO_2 + O + M \rightarrow SO_3 + M \qquad (37)$$
$$SO_3 + H_2O + M \rightarrow H_2SO_4 + M \qquad (38)$$
$$SO + ClO \rightarrow SO_2 + Cl \qquad (7)$$

$$\text{Net: } CO + O_2 + SO_2 + H_2O \rightarrow CO_2 + H_2SO_4 \qquad (H)$$

where $ClC(O)OO$ is the peroxychloroformyl radical and M is a third body (ambient atmosphere). Note that this scheme oxidizes SO_2 to H_2SO_4 while oxidizing CO to CO_2. The relative importance of Scheme (H), when compared to Schemes (G1), (G2), and (G3), determines the efficiency with which the O-O bond in O_2 is broken. Scheme (H) has a solid basis in laboratory kinetic data (Sec. (5)) but likely requires vertical transport of SO because photodissociation of CO_2 occurs predominantly above 75 km altitude while photodissociation of SO_2 occurs predominately below 70 km altitude. Vertical transport of significant amounts of O_2 would likely require violating the observational upper limit on its abundance, and the lifetime of ClO should be short.

Cl-SO_2 interaction can lead to formation of sulfuryl chloride, SO_2Cl_2, [*DeMore et al.*, 1985] via

$$2 (Cl + SO_2 + M \rightarrow ClSO_2 + M) \qquad (39)$$
$$2ClSO_2 \rightarrow SO_2Cl_2 + SO_2 \qquad (40)$$

$$\text{Net: } 2Cl + SO_2 \rightarrow SO_2Cl_2 \qquad (I)$$

The rate for Reaction (39) has been determined in laboratory studies [*Stratton et al.*, 1979; *Eibling and Kaufman*, 1983], although none of the work was conducted in CO_2. The rate for Reaction (40) has not been determined but quantitative evidence of the production of SO_2Cl_2 from photolysis of a mixture of Cl_2 and SO_2 has been found in laboratory experiments [*DeMore et al.*, 1985].

The inclusion of Reaction (39) but not Reaction (40) in the models whose results are presented in Table 5 increases the SO_2 scale height near 70 km altitude from ~ 2.5 km [*Yung and DeMore*, 1981; 1982] to ~ 3.0–3.5 km, both of which are within the 2–4 km observational constraint [*Na et al.*, 1994]. This buffering of SO_2 also contributes to decreasing the H_2SO_4 production rate from 1.4×10^{12} cm^{-2} s^{-1} [*Yung and DeMore*, 1982] to ~ 5×10^{11} cm^{-2} s^{-1}, which is closer to an early observational estimate of the H_2SO_4 production rate, ~ 2×10^{11} cm^{-2} s^{-1} [*Winick and Stewart*, 1980]. All of the production rates for H_2SO_4 from these photochemical models are smaller than those required to match either model from *Krasnopolsky and Pollack* [1994]. In addition, *Krasnopolsky* [2006a] found that inclusion of NO_x in a photochemical model opened an additional important pathway for oxidizing SO

$$NO_2 + SO \rightarrow NO + SO_2 \qquad (41)$$

which may further reduce the modeled H_2SO_4 production rate. If the H_2SO_4 production rate is as large as has been modeled by *Krasnopolsky and Pollack* [1994], then a critical process may be missing from current photochemical models.

One way the rate of formation of sulfuric acid might be enhanced is via $Cl-SO_2$ interactions

$$Cl + SO_2 + M \rightarrow ClSO_2 + M \qquad (39)$$
$$ClSO_2 + O_2 + M \rightarrow ClSO_4 + M \qquad (42)$$
$$ClSO_4 + Cl \rightarrow SO_3 + ClO + Cl \qquad (43)$$
$$SO_2 + h\nu \rightarrow SO + O \qquad (36)$$
$$SO + ClO \rightarrow SO_2 + Cl \qquad (7)$$
$$SO_2 + O + M \rightarrow SO_3 + M \qquad (37)$$
$$2(SO_3 + H_2O + M \rightarrow H_2SO_4 + M) \qquad (38)$$

$$\text{Net: } O_2 + 2SO_2 + 2H_2O \rightarrow 2H_2SO_4 \qquad (J)$$

DeMore et al. [1985] proposed these processes could be quite important in Venus' atmosphere and demonstrated quantitative production of H_2SO_4 from irradiated mixtures of Cl_2, SO_2, and O_2 in which the yield of H_2SO_4 (versus SO_2Cl_2) increased as the partial pressure of O_2 increased, but they have not been developed further. More specific information from laboratory experiments on the reaction mechanism and the rates of the reactions leading to production of H_2SO_4 is needed. The interaction of chlorine and sulfur chemistry in the Venus cloud layers merits further study as it provides key links among the primary chemical cycles and may provide a mechanism to sequester oxygen for transport to the lower atmosphere.

7.2. Nitric Oxide (NO)

Lightning is equivalent to thermodynamic equilibrium chemistry at high temperatures, initially in excess of 30,000

K in the fireball, but as temperature decreases rapidly due to adiabatic expansion, the constituents are quenched at about 2000 K. The net result may be summarized by the chemical scheme [*Yung and McElroy*, 1979],

$$CO_2 \leftrightarrow CO + O \qquad (44)$$
$$O + N_2 \leftrightarrow NO + N \qquad (45)$$
$$N + CO_2 \leftrightarrow NO + CO \qquad (46)$$

$$\text{Net: } 2CO_2 + N_2 \leftrightarrow 2CO + 2NO \qquad (K)$$

This scheme is analogous to that operating in the present terrestrial atmosphere [*Price at al.*, 1997], where NO is made via the chemical scheme,

$$O_2 + N_2 \leftrightarrow 2NO \qquad (47)$$

On Venus, as it was on prebiotic Earth, CO_2 replaces O_2 as the oxidant for N_2. The NO is eventually removed by reactions above the cloud tops,

$$NO + h\nu \rightarrow N + O \qquad (48)$$
$$N + NO \rightarrow N_2 + O \qquad (49)$$

$$\text{Net: } 2NO \rightarrow N_2 + 2O \qquad (L)$$

8. COMPARISONS TO OTHER PLANETS

8.1. Comparisons Among Venus, the Earth, and Mars

Catalytic chemistry plays fundamental roles in the atmospheres of all three planets [see, *e.g.*, *Yung and DeMore*, 1999]. This means trace abundances of highly reactive radicals govern the primary chemical cycles on each planet. On Venus, a small amount of Cl has a major impact on the production of CO_2 from CO and O_2. On the Earth, hydrogen, nitrogen, and halogen radicals play an analogous role. The most prominent example is the regulation of the stratospheric ozone layer. A large number of the catalytic cycles are of the form

$$O_3 + X \rightarrow XO + O_2 \qquad (50)$$
$$O + XO \rightarrow X + O_2 \qquad (51)$$

$$\text{Net: } O_3 + O \rightarrow 2O_2 \qquad (M)$$

where X = H, OH, NO, Cl and Br. The net result of these catalytic cycles is the destruction of O_3. On Mars the key catalytic cycle that recombines CO and O to form CO_2 involves hydrogen radicals [*McElroy and Donahue*, 1972].

$$CO + OH \rightarrow CO_2 + H \qquad (52)$$
$$H + O_2 + M \rightarrow HO_2 + M \qquad (53)$$
$$O + HO_2 \rightarrow OH + O_2 \qquad (54)$$

$$\text{Net: } CO + O \rightarrow CO_2 \qquad (N)$$

A similar scheme by *Parkinson and Hunten* [1972] is also important.

The greatest difference between Venus and Earth concerns the amount of water on these two planets [see, *e.g., Yung and DeMore*, 1999]. The mixing ratio of water vapor in the lower atmosphere of Venus is variable, with a maximum value of 1.5×10^{-4}. This is equivalent to a layer of 2–10 cm of water, uniformly spread over the surface of the planet. For comparison, Earth contains an average layer of 2.7 km of water, residing mostly in the oceans. The lack of an ocean on Venus has at least three dramatic consequences for the atmosphere. First, most of the planet's CO_2 remains in the atmosphere, in contrast to Earth, where most of the 50 bars of CO_2 are sequestered as carbonate rock in the sediments. Second, the atmosphere of Venus contains large quantities of SO_2. On Earth, most of the volatile sulfur resides in the ocean as sulfate ions. The presence of this large amount of SO_2 in the atmosphere is responsible for the production of a dense H_2SO_4 cloud on Venus. Third, the atmosphere of Venus contains large amounts of HCl. On Earth, the bulk of chlorine is in the form of salt (NaCl) in the oceans.

8.2 Atmospheric Chemistry and Atmospheric Evolution

It is now generally accepted that the origin of water in the planets is dehydration of hydrous minerals during formation. The bulk of infalling material has composition similar to that of CI chondrites, which contain up to 3 wt% H_2O. Once the accreting planetary body exceeds a critical size, impact velocities become sufficiently high that devolatilization can start. The critical radii, R_1, for Earth, Venus, and Mars are about 20%, 20%, and 40% of their present radii, respectively. Complete devolatilization can occur when the planets grow to another critical radius, R_2. The values for R_2 are roughly 50% of the present radii for the Earth and Venus; Mars never reached R_2. This implies that the terrestrial planets all should have acquired massive amounts of water of the order of terrestrial oceans. Why, then, is the ocean of water absent from Venus?

In a seminal paper, *Ingersoll* [1969] showed that water vapor could have been a major constituent in the atmosphere of Venus, and that photolysis of H_2O, followed by hydrogen escape could approach the UV photon limit of about 10^{13} cm^{-2} sec^{-1}. Under such conditions, an amount of water equivalent to the terrestrial ocean could be lost in 100 Myr.

Evidence that the atmosphere of Venus has lost most of its water (assuming it had acquired a large quantity) came from the Pioneer Venus measurements of the D/H ratio, which is about 100 times the terrestrial value [*McElroy et al.*, 1982; *Donahue et al.*, 1982] and has been confirmed by Earth-based spectroscopy [*de Bergh et al.*, 1991]. This implies that Venus has likely lost at least the equivalent of 100 times more water than its present reservoir. The loss of hydrogen from Venus is determined by the efficiency of the escape processes and the rate at which hydrogen is transferred from the lower atmosphere to the upper atmosphere where the escape processes operate [*Gurwell and Yung*, 1993]. *Kasting and Pollack* [1983] suggested the early escape of hydrogen from Venus could have been a hydrodynamic process rather than kinetic evaporation from the exosphere, similar to that proposed for Pluto [*Krasnopolsky*, 1999; *Tian and Toon*, 2005]. However, there was no direct observational evidence for hydrodynamic escape from a planetary atmosphere until the recent observation of H atoms escaping from the extra-solar planet HD209458b [*Vidal-Madjar et al.*, 2003]. *Liang et al.* [2003] showed that the source of the H atoms that fuel the hydrodynamic escape on this planet is photolysis of H_2O. This process could have been the same as the one postulated by *Ingersoll* [1969] for the atmosphere of Venus.

The above interpretation is by no means unique. There is an alternative view of the evolutionary history of water on Venus. According to *Grinspoon and Lewis* [1988], Venus lost all its initial inventory of water very early. The current water vapor content is the result of a balance between supply by cometary impact and loss processes. *Grinspoon* [1993] argued that another recent source of water might have been the massive outgassing from catastrophic resurfacing of the planet in the past 0.5–1 Gyr. Determination of noble gas isotopic abundances for Venus may help constrain the relative importance of external versus internal sources of volatiles, but the interpretation of such measurements has often been contentious. Resolution of these two different evolutionary models may depend on an accurate determination of the present escape flux of H and D from Venus [*Donahue*, 1999; *Lammer et al.*, 2006].

The D/H ratio in water vapor on Mars also is enhanced, but by a factor of 6 [*Owen et al.*, 1988; see also *Krasnopolsky et al.*, 1997], which implies that Mars also may have lost large amounts of water in the past [*Yung et al.*, 1988; *Kass and Yung*, 1995; 1999; *Cheng et al.*, 1999; *Miller and Yung*, 2000]. This enrichment of D over H is largely the result of photochemical processes that preferentially dissociate H_2O relative to HDO and the lower efficiency for the escape of D relative to H at the exobase (where the atmosphere merges into space). In addition, *Fouchet and Lellouch* [2000] and *Bertaux and Montmessin* [2001] point out that there is a trap for deuterium on Mars due to water vapor condensation. The latter authors name this the "deuteropause", by analogy with the tropopause in the terrestrial atmosphere [*Moyer et al.*, 1996; *Kuang et al.*, 2003]. One implication of these theories about Mars' evolution is that Mars must have been warmer in the past, which implies it was more likely to have been habitable. However, it is currently impossible to determine the precise amount of water that has escaped in the past. Estimates of the lost water range from lower values of 9 m [*Kass and Yung*, 1999] and 14 m [*Jakosky*, 1990] to higher values of 65 to 120 m [*Krasnopolsky*, 2000].

It is a remarkable fact that the Earth has remained habitable throughout its planetary history even though the solar luminosity has increased substantially. The major energy source for the solar system is the sun. The sun's luminosity has gradually increased by about 40% since the origin of the solar system. The reduced solar constant during the nascent period of the Earth would imply that the planet was completely frozen, a result in conflict with known geological evidence (e.g., sedimentary rocks). A resolution of this paradox is to postulate that the CO_2 content of the atmosphere has been changing with time to compensate for the changing solar constant. According to the Gaia hypothesis, the biosphere may indeed have evolved since its origin to counteract the problem of the increasing luminosity of the sun. However, there is a finite limit to the power of Gaia. The abundance of CO_2 is now very low, while the luminosity of the sun continues to increase. Further decrease of CO_2 by biological activity may be difficult because photosynthesis itself stops when the CO_2 mixing ratio falls below a threshold level of about 150 ppmv. Thus, there is a point beyond which the Gaian control of the global environment would fail and the Earth would become Venus-like in 30–300 Myr [*Lovelock and Whitfield*, 1982]. Venus may be the ultimate graveyard of all terrestrial planets above a certain critical size.

8.3. Implications of Venus Clouds and Chemistry for Studies of Extra-Solar Planets

NASA is undertaking a major effort to find terrestrial planets around other stars. A hope is that they may resemble Earth. A likely possibility from the present state of the solar system, and from the increasing luminosity of the Sun, is that some significant fraction of these terrestrial planets will resemble Venus. Many of the observable characteristics of a terrestrial planet resembling Venus are determined by its atmospheric chemistry and clouds. The first observations will not resolve an exta-solar planet: its global brightness, variability and spectrum will be the first measurements. The mass, radius and albedo can also be inferred from the observations. It is important to remember that the observed temperature may differ from the effective temperature and also from the surface temperature. Venus provides an excellent example in our own solar system.

The distinctive characteristics of Venus would be clear to the first observations from another solar system:
1. High albedo with blue and UV broadband absorption.
2. CO_2 atmosphere, detectable from CO_2 absorption bands and CO emissions.
3. Lack of water, from absence of H_2O absorption bands and OH emission.
4. Clouds of H_2SO_4, inferred from SO_2 absorption bands.
5. In the UV, Venus' brightness will vary over the 4-day super-rotation of the cloud-top markings.

Thus, Venus-like planets could be clearly distinguished from potentially habitable Earth-like planets. The models we have developed for Venus' clouds and chemistry (along with the coupled radiation and dynamics) provide a starting point for understanding those planets. Since our evolutionary models of Venus may encompass Earth's future evolution, this may allow inferences about whether Venus-like planets around other stars ever *were* habitable, and thus to examine the effects of the runaway greenhouse in other solar systems.

9. RECOMMENDATIONS AND FUTURE DIRECTIONS

Better observations of atmospheric composition, including vertical, horizontal, and temporal variations are required to advance our understanding of Venus' clouds and chemistry.
1. The most critical observation required for understanding the chemistry of the middle atmosphere is a determination of the ground-state O_2 abundance. Venus Express will not provide this, but high resolution (resolving power $\sim 400,000$) ground-based spectra might determine its column abundance.
2. Vertical profiles of the time-varying $O_2(^1\Delta)$ airglow emission from Venus Express will constrain models of the oxygen airglow chemistry and limit the range of acceptable models.
3. The most direct observational test of the proposed chlorine chemistry would be detection of ClC(O)OO, which may be possible now that its structure and spectroscopy are known [*Pernice et al.*, 2004]. Observing O_3, $COCl_2$, ClO, or Cl_2 would provide important indirect tests of models. It is not clear whether Venus Express can observe these compounds, but simulations indicate a submillimeter spectrometer, such as that proposed for the *Vesper* mission could measure the abundances of ClO, $O_2(^1\Delta)$, and O_2

even if their concentrations are a factor of 10 smaller than predicted by recent global-average models [*Pernice et al.,* 2004; *Mills et al.,* 2006]. The O_3 abundances predicted by these recent models should also be detectable by *Vesper's* submillimeter spectrometer.

4. Direct observational tests of sulfur chemistry will be provided by three-dimensional maps of SO_2, SO, H_2S (if present and observable), and OCS derived from Venus Express spectroscopy and VMC images.

Better observations of cloud particle properties and their variation also are required. The size distribution, shape, and composition of the majority of the aerosol mass are still open, despite our assurance that "mode 2" (the aerosols visible at the cloud tops) are spherical droplets of concentrated sulfuric acid. Venus Express will make remote measurements of the clouds. Multiple observations will be combined to improve our understanding. Key tasks are to compare, reconcile and consolidate these measurements.

The information expected from Venus Express also should justify more detailed studies of the interactions among radiation, clouds, chemistry and dynamics. For example,

1. Simultaneous three-dimensional maps of time-varying temperatures and the abundances of CO and other species will permit development of multi-dimensional chemical transport models.

2. Spatially-resolved observations of H_2O are needed to understand the large temporal variations reported in ground-based studies [*Sandor and Clancy,* 2005] and to refine our understanding of the cloud layers.

Determination of the mineralogy on the surface of Venus is needed for significant progress on understanding the chemical interaction between the atmosphere and surface and how this interaction may have changed over the course of Venus' evolution.

The most important laboratory measurement needed for understanding the chemistry in the Venus middle atmosphere is a more accurate determination of the equilibrium constant for ClCO and any temperature dependence. Verification of the rate for Reaction (29) which forms ClC(O)OO and an assessment of its uncertainties is also needed. Laboratory measurements are also required to assess whether more speculative schemes for oxidizing CO, such as heterogeneous reactions, are sufficiently rapid. Finally, laboratory spectra and in situ measurements will be the only means by which the unknown UV absorber can be definitively identified. Spectra do not exist for many of the proposed candidates, so it is not possible to compare model calculations with observations.

For the lower atmosphere's chemistry, the validity of reactions (15), (16), (17), and (18) needs to be ascertained in laboratory studies. Laboratory work also is needed to deter-mine whether equivalent heterogeneous reactions are more important than these gas-phase reactions. Further modeling may be required to identify alternative chemical schemes, particularly if any of these reactions is found to not occur or to be too slow to explain observations.

The discovery of NO in the cloud layers presents the exciting possibility that heterogeneous chemistry on the surface of sulfate aerosols may convert the reservoir species of chlorine into active species, which can then participate in catalytic chemistry. The reservoir species (HCl, $ClONO_2$) are chemically unreactive, but the labile species can readily participate in catalysis. These reactions have been shown to play a major role in the destruction of ozone in the Earth's atmosphere [*Solomon et al.,* 1997].

Acknowledgements. Helpful comments were received from two anonymous referees and Dr. Thomas Cravens. YLY was supported in part by NASA grant NNX07AI63G to the California Institute of Technology. This research was partially supported under the Australian Research Council's Discovery Projects funding scheme.

REFERENCES

Allen, D. A., and J. W. Crawford (1984), Cloud structure on the dark side of Venus. *Nature, 307,* 222–224.

Andreichikov, B. M. (1987), Chemical composition and structure of the clouds of Venus inferred from the results of X-ray fluorescent analysis on descent probes VEGA 1 and 2, *Kosmich. Issled., 25,* 737–743 (in Russian).

Andreichikov, B. M., et al. (1987), X-ray radiometric analysis of the cloud amount of Venus by the Vega 1 and 2 probes, *Kosmich. Issled., 25,* 721–736 (in Russian).

Barker, E. S. (1975), Observations of Venus water vapor over the disk of Venus: The 1972–74 data using the H_2O lines at 8197 Å and 8176 Å, *Icarus, 25,* 268–281.

Barker, E. S. (1979), Detection of SO_2 in the UV spectrum of Venus, *Geophys. Res. Lett., 6,* 117–120.

Barker, E. S., et al. (1975) Relative spectrophotometry of Venus from 3067 to 5960 Å, *J. Atmos. Sci., 32,* 1205.

Bertaux, J. L., and F. Montmessin (2001), Isotopic Fractionation through Water Vapor Condensation: The Deuteropause, a Cold Trap for Deuterium in the Atmosphere of Mars., *J. Geophys. Res., 106* (E12), 32879–32884.

Bertaux, J-L, T. Widemann, A. Hauchecorne, V.I. Moroz, and A.P. Ekonomov (1996), VEGA 1 and VEGA 2 entry probes: An investigation of local UV absorption (220–400 nm) in the atmosphere of Venus (SO_2, aerosols, cloud structure), *J. Geophys. Res., 101,* 12709–12745.

Bezard, B., and C. de Bergh (2007), Composition of the atmosphere of Venus below the clouds, *J. Geophys. Res., in press.*

Bezard, B., C. de Bergh, B. Fegley, J-P Maillard, D. Crisp, T. Owen, J. B. Pollack, and D. Grinspoon (1993), The abundance of sulfur-dioxide below the clouds of Venus, *Geophys. Res. Lett., 20,* 1587–1590.

Bjoraker, G. L., H. P. Larson, M. J. Mumma, R. Timmerman, and J.L. Montani (1992), Airborne observations of the gas composition of Venus above the cloud tops: Measurements of H_2O, HDO, HF, and the D/H and $^{18}O/^{16}O$ isotopic ratios, *Bull. Am. Astron. Soc., 24*, 995.

Braak, C. J., J. F. de Haan, J. W. Hovenier, and L. D. Travis (2002), Spatial and temporal variations of Venus haze properties obtained from Pioneer Venus Orbiter polarimetry, *J. Geophys. Res., 107* (E5), 5029.

Butler, B. J., P. G. Steffes, S. H. Suleiman, and J. M. Jenkins (2001), Accurate and consistent microwave observations of Venus and their implications, *Icarus, 154*, 226–238.

Carlson, R. (2007), Clouds and hazes of Venus: Past observations and open questions, presentation at *Venus Express* science workshop, La Thuile, Italy, 23 March.

Cheng, B. M., E. P. Chew, C. P. Liu, M. Bahou, Y. P. Lee, Y. L. Yung and M. F. Gerstell (1999), Photo-induced fractionation of water isotopomers in the Martian atmosphere, *Geophys. Res. Lett., 26*, 3657–3660.

Cimino, J. (1982), The composition and vertical structure of the lower cloud deck on Venus, *Icarus, 51*, 334–357.

Clancy, R. T., and Muhleman, D. O. (1991), Long-term (1979–1990) changes in the thermal dynamical, and compositional structure of the Venus mesosphere as inferred from microwave spectral observations of ^{12}CO, ^{13}CO and $C^{18}O$, *Icarus, 89*, 129–146.

Connes, P., Connes, J., Kaplan, L., and Benedict, W. (1967), Traces of HCl and HF in the atmosphere of Venus, *Ap. J., 147*, 1230–1237.

Connes, P., Noxon, J. F., Traub, W. A., and Carleton, N. P. (1979), $O_2(^1\Delta)$ emission in the day and night airglow of Venus, *Astrophys. J. Lett., 233*, 29–32.

Conway, R. R., McCoy, R. P., Barth, C. A., and Lane, A. L. (1979), IUE detection of sulfur dioxide in the atmosphere of Venus, *Geophys. Res. Lett., 6*, 629–631.

Crisp, D., and D. Titov. (1997), The thermal balance of the Venus atmosphere, in *Venus II* edited by S. W. Bougher, D. M. Hunten, and R. J. Phillips, pp. 353–384, Univ. of Arizona Press, Tucson.

Crisp, D., Meadows, V.S., Bezard, B., de Bergh, C., Maillard, J-P., and Mills, F.P. (1996), Ground-based Near-Infrared Observations of the Venus Night Side: Near-Infrared O Airglow from the Upper Atmosphere, *J. Geophys. Res., 101*, 4577–4593.

Dalton, J. B., J. B. Pollack, D. H. Grinspoon, B. Bezard, and C. de Bergh (2000), Distribution of chlorine in the lower atmosphere of Venus, *Bull. Am. Astron. Soc., 32*, 1120. Presented at the 32nd Annual Meeting of the Division of Planetary Sciences of the American Astronomical Society, Pasadena, CA, October 27, 2000.

de Bergh, C., B. Bezard, T. Owen, D. Crisp, J-P Maillard, and B. L. Lutz (1991), Deuterium on Venus—Observations from Earth, *Science, 251*, 547–549.

de Bergh, et al. (2006) , The composition of the atmosphere of Venus below 100 km altitude: An overview, *Plan. Space Sci., 54*, 1389–1397 (2006).

DeMore, W. B., M-T. Leu, R. H. Smith, and Y. L. Yung (1985), Laboratory studies on the reactions between chlorine, sulfur dioxide, and oxygen: Implications for the Venus stratosphere, *Icarus, 63*, 347–353.

Donahue, T. M. (1999), New analysis of hydrogen and deuterium escape from Venus, *Icarus, 141*, 226–235.

Donahue, T. M., *et al.* (1982), Venus was wet—a measurement of the ratio of deuterium to hydrogen, *Science, 216*, 630–633.

Eibling, R. E., and M. Kaufman (1983), Kinetics studies relevant to possible coupling between the stratospheric chlorine and sulfur cycles, *Atmos. Environ., 17*, 429–431.

Ekonomov, A. P., et al. (1983), UV photometry at the VENERA 13 and 14 landing probes, *Cosmic Res., 21*, 194–206.

Ekonomov, A. P., V. I. Moroz, B. E. Moshkin, V. I. Gnedykh, Yu. M. Golovin, and A. V. Crigoryev (1984), Scattered UV solar radiation within the clouds of Venus, *Nature,, 307*, 345–347.

Esposito, L. W. (1980), Ultraviolet contrasts and the absorbers near the Venus cloud tops. *J. Geophys. Res., 85*, 8151–8157.

Esposito, L. W. (1984), Sulfur dioxide: Episodic injection shows evidence for active Venus volcanism. *Science, 223*, 1072–1074.

Esposito, L. W., and L. D. Travis (1982), Polarization studies of the Venus UV contrasts: Cloud height and haze variability. *Icarus, 51*, 374–390.

Esposito, L. W., R. G. Knollenberg, M. Y. Marov, O. B. Toon, and R. P. Turco (1983), The clouds and hazes of Venus. in *Venus*, edited by D. M. Hunten, L. Colin, T. M. Donahue, and V. I. Moroz, pp. 484–564, Univ. of Arizona Press, Tucson.

Esposito, L. W., M. Copley, R. Eckert, L. Gates, A. I. F. Stewart, and H. Worden (1988), Sulfur dioxide at the Venus cloud tops, 1978–1986. *J. Geophys. Res., 93*, 5267–5276.

Esposito, L. W., J. L. Bertaux, V. Krasnopolsky, V.I. Moroz, and L. V. Zasova (1997), Chemistry of lower atmosphere and clouds, in *Venus II*, edited by S. W. Bougher, D. M. Hunten, and R. J. Phillips, pp. 415–458, Univ. of Arizona Press, Tucson.

Fegley, B., Jr., and A. H. Treiman (1992), Chemistry of atmosphere–surface interactions on Venus and Mars, in *Venus and Mars: Atmospheres, Ionospheres, and Solar Wind Interaction* (J. Luhmann and R.O. Pepin, Eds.), AGU Chapman Conference volume, 7–71.

Fegley, B., Jr., G. Klingelhofer, K. Lodders, and T. Widemann (1997a), Geochemistry of surface-atmosphere interactions on Venus. in *Venus II*, edited by S. W. Bougher, D. M. Hunten, and R.J. Phillips, pp. 591–636, Univ. of Arizona Press, Tucson.

Fegley, B., Jr., M. Y. Zolotov, and K. Lodders (1997b), The oxidation state of the lower atmosphere and surface of Venus, *Icarus, 125*, 416–439.

Fouchet, T,, and E. Lellouch (2000), Vapor pressure isotope fractionation effects in planetary atmospheres: Application to deuterium, *Icarus, 144,* 114–123.

Gel'man, B. G., V. G. Zolotukhin, N. I. Lamonov, B. V. Levchuk, A. N. Lipatov, L. M. Mukhin, D. F. Nenarokov, V. A. Rotin, and B. P. Okhotnikov (1979), An analysis of the chemical composition of the atmosphere of Venus on an AMS of the Venera-12 using a gas chromatograph, *Cosmic Res., 17*, 585–589.

Grinspoon, D. H. (1993), Implications of the high D/H ratio for the sources of water in Venus atmosphere, *Nature, 363*, 428–431.

Grinspoon, D. H., and J. S. Lewis (1988), Cometary water on Venus—implications of stochastic impacts, *Icarus, 74*, 21–35.

Groves, C., and E. Lewars (2000), Dimers, trimers, and oligomers of sulfur oxides: An ab initio and density functional study, *J. Molec. Structure, 530*, 265–279.

Gubenko, V. N., O. I. Yakovlev, and S. S. Matyugov (2001) Radio occultation measurements of the radio wave absorption and the sulfuric acid content in the atmosphere of Venus, *Cosmic Res., 39*, 439–445.

Gurwell, M. A., and Y. L. Yung (1993), Fractionation of hydrogen and deuterium on venus due to collisional ejection, *Planet. Space Sci., 41*, 91–104.

Hansen, J. E., and J. W. Hovenier (1974), Interpretation of the polarization of Venus, *J. Atm. Sci., 31*, 1137–1160.

Hapke, B., and F. Graham (1985), Disulfur monoxide and the spectra of Io and Venus. *Lunar Planet. Sci., XV*, 316–317 (abstract).

Hapke, B. and F. Graham (1989), Spectral Properties of Condensed Phases of Disulfur Monoxide, Polysulfur Oxide, and Irradiated Sulfur. *Icarus, 79*, 47–55.

Hapke, B., and R. Nelson (1975), Evidence for an Elemental Sulfur Component of Clouds from Venus Spectrophotometry, *J. Atmos. Sci., 32*, 1212–1218.

Hartley, K. K., A. R. Wolf, and L. D. Travis (1989), Croconic acid: An absorber in the Venus clouds? *Icarus, 77*, 382–390.

Hoffman, J. H., R. R. Hodges, Jr., T. M. Donahue, and M. B. McElroy (1980), Composition of the Venus lower atmosphere from the Pioneer Venus mass spectrometer. *J. Geophys. Res., 85*, 7882–7890.

Hong, Y., and B. Fegley, Jr. (1997), Formation of carbonyl sulfide (OCS) from carbon monoxide and sulfur vapor and applications to Venus, *Icarus, 130*, 495–504.

Huggins, W., and W. A. Miller (1864), On the spectra of some of the fixed stars. *Phil. Trans. Royal Soc. London, 154*, 413–435.

Ignatiev, N. I., et al. (1997), Water vapour in the lower atmosphere of Venus: A new analysis of optical spectra measured by entry probes, *Adv. Space Res., 19(8)*, 1159–1168.

Ignatiev, N. I., V. I. Moroz, L.V. Zasova, and I.V. Khatuntsev (1999), Water vapour in the middle atmosphere of Venus: An improved treatment of the Venera 15 IR spectra, *Plan. Space Sci., 47*, 1061–1075.

Imamura, T., and G. L. Hashimoto (1998), Venus cloud formation in the meridional circulation, *J. Geophys. Res., 103*, 31349–31366.

Ingersoll, A. P. (1969), Runaway greenhouse: A history of water on Venus, *J. Atmos. Sci., 26(6)*, 1191–1198.

Jenkins, J. M., M. A. Kolodner, B. J. Butler, S. H. Suleiman, and P. G. Steffes (2002), Microwave remote sensing of the temperature and distribution of sulfur compounds in the lower atmosphere of Venus, *Icarus, 158*, 312–328.

Johnson, N. M., and B. Fegley, Jr. (2002), Experimental studies of atmosphere-surface interactions on Venus. *Adv. Space Res., 29(2)*, 233–241.

Kass, D. M., and Y. L. Yung (1995), Loss of Atmosphere from Mars Due to Solar Wind-Induced Sputtering, *Science, 268* (5211), 697–699.

Kass, D. M., and Y. L. Yung (1999), Water on Mars: Isotopic constraints on exchange between the atmosphere and surface, *Geophys. Res. Lett., 26*, 3653–3656.

Kasting, J. F., and J. B. Pollack (1983), Loss of water from venus .1. hydrodynamic escape of hydrogen, *Icarus, 53*, 479–508.

Knollenberg, R. G., and D. M. Hunten (1980), Results of the Pioneer Venus particles and size spectrometer experiment. *J. Geophys. Res., 85*, 8039–8058.

Knollenberg, R., L. Travis, M. Tomasko, P. Smith, B. Ragent, L. Esposito, D. McCleese, J. Martonchik, and R. Beer. (1980), The clouds of Venus: A synthesis report, *J. Geophys. Res., 85*, 8059–8081.

Kondratiev, V. N. (1971), *Rate Coefficients of Gas Phase Reactions,* Nauka Press, Moscow.

Koukouli, M. E., P. G. J. Irwin, and F. W. Taylor (2005), Water vapor abundance in Venus' middle atmosphere from Pioneer Venus OIR and Venera 15 FTS measurements, *Icarus, 173*, 84–99.

Krasnopolsky, V. A. (1985), Chemical composition of Venus' clouds. *Planet. Space Sci., 33*, 109–117.

Krasnopolsky, V. A. (1986), *Photochemistry of the atmosphere of Mars and Venus,* Springer-Verlag, Berlin.

Krasnopolsky, V. A. (1987), S_3 and S_4 absorption cross sections in the range of 340 to 600 nm and evaluation of the S_3 abundance in the lower atmosphere of Venus, *Adv. Space Res., 7(12)*, 25–27.

Krasnopolsky, V. A. (1989), Vega mission results and chemical composition of Venusian clouds. *Icarus, 80*, 202–210.

Krasnopolsky, V. A. (1999), Hydrodynamic flow of N_2 from Pluto, *J. Geophys. Res., 104*, 5955–5962.

Krasnopolsky, V. (2000), On the deuterium abundance on Mars and some related problems, *Icarus, 148*, 597–602.

Krasnopolsky, V. A. (2006a), A sensitive search for nitric oxide in the lower atmospheres of Venus and Mars: Detection on Venus and upper limit for Mars, *Icarus 182*, 80–91.

Krasnopolsky, V. A. (2006b), Chemical composition of Venus atmosphere and clouds: Some unsolved problems, *Planet. Space Sci. 54*, 1352–1359.

Krasnopolsky, V. A. (2007), Chemical kinetic model for the lower atmosphere of Venus, *Icarus,* doi: 10.1016/j.icarus.2007.04.028, in press.

Krasnopolsky V. A., and V. A. Parshev (1979), Chemical composition of Venus' troposphere and cloud layer based on Venera 11, Venera 12, and Pioneer-Venus measurements. *Cosmic Res. 17*, 630–637 (translated from *Kosmich. Issled.17*, 763–771, 1979).

Krasnopolsky, V. A., and V. A. Parshev (1981a), Chemical-Composition of the Atmosphere of Venus, *Nature, 292*, 610–613.

Krasnopolsky, V. A., and V. A. Parshev (1981b), Photochemistry of the atmosphere of Venus at altitudes greater than 50 km II. Calculations. *Cosmic Res. 19*, 176–189 (translated from *Kosmich. Issled. 19*, 261–278, 1981).

Krasnopolsky, V. A., and V. A. Parshev (1981c), Photochemistry of Venus' atmosphere at altitudes over 50 km I. Initial calculations data. *Cosmic Res. 19*, 61–74 (translated from *Kosmich. Issled. 19*, 87–104, 1981).

Krasnopolsky, V. A., and V. A. Parshev (1983), Photochemistry of the Venus atmosphere. in *Venus,* edited by D. M. Hunten, L. Colin, T. M. Donahue, and V. I. Moroz, pp.431–458, Univ. Arizona Press, Tucson.

Krasnopolsky, V. A., and J. B. Pollack (1994), H_2O-H_2SO_4 systen in Venus' clouds and OCS, CO, and H_2SO_4 profiles in Venus' troposphere. *Icarus, 109,* 58–78.

Krasnopolsky, V. A., G. L. Bjoraker, M. J. Mumma, et al. (1997), High-resolution spectroscopy of Mars at 3.7 and 8 µm: A sensitive search for H_2O_2, H_2CO, HCl, and CH_4, and detection of HDO, J. Geophys. Res., 102, 6525–6534.

Kuang, Z. M., G. C. Toon, P. O. Wennberg, and Y. L. Yung (2003), Measured HDO/H_2O ratios across the tropical tropopause, *Geophys. Res. Lett., 30*(7), 1372, doi:10.1029/2003GL017023.

Lammer, H, H. I. M. Lichtenegger, H. K. Biernat, et al. (2006), Loss of hydrogen and oxygen from the upper atmosphere of Venus, *Planet. Sp. Sci.,* 54, 1445–1456.

Lewis, J. S. (1970), Venus: atmospheric and lithospheric composition. *Earth Planet. Sci. Lett., 10,* 73–80.

Liang, M. C., *et al.* (2003), Source of atomic hydrogen in the atmosphere of HD209458b, *Astrophys. J. Lett., 596,* 247–250.

Linkin, V. M., et al. (1985), VENERA 15 and VENERA 16 infrared experiment. 5. Preliminary results of analysis of brightness temperature and thermal flux fields, *Cosmic Res., 23,* 212–221.

Lovelock, J. E., and M. Whitfield (1982), Life-span of the biosphere, *Nature, 296,* 561–563.

Maiorov, B. S., et al. (2005), A new analysis of the spectra obtained by the VENERA missions in the Venusian atmosphere. I. The analysis of the data received from the VENERA-11 Probe at altitudes below 37 km in the 0.44–0.66 µm wavelength range, *Solar System Research, 39,* 267–282.

Mallama, A., D. Wang, and R. A. Howard (2006), Venus phase function and forward scattering from H_2SO_4, *Icarus,* in press.

Marcq, E., Bruno, B., Encrenaz, Th., and Birlan, M. (2005), Latitudinal variations of CO and OCS in the lower atmosphere of Venus from near-infrared nightside spectro-imaging. *Icarus, 179,* 375–386.

Marcq, E., T. Encrenaz, B. Bezard, M. Birlan. (2006), Remote sensing of Venus' lower atmosphere from ground-based IR spectroscopy: Latitudinal and vertical distribution of minor species, *Plan. Space Sci., 54,* 1360–1370.

McElroy, M. B., and T. M. Donahue (1972), Stabiity of the Martian atmosphere, *Science, 177,* 986–988.

McElroy, M. B., *et al.* (1982), Escape of Hydrogen from Venus, *Science, 215,* 1614–1615.

Meadows, V. S., and D. Crisp (1996), Ground-based near-infrared observations of the Venus nightside: The thermal structure and water abundance near the surface, *J. Geophys. Res., 101,* 4595–4622.

Miller, C. E., and Y. L. Yung (2000), Photo-induced Isotopic Fractionation, *J. Geophys. Res. 105,* 29039–29051.

Mills, F. P. (1998), I. Observations and Photochemical Modeling of the Venus Middle Atmosphere. II. Thermal Infrared Spectroscopy of Europa and Callisto, Ph.D. dissertation, 366 pp., California Institute of Technology, Pasadena, Calif.

Mills, F. P. (1999a), Water vapor in the Venus middle atmosphere, *Adv. Space Res., 23*(9), 1573–1576.

Mills, F. P. (1999b), A spectroscopic search for molecular oxygen in the Venus middle atmosphere, *J. Geophys. Res., 104,* 30757–30763.

Mills, F. P., and M. Allen (2007), A review of selected issues concerning the chemistry in Venus' middle atmosphere, *Plan. Space Sci.,* doi: 10.1016/j.pss.2007.01.012, *in press*

Mills, F. P., M. Sundaram, T. G. Slanger, M. Allen, and Y. L. Yung (2006), Oxygen chemistry in the Venus middle atmosphere, in *Advances in Geoscience Volume 3: Planetary Science (PS),* edited by W-H Ip and A. Bhardwaj, pp. 109–117, World Scientific Publishing, Singapore.

Moroz, V. I., N. A. Parfent'ev, and N. F. Sanko (1979), Spectrophotometric experiment on the Venera 11 and Venera 12 descent modules. 2. Analysis of Venera 11 spectra data by the layer-addition method, *Cosmic Res.,* 17, 601–614 (translated from *Kosmich. Issled.,* 17, 727–742).

Moroz, V. I., et al. (1980), Spectrum of the Venus day sky, *Nature, 284,* 243–4.

Moroz, V. I., et al. (1983), The VENERA 13 and 14 spectrophotometric experiment. II. Preliminary analysis of H_2O absorption bands in spectra, *Cosmic Res., 21,* 187–194.

Moroz, V. I., W. Dohler, E. A. Ustinov, K. Schafer, L. V. Zasova, D. Spankuch, A. A. Dyachkov, R. Dubois, V. M. Linkin, D. Oertel, V. V. Kerzhanovich, I. Nopirakovskii, I. A. Matsygorin, H. Becker-Ross, A. A. Shurupov, W. Stadthaus, and A. N. Lipatov (1985), VENERA 15 and VENERA 16 infrared experiment. 4. Preliminary results of spectral analyses in the region of H_2O and SO_2 absorption bands, *Cosmic Res., 23,* N2, 202–211.

Moroz, V. I., D. Spankuch, V. M. Linkin, W. Dohler, I. A. Matsygorin, K. Schafer, L. V. Zasova, D. Oertel, A. V. Dyachkov, R. Schuster, V. V. Kerzhanovich, H. Becker-Ross, E. A. Ustinov, and W. Stadthaus (1986), Venus spacecraft infrared radiance spectra. Some aspects of their interpretation, *Applied Optics, 25,* N10.

Moroz, V. I., D. Spankuch, D. V. Titov, K. Schafer, A. V. Dyachkov, W. Dohler, L. V. Zasova, D. Oertel, V. M. Linkin, and J. Nopirakowski (1990), Water vapor and sulfur dioxide abundances at the Venus cloud tops from the VENERA 15 infrared spectrometry data, *Adv. Space Res., 10,* N5, 77.

Moshkin, B. E., et al. (1983), Spectrophotometric experiment aboard the Venera 13 and Venera 14 probes. 1. Technique, results, and preliminary analysis of the measurements, *Kosm. Issled., 21,* 246.

Moyer, E. J., F. W. Irion, Y. L. Yung, and M. R. Gunson. (1996). ATMOS Stratospheric Deuterated Water and Implications for Troposphere-Stratosphere Transport, *Geophys. Res. Lett., 23,* 2385–2388.

Na, C. Y., and L. W. Esposito, L. W (1995), UV observations of Venus with HST, *Bull. Amer. Astron. Soc., 27,* 1071 (abstract).

Na, C. Y., and L. W. Esposito (1997), Is disulfur monoxide the second absorber on Venus? *Icarus, 125,* 361–368.

Na, C. Y., L. W. Esposito, and T. E. Skinner (1990), International Ultraviolet Explorer observation of Venus SO_2 and SO, *J. Geophys. Res., 95,* 7485–7491.

Na, C. Y., L. W. Esposito, W. E. McClintock, and C. A. Barth (1994), Sulfur dioxide in the atmosphere of Venus, II. Modeling results. *Icarus, 112,* 389–395.

Nicovich, J., K. Kreutter and P. Wine (1990), Kinetics and thermochemistry of ClCO formation from the Cl+CO association reaction, *J. Chem. Phys., 92,* 3539–3544.

Owen, T., and C. Sagan (1972), Minor constituents in planetary atmospheres: Ultraviolet spectroscopy from the Orbiting Astronomical Observatory, *Icarus, 16,* 557–568.

Owen, T., J-P. Maillard, C. de Bergh, and B. L. Lutz (1988), Deuterium on Mars: The Abundance of HDO and the Value of D/H, *Science, 240* (4860), 1767–1770.

Oyama, V. I., G. C. Carle, F. Woeller (1980), Corrections in the Pioneer Venus Sounder probe gas-chromatographic analysis of the lower Venus atmosphere, *Science, 208,* 399–401.

Parkinson, T. D., and D. M. Hunten (1972), Spectroscopy and aeronomy of O_2 on Mars, *J. Atmos. Sci., 29*(7), 1380–1390.

Pernice, H., *et al.* (2004), Laboratory evidence for a key intermediate in the Venus atmosphere: Peroxychloroformyl radical, *Proc. Natl. Acad. Sci. U. S. A., 101,* 14007–14010.

Petryanov, I. V., et al. (1981), Iron in the Venus clouds. *Dokl. AN SSSR, 260,* 834.

Pollack, J. B., *et al.* (1980), Distribution and source of the UV absorption in Venus atmosphere, *J. Geophys, Res, 85,* 8141–8150.

Pollack, J. B., J. B. Dalton, D. Grinspoon, R. B. Wattson, R. Freedman, D. Crisp, D. A. Allen, B. Bezard, C. de Bergh, L. P. Giver, Q. Ma, and R. Tipping (1993), Near-infrared light from Venus' nightside: A spectroscopic analysis, *Icarus, 103,* 1–42.

Porshnev, N. V., L. M. Mukhin, B. G. Gelman, D. F. Nenarokov, V. A. Rotin, A. V. Dyachkov, and V. B. Bondarev (1987), Gas chromatographic analysis of products of thermal reactions of the cloud aerosol of Venus by the Vega 1 and 2 probes, *Cosmic Res., 25,* 715.

Price C, Penner J, Prather M. (1997), NO_x from lightning. 1. Global distribution based on lightning physics, *J. Geophys. Res., 102,* 5929–5941.

Prinn, R. G. (1971), Photochemistry of HCL and Other Minor Constituents in Atmosphere of Venus, *J. Atmos. Sci., 28* (6), 1058–1068.

Prinn, R. G. (1975), Venus: chemical and dynamical processes in the stratosphere and mesosphere, *J. Atmos. Sci., 32,* 1237–1247.

Prinn, R. G. (1978), Venus: chemistry of the lower atmosphere prior to the Pioneer Venus mission, *Geophys. Res. Lett., 5,* 973–976.

Prinn, R. G. (1979), On the possible roles of gaseous sulfur and sulfanes in the atmosphere of Venus, *Geophys. Res. Lett., 6,* 807–810.

Prinn, R. G. (1985), The photochemistry of the atmosphere of Venus, in *The Photochemistry of Atmospheres*, edited by J. S. Levine, pp. 281–336, Academic Press, Orlando.

Rossow, W. B., et al. (1980), Cloud morphology and motions from Pioneer Venus images, *J. Geophys. Res., 85,* 8107–8128.

Sander, S. P., et al. (2002), Chemical Kinetics and Photochemical Data for Use in Stratospheric Modeling Evaluation Number 14, Jet Propul. Lab. Pub. 02–25, Pasadena, Calif. (http://jpldataeval.jpl.nasa.gov/).

Sandor, B. J., and R. T. Clancy (2005), Water vapor variations in the Venus mesosphere from microwave spectra, *Icarus, 177,* 129–143.

Sanko, N. F. 1980. Gaseous sulfur in the atmosphere of Venus. *Kosmich. Issled., 18,* 600.

Schaefer, K., et al (1987), Structure of the middle atmosphere of Venus from analyses of Fourier-spectrometer measurements aboard VENERA-15, *Adv. Space Res., 7,* 17.

Schaefer, K., et al (1990), Infrared Fourier Spectrometer experiment from VENERA-15, *Adv. Space Res., 10,* 57

Sill, G. T. (1983), The clouds of Venus: sulfuric acid by the lead chamber process. *Icarus, 53,* 10.

Slanger, T. G., P. C. Cosby, D. L. Huestis, and T. A. Bida (2001), Discovery of the atomic oxygen green line in the Venus night airglow, *Science, 291,* 463–465.

Slanger, T. G., D. L. Huestis, P. C. Cosby, N. J. Chanover, and T.A. Bida (2006), The Venus nightglow: Ground-based observations and chemical mechanisms, *Icarus,* in press.

Solomon, S., *et al.* (1997), Heterogeneous chlorine chemistry in the tropopause region, *J. Geophys. Res., 102,* 21411–21429.

Spankuch, D., L. V. Zasova, K. Schafer, E. A. Ustinov, J. Guldner, V. I. Moroz, W. Dohler, V. M. Linkin, R. Dubois, A. A. Dyachkov, H. Becker-Ross, A. A. Lipatov, W. Stadthaus, I. A. Matsygorin, D. Oertel, V. V. Kerzhanovich, I. Nopirokovskii, H. Jahn, G. Fellberg, W. Schuster, and A. A. Shurupov (1985), VENERA-15 and VENERA-16 infrared experiment. 2. Preliminary results of temperature profile retrieval, *Cosmic Res., 23,* N2, 176–188.

Spankuch, D., I. A. Matsygorin, R. Dubois, and L. V. Zasova (1990), Venus middle-atmosphere temperature from VENERA-15, *Adv. Space Res., 10,* N5, 67.

Stewart, A. I., D. E. Anderson, L. W. Esposito, and C. A. Barth (1979), Ultraviolet spectroscopy of Venus: Initial results from the Pioneer Venus orbiter. *Science, 203,* 777–779.

Stratton, L. W., R. E. Eibling, and M. Kauffman (1979), Rate constant of the reaction between chlorine atoms and sulfur dioxide and its significance for stratospheric chlorine chemistry, *Atmos. Environ., 13,* 175–177.

Surkov, Yu. A. et al. (1979), Chemical composition of the atmosphere of Venus, *Pisma AZh., 5,* 7.

Surkov, Yu. A., et al (1981), A study of the Venus cloud aerosol by Venera 12 (preliminary data), *Cosmic Res., 20,* 435.

Surkov, Yu. A., V. L. Barsukov, L. P. Moskalyova, V.P. Kharyukova, and A. L. Kemurdzhian, A. L. (1984), New data on the composition, structure and properties of Venus rock obtained by VENERA 13 and VENERA 14, *J. Geophys. Res., 89,* B393–B402.

Surkov, Y. A., L. P. Moskalyova, V.P. Kharyukova, A. D. Dudin, G.G. Smirnov, and S.Ye. Zaitseva (1986), Venus rock composition at the Vega 2 landing site, *J. Geophys. Res. Suppl., 91,* 215–218.

Surkov, Yu. A., V. F. Ivanova, A. N. Pudov, and D. Caramel (1987), Determination of the aerosols chamical composition in the

Venusian clouds by means of the mass-spectrometer MALA-HIT on the VEGA-1 probe, *Kosmich. Issled., 15*, 744–750 (in Russian).

Sze, N. D., and M. B. McElroy (1975), Some problems in Venus' aeronomy, *Planet. Space Sci., 23, 763–786.*

Taylor, F. W., D. Crisp, B. Bezard (1997), Near-infrared sounding of the lower atmosphere of Venus, in *Venus II*, edited by S. W. Bougher, D. M. Hunten, and R. J. Phillips, pp. 325–351, Univ. of Arizona Press, Tucson.

Tian, F., and O. B. Toon (2005), Hydrodynamic escape of nitrogen from Pluto, *Geophys. Res. Lett., 32,* L18201.

Tomasko, M. G., Doose, L. R., Smith, P. H., and Odell, A. P. (1980), Measurements of the flux of sunlight in the atmosphere of Venus, *J. Geophys. Res., 85,* 8167–8186.

Toon, O. B., R. P. Turco, J. B. Pollack (1982), The ultraviolet absorber on Venus: amorphous sulfur, *Icarus, 51,* 358.

Trauger, J. T., and J. I. Lunine (1983), Spectroscopy of Molecular-Oxygen in the Atmospheres of Venus and Mars, *Icarus, 55,* 272–281.

Titov, D. V. (1983), On the possibility of aerosol formation by the reaction between SO_2 and NH_3 in Venus' atmosphere, *Cosmic Res., 21,* 401.

Vidal-Madjar A., *et al.* (2003), An extended upper atmosphere around the extrasolar planet HD209458b, *Nature, 422,*143–146.

von Zahn, U., S. Kumar, H. Niemann, and R. G. Prinn (1983), Composition of the Venus atmosphere, in *Venus*, edited by D. M. Hunten, L. Colin, T. M. Donahue and V. I. Moroz, pp. 299–430, Univ. of Arizona Press, Tucson.

Watson, A. J., et al. (1979), Oxides of nitrogen and the clouds of Venus, *Geophys. Res. Lett., 6,* 743–746.

Widemann, T., J-L. Bertaux, V. I. Moroz, and A.P. Ekonomov (1993), VEGA-1 and VEGA-2 descent modules: In-situ measurements of ultraviolet absorption and relationship with present active volcanism on Venus, *Bull. Amer. Astron. Soc., 25,* 1094 (abstract).

Winick, J. R., and A.I. Stewart (1980), Photochemistry of SO_2 in Venus' upper cloud layers, *J. Geophys. Res., 85,* 7849–7860.

Young, A. T. (1977), An improved Venus cloud model, *Icarus, 32,* 1–26.

Young, A. T. (1983), Venus cloud microphysics. *Icarus 56,* 568.

Young, L. D. G. (1972), High resolution spectrum of Venus—A review, *Icarus, 17,* 632–658.

Young, L. D. G., A. T. Young, and L. V. Zasova (1984), A new interpretation of the Venera 11 spectra of Venus, *Icarus, 60,* 138.

Yung, Y. L., and W. B. Demore (1981), Photochemical models for the stratosphere of Venus, in *An International Conference on the Venus Environment*, p. 26, NASA Ames Research Center.

Yung, Y. L., and W. B. Demore (1982), Photochemistry of the Stratosphere of Venus—Implications for Atmospheric Evolution, *Icarus, 51* (2), 199–247.

Yung, Y. L., and W. B. DeMore (1999), *Photochemistry of Planetary Atmospheres*, Oxford University Press, New York.

Yung, Y. L. and M. B. McElroy (1979) Fixation of Nitrogen in the Prebiotic Atmosphere. *Science, 203*(4384), 1002–1004.

Yung, Y. L., J. S. Wen, J. P. Pinto, M. Allen, K. K. Pierce, and S. Paulson (1988), HDO in the Martian Atmosphere—Implications for the Abundance of Crustal Water, *Icarus, 76* (1), 146–159.

Zasova, L. V., V. A. Krasnopolsky and V. I. Moroz (1981), Vertical distribution of SO_2 in the upper cloud layer of Venus and origin of UV absorption, *Adv. Space Res., 1*(9), 13–16.

Zasova, L. V. et al (1985), Infrared experiment on VENERA 15 and VENERA 16 spacecraft, 3: Some conclusions on the structure of clouds based on the spectra analysis, *Kosmich. Issled., 23,* 221–235.

Zasova, L. V., et al. (1989), Venusian clouds from VENERA-15 data, *Veroffenlichungen Forschungsbereichs Geo-Kosmoswissenschaften, 18,* Academie-Verlag, Berlin.

Zasova, L. V., V. I. Moroz, L. W. Esposito, and C. Y. Na (1993), SO_2 in the middle atmosphere of Venus: IR measurements from Venera-15 and comparison to UV data, *Icarus, 105,* 92–109.

Venus Atmosphere Dynamics: A Continuing Enigma

G. Schubert[1], S. W. Bougher[2] , C. C. Covey[3] , A. D. Del Genio[4] ,
A. S. Grossman[3], J. L. Hollingsworth[5] , S. S. Limaye[6], and R. E. Young[7]

1. INTRODUCTION AND FOCUSED LITERATURE REVIEW

The dynamics of the Venus atmosphere presents a major unsolved problem in planetary science: the so-called super-rotation of the lower atmosphere and its transition to a sub-solar-to-antisolar circulation in the upper atmosphere. (In this paper we place the dividing line between the lower and upper atmosphere at 90–100 km altitude (pressure 0.39 to 0.028 mbar), the base of the day-side thermosphere.) Superrotation has also been observed in the atmosphere of Titan, the only other slowly rotating world with a substantial atmosphere known at present; in this case also the transition to a different circulation in the upper atmosphere is also apparent but not well understood. Thus, the issues discussed below may be generic to any slowly rotating terrestrial planet's atmosphere.

[1] Department of Earth and Space Sciences and Institute of Geophysics and Planetary Physics, University of California, Los Angeles, California, USA

[2] Space Physics Research Laboratory, AOSS Department, University of Michigan, Ann Arbor, Michigan, USA

[3] Lawrence Livermore National Laboratory, Livermore, California, USA

[4] NASA Goddard Institute for Space Studies, New York, New York, USA

[5] NASA Ames Research Center/SJSUF, Moffett Field, California 94035, USA

[6] Space Science and Engineering Center, University of Wisconsin-Madison, Madison, Wisconsin, USA

[7] NASA Ames Research Center, Moffett Field, California, USA

Exploring Venus as a Terrestrial Planet
Geophysical Monograph Series 176
Copyright 2007 by the American Geophysical Union.
10.1029/176GM07

1.1 Lower Atmosphere

The lower atmosphere of Venus is observed to rotate in the same retrograde (westward) direction as the solid planet, to first approximation at constant angular velocity at any given altitude. The maximum superrotation rate, at cloud-top levels, is about once every 4 Earth-days, 60 times faster than that of Venus itself. This retrograde superrotating zonal (RSZ) flow has been known for over three decades, but at present there is still no adequate explanation. Not a single theoretical model using realistic formulations of the Venus atmosphere has been able to quantitatively reproduce this fundamental characteristic of the atmosphere of the nearest planet to Earth.

Whereas the RSZ circulation is not understood theoretically, the meridional (north-south) circulation of the lower atmosphere is even less well-known. A thermally direct Hadley cell (upwelling at the equator, poleward flow aloft, descending flow at high latitude, and equatorward flow at lower altitude) is expected at cloud levels where most of the solar energy absorption takes place in the Venus atmosphere. However, it has not been observed except for an indication of poleward motion in low-latitude cloud-tracked winds (*Schubert*, 1983). The latitudinal and vertical extent of such a circulation is unknown as well as whether two or more Hadley cells lie "stacked" above and below each other. Intriguing features of the polar circulation and thermal structure such as the polar vortex, polar dipole, cold collar, and reversed equator-pole temperature gradient (*Taylor et al.*, 1983) above the cloud tops are probably related to the Hadley cells.

The most recent comprehensive reviews of the thermal structure and dynamics of the Venus lower atmosphere have been presented by *Crisp and Titov* (1997) and *Gierasch et al.* (1997), respectively. A number of studies (*Fels and Lindzen*, 1974; *Pechman and Ingersoll* , 1984; *Fels* , 1986; *Leovy and Baker*, 1987; *Hou et al.*, 1990; *Newman and*

Leovy, 1992) indicate that momentum fluxes from thermal tides maintain superrotation in the altitude range of 60–90 km (pressure ~ 0.21 bar to 0.39 mbar), at and above the level of Venus' planetwide cloud cover. The tides might act in combination with a direct equator-to-pole Hadley cell, which would transport retrograde momentum upward from the solid planet (*Gierasch*, 1975). However, if thermal tide amplitudes are small below the clouds (*Gierasch et al.*, 1997), then middle atmosphere dynamics could be decoupled from lower atmosphere dynamics. In that case, the key to unlocking the mechanisms generating superrotation would be contained in the deep atmosphere, especially in the lowest atmospheric scale height.

Two observed features of the superrotation illustrate why this should be the case. First, the RSZ winds are generally in the same direction all the way from the surface to well above the clouds (*Counselmann et al.*, 1980; *Gierasch et al.*, 1997). Second, as *Schubert* (1983) first pointed out, the product of atmospheric density times the mean zonal wind, i.e., the momentum per unit volume of the RSZ flow, is a maximum near 20 km altitude, well below cloud levels, but just above the lowest atmospheric scale height. The only plausible momentum source for producing a net zonal momentum in the atmosphere is the solid planet itself. Therefore, in attempting to understand superrotation, emphasis must be placed on the deep atmosphere, in particular the lowest scale height.

In this section we concentrate on developments in understanding the dynamics of the Venus lower atmosphere that have occurred since the publication of *Gierasch et al.* (1997). The reader is referred to that paper and references therein for extensive discussion of many of the fundamental aspects of Venus atmospheric dynamics. There have been few recent observational studies of significance for modeling the circulation of the atmosphere. This will change as two missions, Venus Express and Venus Climate Orbiter, both designed to make extensive remote sensing observations of the atmosphere of Venus, return data over the next several years.

Probably the most significant new developments regarding Venus lower atmosphere dynamics have occurred in theoretical models. *Young and Pollack* (1977) reported the first—and for many years the only—successful attempt to simulate RSZ with a three-dimensional general circulation model (GCM). This model included the atmosphere from the surface to the cloud tops, albeit at low resolution by today's standards (16 levels in the vertical dimension and wavenumber 4 in the horizontal). The model evolved cloud-top superrotation at all latitudes with maximum wind velocities of 30–85 m s^{-1} depending on initial conditions. The way the model did this seemed to involve a weak *Gierasch* (1975) mechanism (transport of retrograde zonal momentum upward at the

equator and then poleward by a Hadley cell) amplified by *Thompson*'s (1970) suggested instability process. However, controversy arose about the model's vertical diffusion formulation, which may have exaggerated the RSZ flow (*Rossow et al.*, 1980; *Young and Pollack*, 1980), and in any case the model failed to produce any superrotation below 30 km altitude (~9.6 bar), where observations show that angular momentum per unit volume actually reaches its greatest value (*Schubert*, 1983).

Two decades later, *Del Genio et al.* (1993) and *Del Genio and Zhou* (1996) reported a series of simulations with Earth-like GCMs in which the assumed planetary rotation period was increased to that of either Titan (16 Earth days) or Venus (243 Earth days). These simulations retained the mass of Earth's atmosphere—about the same as Titan's but nearly two orders of magnitude less than Venus' atmospheric mass. Nevertheless, the results suggested ways the real Venus atmosphere could maintain the observed dynamical state. For Venus' planetary rotation rate, *Del Genio et al.* (1993) and *Del Genio and Zhou* (1996) obtained superrotation at virtually all latitudes and altitudes, provided that (1) higher levels of the atmosphere were decoupled from lower levels by a layer of high static stability and (2) care was taken to have the model conserve momentum to high accuracy. The first requirement seems to be satisfied by the actual Venus atmosphere (*Schubert*, 1983). The second requirement is understandable since, dynamically speaking, Venus' atmosphere is a weakly forced/weakly dissipative system, so a small numerical inaccuracy can ruin its simulation.

The simulations of *Del Genio et al.* (1993) and *Del Genio and Zhou* (1996) achieved superrotation by a strong *Gierasch* (1975) mechanism in which the requisite equatorward eddy momentum transport was accomplished by barotropic instability, as suggested originally by *Rossow and Williams* (1979). Similar results were obtained by *Hourdin et al.* (1992, 1995, 1996) and *Rannou et al.* (2006) in their Titan GCM simulations. The Gierasch mechanism emerged thereby as the favored explanation for Venus' superrotation. None of these simulations, however, was able to completely replicate the observed data. *Del Genio and Zhou*'s (1996) results, for example, fell an order of magnitude short of the real Venus in terms of cloud-top equatorial wind speed. These model simulations also assumed axisymmetric heating, thus excluding the day-night cycle and phenomena such as atmospheric tides and the "moving flame" (*Schubert and Whitehead*, 1969).

Yamamoto and Takahashi (2003a,b, 2004, 2006a) have taken the next logical steps of including the full mass of Venus' atmosphere and the day-night cycle in their GCM. They found that RSZ flow evolved by a strong Gierasch mechanism involving a single surface-to-cloud-top and equator-to-pole Hadley circulation, with a variety of waves transporting angular momentum equatorward. Unfortunately, it seemed necessary to alter the assumed thermal forcing of the atmosphere beyond the limits allowed by observation in order to establish a strong enough Hadley cell to make superrotation occur at the observed magnitude and extent—in short, to obtain the right answer for at least partly the wrong reason. Either a lower atmospheric heating rate orders of magnitude greater than observed (*Yamamoto and Takahashi*, 2003a,b), or an equator-to-pole surface temperature gradient far greater than expected to exist (*Yamamoto and Takahashi*, 2004, 2006a) appeared necessary for the model to "work."

Independent GCM simulations by ourselves (*Hollingsworth et al.*, 2007), *Lee* (2006), and *Lee et al.* (2006) are consistent with Yamamoto and Takahashi's experience. Our model, a full 3D global circulation model, is adapted from the NASA Goddard Space Flight Center ARIES/GEOS 'dynamical core' (*Suarez and Takacs*, 1995), which is based on the meteorological primitive equations. In contrast to a complex atmospheric general circulation model (AGCM), we utilize simplified forcing terms on the right sides of the governing equations, i.e., the net diabatic heating is expressed analytically, in addition to a Newtonian cooling term. Further, momentum drag is specified in terms of a Rayleigh friction, and a simple boundary layer scheme is imposed. Further details are provided in Hollingsworth et al. (2007). With unrealistically strong heating in the lower atmosphere, we found significant superrotation in our GCM simulations (Plate 1). RSZ flow occurred throughout the atmosphere except for very weak prograde flow at low altitudes and latitudes between $\pm 30°$, and maximum mean zonal westward winds of $O(90 \text{ m s}^{-1})$ occurred at cloud-top levels, as observed. The bulk of the atmosphere below the clouds ($p > 1$ bar; $z/H \lesssim 5$) showed very weak north-south temperature contrasts and wind velocity of order (10 m s^{-1}). The zonal wind and temperature fields were associated with a deep, equatorially-symmetric mean meridional overturning (i.e., Hadley circulation) with rising motion at the equator, poleward motion aloft and sinking motion in high latitudes (see Figure 2 in *Hollingsworth et al.*, 2007). However, when lower atmospheric heating was consistent with Pioneer Venus observations (*Tomasko et al.*, 1980), we found only extremely weak $O(1 \text{ m s}^{-1})$ superrotation in the upper tropical atmosphere; somewhat stronger winds $O(10 \text{ m s}^{-1})$ were found in middle/high latitudes at high altitudes (Plate 2). Pole-to-equator temperature contrast, mean zonal flow, wave activity and associated equatorward

eddy momentum transports—all prominent in the case of unrealistic heating—were dramatically reduced with realistic heating.

In agreement with our results, *Yamamoto and Takahashi* (2006a) state: "Although the superrotation is produced in the simplified AGCMs, a real Venusian superrotation mechanism is still unknown at the present stage. In addition to the further observations, the further improvements of AGCMs are needed in order to elucidate the real superrotation mechanism. Together with the improvement of the radiation code, more realistic surface processes should be incorporated into Venus atmosphere AGCM."

In assessing Venus lower atmosphere modeling efforts, some guidance may be obtained from current assessments of Earth climate models, which include both comparison of different models with different underlying assumptions (*Phillips et al.*, 2006) and a thorough exploration of "adjustable parameter space" (http://www.climateprediction.net). The stunning enhancement of computer capability since Venus GCMs were first developed certainly offers many opportunities along these lines. However, the question remains: What particular ingredient(s) are missing from Venus GCMs that prevent them from fully explaining superrotation?

The most recent version of Yamamoto and Takahashi's model, with day-night variations included, should be capable of simulating most of the superrotation mechanisms suggested since the phenomenon was observed, including the meridional circulation and non-axisymmetric eddies (*Gierasch*, 1975; *Rossow and Williams*, 1979), instability (*Thompson*, 1970) and diurnal solar heating variations such as thermal tides (*Schubert and Whitehead*, 1969; *Newman and Leovy*, 1992). A prominent exception is the class of theories involving surface-atmosphere interaction. Thermally excited gravity waves (*Fels and Lindzen*, 1974) originating near the surface might help to maintain superrotation (*Hou and Farrell*, 1987). Gravity waves, perhaps excited by the Venusian topography, were apparently detected by the Venera balloons (*Young et al.*, 1987). "One usually expects waves generated in the boundary layer to have small horizontal phase velocities and to contribute to drag rather than acceleration" (*Gierasch et al.*, 1997) and indeed the representation of so-called gravity wave drag is an important feature of modern Earth atmosphere GCMs, but these matters are only beginning to be explored in Venus lower atmosphere GCMs (*Herrnstein and Dowling*, 2007). Surface-atmosphere interaction is one of many issues that need to be addressed in order to simulate Venus' atmospheric circulation in a fully satisfactory manner.

---

[1] The tropopause (base of the mesosphere) is conventionally placed at an altitude of 60 ± 5 km, where the near-adiabatic lapse rate in the lowest levels of the atmosphere becomes less steep; the mesopause (base of the thermosphere) is at 90–100 km altitude.

1.2 Upper Atmosphere

The large-scale circulation of the upper atmosphere from ~90 to ~200 km altitude (upper mesosphere and thermosphere[1]) is a combination of two distinct flow patterns: (1) a relatively stable subsolar-to-antisolar (SS-AS) circulation cell driven by solar (EUV- UV) and IR heating, and (2) a highly variable retrograde superrotating zonal (RSZ) flow, in part a continuation of the lower-atmosphere RSZ flow discussed above (see Figure 1). The effects of the superposition of these wind components in the Venus upper atmosphere are: (a) to shift the divergence of the flow from the subsolar point toward the evening terminator (ET), (b) to generate larger evening terminator winds than those along the morning terminator (MT), and (c) to shift the convergence of the flow away from midnight and toward the morning terminator. These components also vary as a function of altitude and

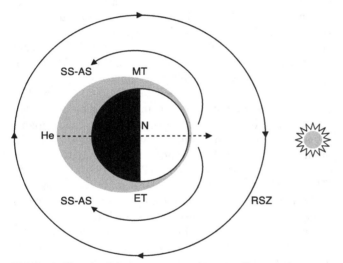

Figure 1. Sketch of the components that contribute to the upper atmospheric circulation on Venus. The sketch looks down on the north pole of the planet. The daylight hemisphere is light and the nightside hemisphere is dark. The overhead motion of the Sun is counterclockwise. SS-AS is the subsolar-antisolar circulation with upward motion and divergence under the Sun and sinking flow and convergence on the night side. RSZ is the retrograde superrotating zonal circulation which is clockwise, opposite to the overhead motion of the Sun. The lightly shaded region represents the variation in the helium content of the upper atmosphere. The SS-AS flow transports helium from the day side to the night side resulting in a helium bulge on the night side. The RSZ and SS-AS winds are in the same direction at the evening terminator ET but in opposite directions at the morning terminator MT resulting in higher zonal wind velocities at the ET compared with the MT. The effect of the superposition of the RSZ and SS-AS winds is to shift the center of dayside divergence toward the ET and shift the center of nightside convergence and the center of the helium bulge toward the MT.

reflect the changing importance of underlying drivers and their day-to-day and solar cycle variations (see Table 1).

This picture of Venus upper atmosphere winds has been gleaned from a number of remote and in-situ datasets collected at the planet. A thorough examination of Pioneer Venus Orbiter neutral density (e.g., CO_2, O, He, and H) and temperature distributions above 130 km (Figure 2) has been used to constrain general circulation model simulations, from which SS-AS and RSZ wind magnitudes can be extracted (see review by *Bougher et al.*, 1997). The spatial distribution of ultraviolet night airglow emissions (e.g., NO^*) and dayglow emissions (e.g., atomic O), and H-Lyman-alpha emissions, have all been used to trace the circulation patterns at thermospheric altitudes above ~115 km (see Table 1). Finally, visible and infrared O_2 nightglow distributions from Veneras 9 and 10, Pioneer Venus, Galileo, and the ground, along with minor species distributions (e.g., CO) have also been used to constrain upper mesospheric wind patterns (~90–110 km altitude). In addition, direct wind velocities were also obtained from CO mm observations (see review by *Lellouch et al.*, 1997). Many of these datasets are discussed in detail in *Bougher et al.* (1997) and *Lellouch et al.* (1997), and are summarized in Table 1.

The processes responsible for maintaining (and driving variations in) the SS-AS and RSZ winds in the Venus upper atmosphere are still not well understood or quantified. For example, it is apparent that some type of deceleration mechanism is necessary to slow the upper atmosphere winds (e.g., *Alexander*, 1992; *Zhang et al.*, 1996; *Bougher et al.*, 1997). This is evident since the day-to-night upper atmosphere circulation results in downwelling and adiabatic heating on the nightside, which must be limited to maintain the observed cold nightside temperatures (Figure 2) and density structure. This deceleration is not symmetric in local time, because as noted above the net zonal winds appear stronger at the dusk versus the dawn terminator. The mechanism responsible for this deceleration and asymmetry is thought to be gravity wave breaking and subsequent momentum and energy deposition in the Venus thermosphere (*Alexander*, 1992; *Zhang et al.*, 1996; *Bougher et al.*, 1997). The large variability of the superrotating zonal winds between 90 and 110 km altitude (*Bougher et al.*, 1997; *Lellouch et al.*, 1997) may arise from the changing nature of this wave breaking. However, only limited observations of gravity wave features are available to constrain the sources, their vertical propagation, and impacts at thermospheric heights (e.g; *Bougher et al.*, 1997).

Gravity waves are thus believed to play a major role in Venus upper atmosphere dynamics, in addition to their possible role (discussed above) in the lower atmosphere's RSZ flow. Wavelike density perturbations with horizontal scales of 100–600 km have been observed in the Venus thermosphere by the Pioneer

Table 1. Upper Atmosphere Wind Constraints

Species/Emissions/Temps.	Altitude Range (km)	SS-AS Winds (m s^{-1})	RSZ Winds (m s^{-1})
Temps. (IR and Occultation)[a]	70–90		variable (weak)
CO mm, CO distribution[b]	90–105	present	variable (weak, occas. strong)
CO mm, winds[c]	90–105	≤ 40–110 ± 20	35–132 ± 10 (variable)
10-micron, CO_2 heterodyne[d]	109 ± 10	120 ± 20	25 ± 15 (weak)
O_2 IR (1.27-microns)[e]	95–110		highly variable (~ 10–50)
CO 4.7-micron, winds[g]	100–110		sum $= 140 \pm 45$
O_2 visible (400–800 nm)[f]	100–130		weak (≤ 30)
NO nightglow (UV)[g]	115–150	~ 200	40–60
O dayglow (130 nm)[h]	130–250		eddy diffusion
Temps. (night)[i]	above 150	~ 200	~ 50–100
H dayglow (121.6-nm)[j]	above 150		~ 45–90
H and He densities[k]	above 150		~ 45–90

[a] *Taylor et al. (1980), Kliore (1985), Schäfer et al. (1990), Roos-Serote et al. (1995)*

[b] *Gulkis et al. (1977), Schloerb et al. (1980), Clancy and Muhleman (1985)*
   Clancy and Muhleman (1991), Lellouch et al. (1994), Gurwell et al. (1995),
   Rosenqvist et al. (1995), Lellouch et al. (1997)

[c] *Shah et al. (1991), Rosenqvist et al. (1995), Lellouch et al. (1997)*

[d] *Goldstein et al. (1991)*

[e] *Crisp et al. (1996), Bougher et al. (1997),*

[f] *Krasnopolsky (1983), Bougher and Borucki (1994)*

[g] *Stewart et al. (1980), Gerard et al. (1981), Bougher et al. (1990)*

[h] *Alexander et al. (1993)*

[i] *Keating et al. (1980), Mayr et al. (1980), Niemann et al. (1980)*
   Bougher et al. (1986, 1997), Mengel et al. (1989)

[j] *Paxton et al. (1985, 1988a,b)*

[k] *Niemann et al. (1979), Brinton et al. (1980)*

Venus Orbiter Neutral Mass Spectrometer (*Kasprzak et al.,* 1988, 1993). These perturbations have been shown to be consistent with vertically propagating gravity waves from a source region at or below ~80 km altitude, well below thermospheric altitudes (*Mayr et al.,* 1988). These thermo- spheric gravity waves may be launched from the Venus cloud region (50–70 km) (*Baker et al.,* 2000a,b) propagate vertically, and break in the thermosphere, thereby providing a significant local source of momentum and energy (see above). Recall from the previous section that vertically propagating gravity waves of long horizontal wavelength driven by solar heating—i.e., atmospheric tides—have been suggested to accelerate the cloud-level atmosphere by their upward transport of prograde momentum; if so they would decelerate the upper atmosphere's RSZ flow. Gravity waves of short horizontal wavelength might also generate small scale vertical mixing (i.e., eddy diffusion) in the Venus upper atmosphere.

A number of circulation models have been constructed to address the Venus upper atmosphere circulation (see review by *Bougher et al.,* 1997). These multi-dimensional models serve both to reproduce the global wind tracers defined above (thereby constraining the wind magnitudes) and to characterize the day-to-day variability of these circulation patterns. Gravity wave breaking has been implemented in some of these model simulations to slow large scale winds. However, it has been difficult for 3-D thermospheric circulation models to reproduce observed diurnal density, temperature, and airglow variations with a unique set of input parameters and wind fields (*Bougher et al.,* 1997).

2. MAJOR UNRESOLVED QUESTIONS

2.1 Lower Atmosphere

2.1.1 Superrotation. The deep atmosphere might hold the key to the driving mechanism of the superrotation. However, it is this region of the atmosphere that we know the least

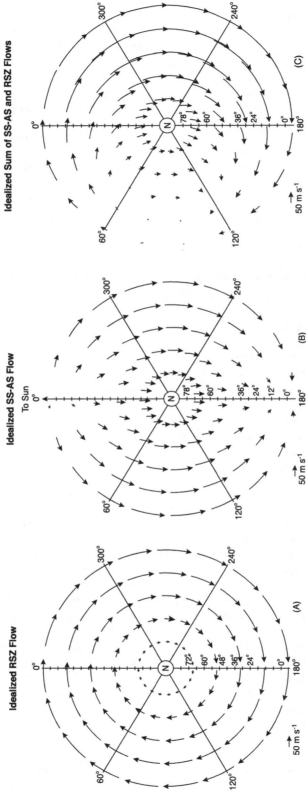

Figure 2. Retrograde superrotating zonal (RSZ) flow (a), idealized subsolar–antisolar (SS–AS) flow (b), and the sum of the two flows (c). The SS–AS flow varies as the sine of the solar zenith angle: $u = u_{max}\sqrt{1 - \cos^2 \lambda \cos^2 \phi}$, where ϕ = north latitude, λ = east longitude, and $(\phi, \lambda) = (0, 0)$ at the subsolar point. The direction of the flow is $\hat{r} \times (\hat{r} \times \underline{\hat{x}})$ where $\hat{r}$ is a unit vector pointing from the center of the planet to the field point and $\underline{\hat{x}}$ is a unit vector pointing from the center of the planet to the Sun (so that the cosine of the solar zenith angle is given by $\hat{r} \cdot \underline{\hat{x}} = \cos \lambda \cos \phi$. This flow direction has (un-normalized) vector components $\{0, \cos \lambda \sin \phi, \sin \lambda\}$ in the $\{\hat{r}, \underline{\hat{\phi}}, \underline{\hat{\lambda}}\}$ basis. The idealized RSZ flow is given by $u = -u_{max} \cos \phi \, \underline{\hat{\lambda}}$. The maximum values of the SS–AS and RSZ flows are set equal.

about. The stratification in the lowest scale height is basically undetermined. All four Pioneer Venus probes failed to obtain thermal structure measurements in this region, and the only profile with sufficient vertical resolution to investigate static stability is that obtained by the VEGA 2 lander (*Crisp and Titov*, 1997). Thus, the latitude-longitude variation of thermal structure and static stability in the lowest scale height are unknown. Dynamical processes affected by lower atmosphere thermal structure include the strength of surface stresses that transfer angular momentum between surface and atmosphere, boundary layer convection, generation and propagation of atmospheric waves, and slope winds (*Gierasch et al.*, 1997). Detailed modeling of atmosphere-surface interactions should at least shed light on which of these interactions would be the most significant. Modeling of these dynamical processes could guide measurement objectives in future Venus missions. Even the basic balances determining the structure of the planetary boundary layer are undetermined for slowly rotating planets. *Allison* (1992) suggests that Titan's rotation is strong enough and its near-surface winds are weak enough to anticipate a geostrophic sublayer and thus a classical Ekman wind spiral structure, and the Huygens Doppler Wind data appear to bear this out (*Bird et al.*, 2005), but the existence of such a regime is problematic for the more slowly rotating Venus. Existing GCMs applied to Venus typically simplify the surface turbulent transfer using constant transfer coefficients, i.e., they assume that surface transfer of heat is proportional to the difference in temperature between the surface and the immediately overlying atmosphere, together with an analogous formulation of surface frictional drag on wind. Nonlocal turbulence schemes, which have improved several terrestrial climate GCMs, may be needed to allow convectively generated near-surface turbulence to efficiently interact with the free-atmosphere Hadley circulation. The altitudes at which GCM-derived winds diverge from the observed zonal wind profile (in the lowest scale height, or above, or at all heights) could be indicative of whether the simulation of the surface-atmosphere interaction or the larger-scale dynamical transport mechanisms at higher altitude is the primary problem. Although measured winds on Venus are retrograde at all altitudes, prograde surface winds are required at some latitudes (if the surface and atmosphere are in equilibrium) to guarantee zero net surface torque and provide a means for angular momentum to be supplied to the atmosphere. The latitudinal profile of surface wind may thus be a useful diagnostic for evaluating the superrotation mechanisms in different GCMs.

Not only 3-D circulation but also solar and infrared radiative transfer, cloud dynamics, and chemistry must eventually be included in models of the Venus atmosphere.

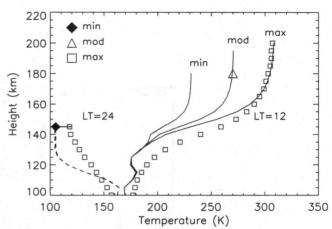

Figure 3. Venus Thermospheric General Circulation Model (VTGCM) temperatures (K) and their variations over the solar cycle. Simulated equatorial profiles are presented for noon (LT = 1200) and midnight (LT = 2400) conditions over 100 to 200 km altitude. *Hedin et al.* (1983) empirical model temperatures (solar maximum conditions) are indicated by squares. Magellan (drag) inferred temperatures (solar moderate conditions) are shown by triangles. Taken from *Bougher et al.* (2002).

(The latter two and near-surface processes are the current emphasis in Earth atmosphere GCM development.) No Venus atmospheric circulation model yet incorporates realistic radiative transfer schemes or attempts to account interactively for the effects of the ubiquitous clouds. Clearly, the heating of the atmosphere by radiative transfer and the influence of the clouds on this process drive the atmospheric circulation. Realistic radiative driving is particularly important to simulate day-night variations (which, as discussed above, are probably important in cloud-top superrotation) and deep atmosphere forcing of the Hadley circulation.

2.1.2 Mean meridional (Hadley) circulation and eddies. We need to characterize how many thermally direct Hadley cells exist between the surface and clouds and to what latitudes they extend (*Schubert*, 1983). This bears directly on how momentum is transported vertically from the surface to accelerate higher levels of the atmosphere. It is equally important to identify dominant eddy (i.e., nonaxisymmetric) components in the atmospheric flow, and to determine how they are generated. It is unlikely that any future Venus mission would be capable of measuring eddies in the deep atmosphere because of their small magnitude and significant space-time variations. Therefore, general circulation modeling would be valuable in providing an idea of what could exist, and what regimes are most capable of affecting the superrotation.

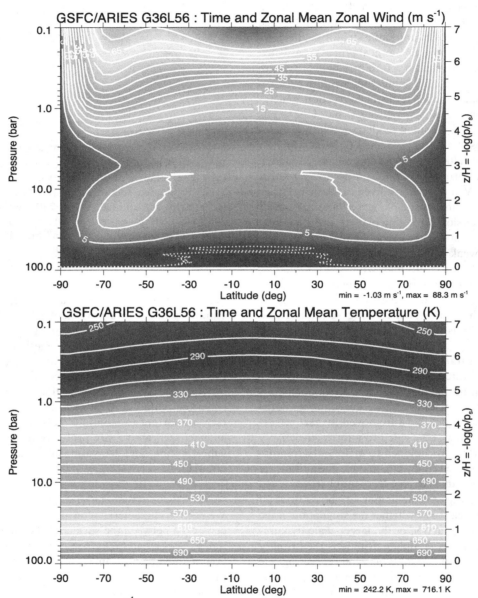

Plate 1. The mean zonal wind (m s^{-1}) and temperature (K) from a VSGCM simulation using diabatic heating similar to Yamamoto and Takahashi (2003a, 2004). Time averages are computed over the last 1000 days from a 10,000 day simulation. The contour interval is 5 m s^{-1} and 20 K in the top and bottom panels, respectively.

A related issue involves the magnitude of thermal contrasts on constant pressure surfaces in the deep atmosphere. Theoretical considerations lead to the expectation of temperature variations of O(0.1 K) with respect to both latitude and longitude in the lowest scale height (*Stone*, 1975); simple order of magnitude estimates of day-night thermal constrasts based on solar flux measurements from Pioneer Venus (*Tomasko et al.*, 1980) yield similar values. Yet the Pioneer Venus probes indicate thermal contrasts of several degrees all the way down to near 12 km altitude, below which the probes ceased making thermal measurements (*Seiff et al.*, 1980; *Seiff*, 1983). Cyclostrophic balance above the lowest scale height is consistent with meridional temperature contrasts $\geq$ 1 K, but as the surface is approached the zonal wind decreases to zero, and cyclostrophic balance is no longer a good approximation. It is not clear whether RSZ flow induces measurable thermal contrasts via cyclostrophic balance, or whether the thermal contrasts are produced by some other process and subsequently induce the RSZ flow and cyclostophic balance.

Given better constraints on the latitudinal temperature contrast (and thus the vertical profile of cyclostrophically-balanced winds) and the static stability in the free troposphere, it would be possible to map the potential vorticity distribution in the Venus troposphere. Potential vorticity essentially maps the degree to which angular momentum and potential temperature surfaces intersect. Stable planetary circulations have zero potential vorticity at the equator and potential vorticity of the same sign as the Coriolis parameter in each hemisphere (*Stevens*, 1983). Baroclinic/barotropic instabilities attempt to homogenize latitudinal potential vorticity gradients equatorward of the jet regions that give rise to the instabilities, just as convection attempts to homogenize potential temperature in the vertical. Thus, depending on the efficiency of the eddies that do the mixing, near-zero values of potential vorticity may extend from the equator poleward as far as the jets. *Allison et al.* (1994) derive zero potential vorticity latitudinal profile envelopes for the zonal flow as a function of Richardson number. These represent the most efficiently latitudinally mixed state possible for a given stratification. *Allison et al.* (1994) show that the Pioneer Venus probe winds at 25 km altitude, the UV cloud-tracked winds at 62 km altitude, and the cyclostrophic thermal winds above cloud top all resemble a zero potential vorticity latitudinal profile, although our limited ability to measure the Richardson number (*Schubert*, 1983; *Gierasch et al.*, 1997) makes the inference inconclusive. This suggests that whatever processes control the Venus superrotation effectively mix potential vorticity. Potential vorticity mixing appears to operate efficiently in GCM simulations of Mars (*Barnes and Haberle*, 1996) and at Titan's rotation rate (*Allison et al.*,

1994), as well as in Earth's atmosphere (*Lindzen*, 1990). This may or may not be the case for Venus models, however, and so comparisons with zero potential vorticity profiles may be diagnostic of whether the dynamical modes that transport angular momentum equatorward are operating at maximum efficiency. This in turn may shed light on why most Venus atmosphere GCMs underestimate superrotation.

2.1.3 Vortex organization of the circulation. The vortex organization of the Venus atmosphere inferred from Mariner 10 ultraviolet images of Venus taken in 1974 (*Suomi and Limaye,* 1978) was also seen from Pioneer Venus images acquired during 1978 through 1983 (*Limaye*, 1990). The global atmosphere is organized as two hemispheric vortices centered over each pole, each exhibiting a broad mid-latitude jet and poleward flow at the ultraviolet cloud level. The vertical extent of these two vortices could not be determined from the reflected ultraviolet images. Early VIRTIS observations from the Venus Express mission indicate that the vortex organization exists deep into the Venus atmosphere. These data show the same spiral bands on the night side in the emitted infrared radiation as those seen on the day side in the ultraviolet images. Further, they also show an "S" shaped feature, akin to the features seen in the "eye" or the inner region of many tropical cyclones. This feature was first detected in the Pioneer Venus thermal infrared observations (*Taylor et al.*, 1980), and was inferred as a warm feature surrounded by a cold collar. Whorls seen in the inner core region of tropical cyclones are a result of potential vorticity mixing (*Kossin et al.*, 2002), and it is possible that the "S" shaped feature seen in the core of the vortices in the north and the south polar regions are similarly produced. Prior results from tracking of the night side features seen in the Galileo near infrared images suggested poleward flow (*Carlson et al.*, 1991) at ~53 km altitude, consistent with the VIRTIS data. Thus we can surmise that the vortex circulation is deep, extending down to this level, and perhaps even deeper. We need more observations, deeper in the atmosphere and near the surface in polar regions to fully understand the vortex circulation on Venus.

Limaye (2007) has noted the similarities between a tropical cyclone and the hemispheric vortex circulation on Venus. One difference seems to be in the longevity—on Venus the vortex circulation has been observed over three decades. Although we do not have continuous spacecraft data, the vortex organization appears to have persisted throughout this period, and has existed for a long time. Tropical cyclones, on the other hand, decay when their energy source is shut down once the cyclone makes landfall. Over vast reaches of the southern oceans, cyclones generally last longer, since they can travel long distances without landfall. On Venus the vortex, being situated over the pole, does not of course "travel" with the steering winds, and the vortex never loses the energy source.

2.1.4 Thermal tides, planetary scale waves. Thermal tides have been suggested to play a role in the maintenance of the atmospheric circulation of Venus. They were detected in the thermal structure data by *Schofield and Taylor* (1983) and also in the atmospheric circulation by *Limaye* (1990). Pechman and Ingersoll (1984) and *Newman and Leovy* (1992) investigated dynamical implications, but the structure (phase and amplitude relationships) of the tides has not yet been adequately determined because of limited observations. (Thermal structure observations by the Pioneer Venus Orbiter Infrared Radiometer cover much less than one solar day, and the solar-locked structure in the cloud motions can be determined only from day-side images.) Measured cloud motions reveal inconsistencies with the theoretical calculations referenced above. Further work is needed to explore the structure of the thermal tides by both comprehensive atmospheric circulation modeling and observational data analysis (the latter incorporating the Pioneer Venus results as well as those from the Venus Express mission).

Many investigators over the years have pointed out that planetary-scale waves are suggested in whole-disk ultraviolet images of the Venus cloud tops (*Covey and Schubert*, 1981, 1982, 1983) and in the zonal winds inferred from both Mariner 10 and Pioneer Venus cloud motions (*Limaye and Suomi*, 1978; *Rossow et al.*, 1990). These waves are believed to arise from resonant oscillations and instabilities of the atmosphere, in contrast to thermal tides forced by solar heating. As with the tides, their role in the maintenance of the global circulation is suggested by theoretical models but unconfirmed due to limitations of the observations. Venus Express, with more capable instruments (see Section 4), will yield a better determination of the structure of the large scale waves.

2.2 Upper Atmosphere

The processes responsible for maintaining and driving variations in the SS-AS and superrotating wind components in the Venus upper atmosphere are not well understood or quantified. In particular, what causes the rapid and spectacular variations observed in the magnitude of the RSZ flow, especially in the upper mesosphere and lower thermosphere regions (~90–130 km altitude)? Why is the structure of the NO (UV) and O_2 (visible and IR) nightglow emissions so complex and variable with time on diurnal as well as solar cycle timescales? Why do hemispheric asymmetries in tracer emissions/species occur in the Venus upper atmosphere? Answers to these questions require many more observations

to establish a comprehensive climatology of the Venus atmospheric structure and wind components over ~90–200 km altitude. A statistical assessment of the relative importance of the separate SS-AS and RSZ wind components during different atmospheric conditions can then be made.

The role of wave activity in driving these large variations in Venus upper atmosphere wind components also remains to be addressed. Gravity wave breaking is one mechanism thought to be responsible for SS-AS wind deceleration and the asymmetries resulting in the observed RSZ flow in the upper atmosphere. Constraints for gravity wave parameters at the cloud tops and above are quite limited at present. Measurements of gravity wave phase speeds and amplitudes at various altitudes are crucial to the construction and validation of gravity wave breaking models. Confirmation of this mechanism will likely require the implementation of a self-consistent gravity wave breaking formulation into three-dimensional circulation models (e.g. *Zhang et al.*, 1996; *Bougher et al.*, 1997). *Fritts and Lu* (1993) developed a gravity wave parameterization scheme for use in terrestrial GCMs based on the spectral characteristics of observed gravity waves on Earth. It has been utilized in Venus GCMs, but requires further knowledge of the gravity wave spectrum in the Venusian atmosphere (*Zhang et al.*, 1996). Other gravity wave parameterization schemes are available for possible incorporation into Venus GCMs (*Lindzen*, 1981; *Hines*, 1997; *Lawrence*, 1997; *Alexander and Dunkerton*, 1999; *Warner and McIntyre*, 2001). In short, substantial unknowns remain about the true nature of gravity waves and their impact on the dynamics of the upper atmosphere of Venus.

3. PRIORITIES FOR GENERAL CIRCULATION MODELING

3.1 Lower Atmosphere

The modeling of the general circulation of Venus' lower atmosphere has reached a peculiar state. We seem to be able to get the "right answer" (dominant RSZ flow) only if we input the "wrong reasons" (heating rates that lie outside the error margins of observation). What improvements to GCMs are needed in order to get the right answer for the right reasons? The following considerations suggest the top priorities:

1. Whatever the number and type of future missions to Venus, they will never in the foreseeable future approach the coverage that Earth's troposphere has enjoyed since the advent of routine global weather balloon observations in the mid-20th century. Accordingly, theoretical general circulation modeling will be essential both to

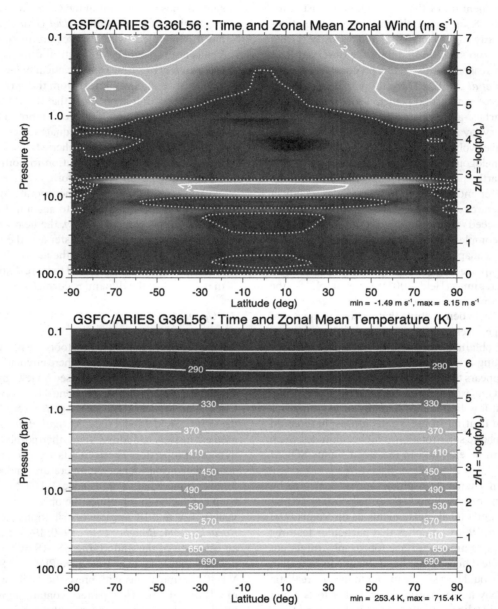

Plate 2. As in Plate 1 but for a simulation that uses a more realistic diabatic heating in the lower atmosphere (i.e., below 35 km). Contour interval for zonal wind plot is 2 m s⁻¹.

explore mechanistic ideas and to make sense of whatever incomplete observations exist. The GCMs will necessarily include poorly constrained empirical parameters, which should be varied over their possible ranges.

2. The angular momentum of the superrotation must come from somewhere. Solar gravitational torque on the atmosphere was an early suggestion (*Gold and Soter*, 1971), but its magnitude is too small to maintain the observed RSZ flow against any reasonable dissipative processes (*Schubert*, 1983; *Gierasch et al.*, 1997). The only remaining angular momentum source is the solid planet. Thus, more precise simulation of surface-atmosphere interactions and the near-surface boundary layer is necessary. As discussed above, Earth GCMs have evolved to include detailed parameterizations of both. Analogous parameterizations for Venus would need to avoid Earth-specific assumptions like the existence of an Ekman layer and start from first principles wherever possible. A "hierarchy of models" strategy (e.g., *Andrews et al.*, 1987) proceeds from simplified to intermediate and finally to fully complex atmospheric models. Global circulation models that adapt highly simplified physics, so-called SGCMs, can help shed light on the parameters that regulate circulation regimes (Held, 2005). For example, for the terrestrial troposphere, *Kraucunas and Hartmann* (2005) utilized an SGCM to demonstrate that persistent equatorial superrotation (i.e., west-to-east mean zonal flow over the equator) can be obtained when steady east-west variations in diabatic heating are imposed within low latitudes. The superrotation appears to be driven by quasi-stationary planetary waves that are excited by the imposed tropical heating. It might be that low-latitude topography or asymmetries in near-surface heating in the Venus atmosphere could enhance atmospheric superrotation at higher altitudes. Such mechanisms can be explored in idealized Venus SGCM simulations.

3. Superrotation in Titan's atmosphere—now confirmed by Cassini/Huygens observations, and quite possibly arising from the same dynamical mechanisms as on Venus (*Del Genio and Zhou*, 1996)—implies that particular features of Venus such as its unique surface topography might not play a critical role. On the other hand, thanks to Magellan observations, it would be a simple matter to include realistic Venus topography as a sensitivity test. Other tests could include various idealized topography boundary conditions including, of course, perfectly smooth topography. These would define what types of topography (if any) are required to efficiently transfer angular momentum from the solid planet to the atmosphere.

4. Higher horizontal resolution in Earth GCMs, up to at least 2-degree latitude/longitude grid point spacing, leads to substantially better agreement not only with small-scale details but also with large-scale flow observations (*Duffy et al.*, 2003). Venus GCMs have never had less than about 5-degree grid point spacing. This relatively coarse resolution is adequate to represent large-scale horizontal waves such as those important in the Gierasch-Rossow-Williams mechanism of superrotation (*Del Genio and Zhou*, 1996), but it is not clear which waves are important in the real Venus atmosphere. In the vertical dimension, both theory (*Gierasch et al.*, 1997) and practical experience imply that at least several dozen levels from the surface to the model top are needed. It is not clear that the current state-of-the-art vertical resolution (50 levels) represents convergence, however, or even that it is adequate to accurately represent vertically propagating atmospheric tides. Evidently experimentation with higher resolution in both horizontal and vertical dimensions is called for.

5. No Venus GCM models yet incorporate realistic radiative transfer schemes or attempt to account for the effects of the ubiquitous clouds. Clearly, the heating/cooling of the atmosphere by radiative transfer and the influence of the clouds on this process drive the atmospheric circulation. These phenomena might need to be realistically simulated in order to get the dynamics correctly.

3.2 Upper Atmosphere

Three-dimensional modeling tools are presently being used to study the Venus upper atmosphere circulation and the tracer emissions/species distributions (see reviews by *Bougher et al.*, 1997, 2002). Both the NCAR Venus Thermospheric General Circulation Model (VTGCM) (e.g., *Bougher et al.*, 1988) and the GSFC SRM Model (e.g., *Mengel et al.*, 1989) have been used extensively to address Venus thermosphere structural and dynamic properties. These GCMs are also crucial to synthesizing the various density, temperature, and airglow datasets for extracting winds. For instance, VTGCM simulations have been conducted to reproduce the observations of the various visible, UV and IR airglow distributions (e.g. *Bougher and Borucki*, 1994; *Bougher et al.*, 1990, 1997) thereby providing an estimate of the underlying SS-AS and RSZ wind components that vary over time (see Plate 3). Solar maximum VTGCM simulations with prescribed RSZ winds (up to ~50 m s^{-1}) reveal a net flow pattern containing mean horizontal winds which are strongest across the evening terminator and converge after midnight, near the equator. However, all existing GCMs are presently unable to reproduce the observed diurnal density, temperature, and airglow variations utilizing a unique set of wind fields, eddy diffusion coefficients, and wave drag parameters (*Bougher et al.*, 1997). This may reflect missing physical processes or inputs, e.g., exospheric transport above 180–200 km altitude, planetary waves, and limited

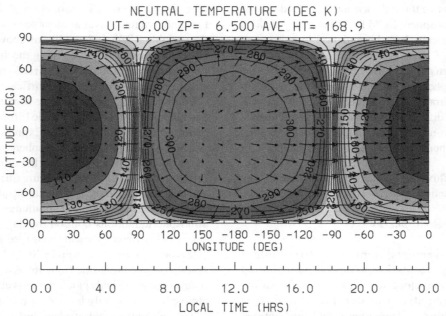

NEUTRAL TEMPERATURE (DEG K)
UT= 0.00 ZP= 6.500 AVE HT= 168.9

Plate 3. Venus Thermospheric General Circulation Model (VTGCM) exobase temperatures (K) and superimposed horizontal wind vectors at the ~7.5 × 10⁻³ nanobar level.Solar maximum conditions are illustrated, consistent with early Pioneer Venus observations. Maximum horizontal wind vectors are indicated across the evening terminator (~260 m s⁻¹). Convergence of the flow is seen after midnight, toward the morning terminator. RSZ wind speeds of ~50 m s⁻¹ are prescribed at this level. Exospheric temperatures range from a nightside minimum (104 K) to a dayside maximum (308 K) yielding a day-night contrast of about 200 K, similar to that observed by Pioneer Venus instruments.

gravity wave constraints for wave breaking. In general, too many free parameters presently exist.

The upper atmosphere modeling task would be greatly improved if simultaneous wind, temperature, and density measurements could be made over ~65–200 km altitude and gravity wave parameters could be quantified. Specific upgrades needed for Venus upper atmosphere models include the following:

a Extend circulation models to cover the region above the cloud tops to the exobase (~65–200 km altitude). This extension permits a likely gravity wave source region (e.g., cloud tops), propagation region (above the cloud tops), and dissipation region (upper mesosphere and lower thermosphere) to be captured in the same three-dimensional model domain.

b Drive the upper atmosphere GCM lower boundary at cloud tops with an empirically based circulation (winds specified).

c Implement a modern gravity wave breaking model into the Venus upper atmosphere GCMs. Make use of the best available gravity wave parameters. Internally consistent momentum drag and eddy diffusion terms can then be simulated.

d Utilize the latest ion-neutral reaction rates of *Fox and Sung* (2001) for photochemistry in upper atmosphere GCM codes.

e Use SOLAR2000 fluxes to calculate the solar heating, dissociation, and ionization rates (*Tobiska et al.*, 2000)

3.3 Lower and Upper Atmosphere

The model improvements summarized in Table 2 would produce a high-resolution Venus GCM including atmospheric circulation, cloud formation and dissipation, solar and infrared radiative transfer, the near-surface boundary layer, the upper atmosphere, and interactions among these components (chemical effects have not been included). All of these improvements need to be tested, in order to determine the sensitivity of "the real superrotation mechanism" and other features of the Venus atmosphere circulation (e.g., SS-AS flow, Hadley circulation and day-night variations) to parameters that observations cannot fully determine. Such a program of model development and testing implies substantial use of computer resources.

4. IMPACT OF ONGOING AND FUTURE OBSERVATIONS

In 2005 the European Space Agency's Venus Express (VEX) mission began observing the planet from a highly elliptic polar orbit. A major goal of VEX is to increase our understanding of the circulation and dynamics of the atmosphere from the surface up to the exobase (*Titov et al.*, 2001,

2006). VEX remote sensing measurements complement and extend those obtained by the Pioneer Venus Orbiter (PVO) between 1978 and 1992 and the Galileo flyby in 1990. VEX will permit the first detailed study of middle and upper atmosphere dynamics using O_2 (visible and IR nightglow), NO (UV nightglow), O (UV dayglow) and H (UV dayglow) emissions as tracers of the circulation. In addition, thermal mapping at and above cloud levels (~60–90 km altitude) can be used to derive thermal wind fields for comparison with the spatial and temporal variations in these emission features. Also, global wind fields will be obtained from cloud feature tracking. VEX measurements will thus provide a more comprehensive investigation of the atmosphere structure and dynamics over another period in the solar cycle and for variable RSZ wind conditions above the cloud tops. Our understanding of the underlying mechanism(s) responsible for maintaining and driving variations in the Venus atmosphere wind system will be significantly advanced.

Equipped with imaging (Venus Monitoring Camera, VMC) and mapping instruments (Visible InfraRed Thermal Imaging Spectrometer, VIRTIS), Venus Express probes the equatorial and southern hemisphere atmosphere from high altitudes to the surface at wavelengths from the ultraviolet to the thermal infrared (*Titov et al.*, 2006). Feature tracking in the ultraviolet and near infrared images of Venus (acquired daily by VMC at 365, 513, 965, and 1000 nm simultaneously), determines the atmospheric motions both on the day side and on the night side at two different vertical levels (~65 km altitude on the day side from 365 nm images and ~50–55 km altitude on the day side and on the night side from 965 nm images). The VIRTIS instrument also yields cloud motions at multiple levels of the atmosphere on a daily basis over a limited region of Venus (due to a smaller field of view) but at multiple wavelengths that complement the VMC cloud tracking results.

In addition to its low resolution spectral mapping (imaging) capability (VIRTIS-M channel, 0.25–5.0 microns), the instrument provides a high spectral resolution channel (VIRTIS-H, 2–5 microns) without imaging capabilities. The latter addresses atmosphere dynamics by: (1) measuring 3-D temperature and derived thermal wind fields at 60–90 km altitude on the nightside and (2) mapping O_2 infrared and visible airglow as a tracer of the dynamics of the upper mesosphere and lower thermosphere (~95–130 km altitude). Nightside VIRTIS global maps of the strong 1.27 micron O_2 emission will also be obtained; repeated observations during several orbits of Venus Express will monitor its variability at time scales from hours to days. Finally, limb observations of 4.3-micron NLTE CO_2 emissions provide a sounding of the upper atmosphere above 100 km. These emissions could be sensitive to atmospheric perturbations produced by gravity-wave breaking.

The Venus Radio Occultation experiment (VeRa) on Venus Express measures the detailed vertical thermal structure of the atmosphere. VeRa profiles are obtained only during occultation opportunities but at high vertical resolution. They enable a composite view of the meridional thermal structure of the atmosphere from which the zonal circulation can be inferred through the assumption of cyclostrophic balance (Newman et al., 1984; *Limaye*, 1985). These inferred winds can be compared with similarly derived winds from the VIRTIS instrument (~ 60–90 km altitude on the nightside). In addition, the O_2 airglow at 1.27 microns will be monitored by limb observations, thereby characterizing the variations in the airglow layer over ~ 95–110 km altitude (see Table 1). Finally, the VMC utilizes filters at 356 and 376-nm to image the O_2 visible nightglow. VMC O_2 visible nightglow distribution maps can be coordinated and compared with corresponding VIRTIS O_2 IR nightglow maps in order to characterize the changing RSZ wind component for overlapping layers spanning ~ 95 to 130 km (see Table 1).

The global wind fields obtained from cloud feature tracking enable the determination of circulation for comparison with

Table 2. Atmosphere-Related Features in Current Models and Possible Future Models.

	2006-vintage GCMs	Available Observations	Model Requirements*
Topography	Flat**	Height variations at 20/1 km resolution (horizontal/vertical)	Very fine horizontal/vertical resolution near surface
Surface and lowest-level atmosphere properties	Highly idealized surface interaction	Temperatures and images from surface in the visible	Direct surface-atmosphere coupling; boundary layer sub-model
Horizontal variations in atmosphere	~ 500 km grid-point spacing	~ 5—20 km (T) and ~ 100 km (wind) at and above clouds	≤ 100 km grid-point spacing
Vertical variations in atmosphere	~ 2 km vertical-level spacing	~ 0.1 km (T) and ~ 1 km (wind) at select locations	≥ 100 levels between surface and 100 km altitude
Clouds	Omitted***	Vertical profiles of particle sizes; horizontal "slices" of cloud properties	Interactive cloud simulation
Heating/cooling rates	Assumed	Vertical profiles of temperatures, fluxes at select locations	Interactive radiative transfer
Upper atmosphere	Upper atmosphere model coverage ≥ 100 km altitude****	Up to ~ 200 km altitude	Single GCM (0–200 km). Extend upper boundary; add gravity wave parameterization and non-LTE physics

*What is needed for models to achieve the level of detail in the observations
**Except for *Herrnstein and Dowling* (2007) and *Yamamoto and Takahashi* (2006b).
***Models generally impose radiative effects of clouds directly, without explicitly including the clouds.
**** Separate upper (90–200 km altitude) and lower atmosphere models are presently used.

GCM results (e.g., meridional profile of zonal and meridional flow, presence of large scale waves). The balanced flow results from the zonal average thermal structure provide a check on the cloud feature results for the average zonal flow, particularly the strength of the mid-latitude jet at the cloud level. Analysis of the cloud motions in solar-fixed coordinates yields information about the magnitude and phases of the diurnal and the semi-diurnal solar thermal tides. The cloud motions also provide estimates of the transport of angular momentum at a minimum of two different vertical levels (~ 65 km and ~ 50–55 km altitude). These results should be particularly useful for evaluating GCM simulations and the Gierasch mechanism for the maintenance of the angular momentum budget. They will also provide diagnostic information about the organization of the global vortex pattern seen in Mariner 10 and Pioneer Venus ultraviolet images (*Suomi and Limaye*, 1978; *Limaye*, 1985), and the stability of the smaller polar vortices observed at thermal infrared wavelengths by Pioneer Venus (*Taylor et al.*, 1980).

The SPICAV (Spectroscopy for Investigation of Characteristics of the Atmosphere of Venus) instrument is a versatile atmospheric spectrometer that consists of both UV (110–310 nm) and IR (0.7–1.7 micron) channels (*Titov et al.*, 2001, 2006). SPICAV (UV) makes nadir, limb, and solar/stellar occultation measurements. It provides measurements of the vertical profiles of atmospheric density (and inferred temperatures) over 80–180 km altitude (dayside) and 80–150 km altitude (nightside) via stellar occultations. These day-to-night density and temperature variations can be used to validate GCM simulations and thereby extract thermospheric SS-AS wind magnitudes. Airglow (nadir and limb) observations of NO (190–270 nm), O (130.4 nm), H (121.6 nm), and CO (Cameron band) emissions are also made. Near apoapsis viewing of the nightside will provide global spectral imaging sequences of these airglow features, enabling maps to be constructed, from which estimates of the changing thermospheric RSZ wind component (above ~ 115 km altitude) can be made. Periapsis limb viewing (e.g., the NO emission layer), at high vertical resolution, will provide further constraints on the changing nature of the local chemical and eddy diffusion processes on the nightside that regulate the nightglow intensity and its vertical distribution.

SPICAV (IR) is a miniature IR spectrometer (with 0.5–1.0 nm spectral resolution). This instrument monitors the 1.27 micron emission with limb viewing. This high vertical resolution profile of the airglow layer, when combined with corresponding O_2 nightglow maps from VIRTIS provides a more comprehensive picture of the O_2 nightglow's vertical and horizontal distributions.

VEX datasets can be assimilated to constrain GCMs, from which separate SS-AS and RSZ wind components can

be extracted. A greatly expanded climatology of the Venus upper atmosphere structure and wind components will thus be compiled. In addition, gravity wave parameters above the cloud tops will be measured (or inferred) and used to constrain the gravity wave breaking model proposed by Alexander (1992). The feasibility of the gravity wave mechanism for regulating the SS-AS and RSZ wind components can then be tested.

The Venus Climate Orbiter (VCO) is a mission proposed to be launched by Japan in 2010 (*Nakamura et al.*, 2007). By making observations of Venus at ultraviolet and near infrared wavelengths from an equatorial, elliptic orbit, VCO will provide observations complementary to those from Venus Express. By the choice of the VCO orbit, with the spacecraft moving in the same direction in the equatorial plane as Venus, VCO will be able to focus more on the short term evolution of the clouds and the concurrent atmospheric circulation and thus provide more information on small scale and short term processes not possible from previous missions. It will provide a good picture of the atmospheric circulation at low and middle latitudes, but it will miss the polar latitudes. Since VCO is planned to arrive at Venus just as the Venus Express mission will be nearing its end, we will have observations of the Venus atmospheric circulation over a long period of time, from 2006 through perhaps ~ 2013, yet the lack of overlap is disappointing.

The Messenger spacecraft flies by Venus in October 2006 and June 2007 on its way to Mercury. It is planned to image the upper cloud layers at visible and near-infrared wavelengths to compare with previous observations. Coordination with VEX could enable some unique simultaneous measurements.

Venus Express, the Messenger flybys, and Venus Climate Orbiter will provide a wealth of new observations to both constrain and evaluate GCM models of atmospheric structure and dynamics. The new data will restore balance between observation and theory in our quest to understand the workings of the Venus atmosphere.

5. CONCLUDING REMARKS

Cloud-top superrotation surprised Venus observers nearly half a century ago. Explaining it—and how it transitions to SS-AS flow in the upper atmosphere—would clear up a long-standing mystery of atmospheric dynamics, and lead to a comprehensive, self-consistent picture of Venus' atmospheric circulation and how it is forced. Not only solar/infrared radiative transfer and 3-D circulation but also cloud dynamics, near-surface processes, and—eventually—chemistry at all levels must be included in models of the Venus atmosphere. (The latter three are the current emphasis in Earth atmosphere GCM development.) Accomplishing this goal is a necessary condition for tackling the "big questions" about how Venus' climate evolved so differently from Earth's. Understanding terrestrial planets and their climatology has become a principal focus of NASA's Astrobiology program through that program's emphasis on studying habitable planets. The Kepler Mission to be launched in 2008 will search over 100,000 stars in an attempt to determine the frequency of existence of Earth-sized planets. A major scientific goal will be to determine whether those planets lie in the habitable zone of the parent star. A planet's atmospheric circulation could be a critical factor in determining a planet's habitability (Joshi et al., 1997). In size and mass Venus is very similar to Earth. Not to understand Venus' atmospheric superrotation, which is such a fundamental feature of this terrestrial planet's dynamic circulation, casts doubt on how well we can assess planet habitability in general, and to what extent we understand dynamic meteorology for situations other than our own planet.

Acknowledgments. S. Bougher was supported by NSF-AST grant #0406650 to the University of Michigan. C. Covey's work was performed under auspices of the Office of Science, U.S. Department of Energy by the University of California, Lawrence Livermore Laboratory under contract W–7405–Eng–48. S. S. Limaye acknowledges the support provided by NASA grant #NNG06GC68G. G. Schubert was supported by NASA Planetary Atmospheres grant NASA NNG04GQ72G and by a grant from the Systemwide Institute of Geophysics and Planetary Physics, University of California.

REFERENCES

Alexander, M. J. (1992), A mechanism for the Venus thermospheric superrotation, *Geophys. Res. Lett., 19,* 2207–2210.

Alexander, M. J., and T. J. Dunkerton (1999), A spectral parameterization of mean-flow forcing due to breaking gravity waves, *J. Atmos. Sci., 56,* 4167–4182.

Alexander, M. J., A. I. F. Stewart, S. C. Solomon, and S. W. Bougher (1993), Local time asymmetries in the Venus thermosphere, *J. Geophys. Res., 98,* 10,849–10,871.

Allison, M. (1992), A preliminary assessment of the Titan planetary boundary layer, in *Proceedings, Symposium on Titan, Toulouse, France, 9-12 September 1991,* pp. 113–118, ESA Publications Division, ESTEC, Noordwijk, The Netherlands.

Allison, M., A. D. Del Genio, and W. Zhou (1994), Zero potential vorticity envelopes for the zonal-mean velocity of the Venus/Titan atmospheres, *J. Atmos. Sci., 51,* 694–702.

Andrews, D. G., J. R. Holton, and C. B. Leovy (1987), *Middle Atmosphere Dynamics,* Academic Press, Orlando, 489 pp.

Baker, R. D., G. Schubert, and P. W. Jones (2000a), Convectively generated internal gravity waves in the lower atmosphere of Venus. Part I: No wind shear, *J. Atmos. Sci., 57,* 184–199.

Baker, R. D., G. Schubert, and P. W. Jones (2000b), Convectively generated internal gravity waves in the lower atmosphere of Venus. Part II: Mean wind shear and wave- mean flow interaction, *J. Atmos. Sci., 57,* 200–215.

Barnes, J. R., and R. M. Haberle (1996), The Martian zonal-mean circulation: Angular momentum and potential vorticity structure in GCM simulations, *J. Atmos. Sci., 53,* 3143–3156.

Bird, M. K., M. Allison, S. W. Asmar, D. H. Atkinson, I. M. Avruch, R. Dutta-Roy, Y. Dzierma, P. Edenhofer, W. M. Folkner, L. I. Gurvits, D. V. Johnston, D. Plettemeier, S. V. Pogrebenko, R. A. Preston, and G. L. Tyler (2005), The vertical profile of winds on Titan, *Nature, 438,* 800–802.

Bougher, S. W., and W. J. Borucki (1994), Venus O_2 visible and IR nightglow: Implications for lower thermosphere dynamics and chemistry, *J. Geophys. Res., 99,* 3759–3776.

Bougher, S. W., R. E. Dickinson, E. C. Ridley, R. G. Roble, A. F. Nagy, and T. E. Cravens (1986), Venus mesosphere and thermosphere: II. Global circulation, temperature, and density variations, *Icarus, 68,* 284–312.

Bougher, S. W., R. E. Dickinson, E. C. Ridley, and R. G. Roble (1988), Venus mesosphere and thermosphere: III. Three-dimensional general circulation with coupled dynamics and composition, *Icarus, 73,* 545–573.

Bougher, S. W., J. C. Gérard, A. I. F. Stewart, and C. G. Fesen (1990), The Venus nitric oxide night airglow: model calculations based on the Venus thermospheric general circulation model, *J. Geophys. Res., 95,* 6271–6284.

Bougher, S. W., M. J. Alexander, and H. G. Mayr (1997), Upper atmosphere dynamics: Global circulation and gravity waves, in *Venus II: Geology, Geophysics, Atmosphere, and Solar Wind Environment,* edited by S. W. Bougher, D. M. Hunten, and R. J. Phillips, pp. 259–291, The University of Arizona Press, Tucson.

Bougher, S. W., R. G. Roble, and T. J. Fuller-Rowell (2002), Simulations of the upper atmospheres of the terrestrial planets, in *Comparative Aeronomy in the Solar System,* edited by M. Mendillo, A. Nagy, and H. Waite, AGU Monograph, pp. 261–288, American Geophysical Union, Washington, DC.

Brinton, H. C., H. A. Taylor, H. B. Niemann, H. G. Mayr, A. F. Nagy, T. E. Cravens, and D. F. Strobel (1980), Venus nighttime hydrogen bulge, *Geophys. Res. Lett., 7,* 865–868.

Carlson, R. W., K. H. Baines, T. Encrenaz, F. W. Taylor, P. Drossart, L. W. Kamp, J. B. Pollack, E. Lellouch, A. D. Collard, S. B. Calcutt, D. Grinspoon, P. R. Weissman, W. D. Smythe, A. C. Ocampo, G. E. Danielson, F. P. Fanale, T. V. Johnson, H. H. Kieffer, D. L. Matson, T. B. McCord, and L. A. Soderblum (1991), Galileo infrared imaging spectroscopy measurements at Venus, *Science, 253,* 1541–1548.

Clancy, R. T., and D. O. Muhleman (1985), Diurnal CO variations in the Venus mesosphere from CO microwave spectra, *Icarus, 64,* 157–182.

Clancy, R. T., and D. O. Muhleman (1991), Long-term (1979–1990) changes in the thermal, dynamical, and compositional structure of the Venus Mesosphere as inferred from microwave spectral line observations of ^{12}CO, ^{13}CO, and $C^{18}O$, *Icarus, 89,* 129–146.

Counselmann, C. C., S. A. Gourevitch, R. W. King, G. B. Loriot, and E. S. Ginsberg (1980), Zonal and meridional circulation of the lower atmosphere of Venus determined by radio inteferometry, *J. Geophys. Res., 85,* 8026–8030.

Covey, C., and G. Schubert (1981), 4-day waves in the Venus atmosphere, *Icarus, 47*, 130–138.

Covey, C., and G. Schubert (1982), Planetary-scale waves in the Venus atmosphere, *J. Atmos. Sci., 39*, 2397–2413.

Covey, C. C., and G. Schubert (1983), Venus atmospheric waves: A challenge for nonlinear dynamics, *Physica D, 6*, 241–248.

Crisp, D., and D. Titov (1997), The thermal balance of the Venus atmosphere, in *Venus II: Geology, Geophysics, Atmosphere, and Solar Wind Environment*, edited by S. Bougher, D. Hunten, and R. Phillips, pp. 353–384, University of Arizona Press, Tucson.

Crisp, D., V. S. Meadows, B. B'ezard, C. de Bergh, J.-P. Maillard, and F. P. Mills (1996), Ground-based near-infrared observations of the Venus nightside: 1.27-μm O_2 ($a^1 \Delta_g$) airglow from the upper atmosphere, *J. Geophys. Res., 101*, 4577–4593.

Del Genio, A. D., and W. Zhou (1996), Simulations of superrotation on slowly rotating planets: Sensitivity to rotation and initial condition, *Icarus, 120*, 332–343.

Del Genio, A. D., W. Zhou, and T. P. Eichler (1993), Equatorial superrotation in a slowly rotating GCM: Implications for Titan and Venus, *Icarus, 101*, 1–17.

Duffy, P. B., B. Govindasay, J. P. Iorio, J. Milovich, K. R. Sperber, K. E. Taylor, M. F. Wehner, and S. L. Thompson (2003), High-resolution simulations of global climate. Part 1. Present climate, *Clim. Dyn., 21*, 371–390.

Fels, S. B. (1986), An approximate analytical method for calculating tides in the atmosphere of Venus, *J. Atmos. Sci., 43*, 2757–2772.

Fels, S. B., and R. S. Lindzen (1974), The interaction of thermally excited gravity waves with mean flows, *Geophys. Fluid Dyn., 6*, 149–191.

Fox, J. L., and K. Y. Sung (2001), Solar activity variations of the Venus thermosphere/ionosphere, *J. Geophys. Res., 106*, 21,305–21,336.

Fritts, D. C., and W. Lu (1993), Spectral estimates of gravity wave energy and momentum fluxes. Part II: Parameterization of wave forcing and variability, *J. Atmos. Sci., 50*, 3695–3713.

Gerard, J. C., A. I. F. Stewart, and S. W. Bougher (1981), The altitude distribution of the Venus ultraviolet nightglow and implications on vertical transport, *Geophys. Res. Lett., 8*, 633–636.

Gierasch, P. J. (1975), Meridional circulation and the maintenance of the Venus atmospheric rotation, *J. Atmos. Sci., 32*, 1038–1044.

Gierasch, P. J., R. M. Goody, R. E. Young, D. Crisp, C. Edwards, R. Kahn, D. McCleese, D. Rider, A. Del Genio, R. Greeley, A. Hou, C. B. Leovy, and M. Newman (1997), The general circulation of the Venus atmosphere: An assessment, in *Venus II: Geology, Geophysics, Atmosphere, and Solar Wind Environment*, edited by S. Bougher, D. Hunten, and R. Phillips, pp. 459–500, University of Arizona Press, Tucson.

Gold, T., and S. Soter (1971), Atmospheric tides and the 4-day circulation on Venus, *Icarus, 14*, 16–20.

Goldstein, J. J., M. J. Mumma, T. Kostiuk, and F. Espenak (1991), A self-consistent picture of circulation in Venus' atmosphere from 70 to 200 km altitude, *Icarus, 94*, 45–63.

Gulkis, S., R. K. Kakar, M. J. Klein, , and E. Olson (1977), Detection of variations in stratospheric carbon monoxide, in Proceedings of the *Symposium on Planetary Atmospheres*, edited by A. V. Jones, pp. 61–65, Royal Soc. of Canada, Ottawa.

Gurwell, M. A., D. O. Muhleman, K. P. Shah, G. L. Berge, D. J. Rudy, and A. W. Grossman (1995), Observations of the CO bulge on Venus and implications for mesospheric winds, *Icarus, 115*, 141–158.

Hedin, A. E., H. B. Niemann, W. T. Kasprzak, and A. Seiff (1983), Global empirical model of the Venus thermosphere, *J. Geophys. Res., 88*, 73–83.

Held, I. M. (2005), The gap between simulation and understanding in climate modeling, *Bull. Amer. Meteor. Soc., 86*, 1609–1614.

Herrnstein, A., and T. E. Dowling (2007), Effects of topography on the spinup of a Venus atmospheric model, *J. Geophys. Res.*, in press.

Hines, C. O. (1997), Doppler-spread parameterization of gravity-wave momentum deposition in the middle atmosphere. 1. Basic formulation, *J. Atmos. Sol.-Terr. Phys., 59*, 371–386.

Hollingsworth, J. L., R. E. Young, G. Schubert, C. C. Covey, and A. S. Grossman (2007), A simple-physics global circulation model for Venus: Sensitivity assessments of atmospheric superrotation, *Geophys. Res. Lett., 34*, L05202, doi:10.1029/2006GL028567.

Hou, A. Y., and B. F. Farrell (1987), Superrotation induced by critical-level absorption of gravity waves on Venus: An assessment, *J. Atmos. Sci., 44*, 1049–1061.

Hou, A. Y., S. B. Fels, and R. M. Goody (1990), Zonal superrotation above Venus' cloud base induced by the semidiurnal tide and the mean meridional circulation, *J. Atmos. Sci., 47*, 1894–1901.

Hourdin, F., P. Levan, O. Talagrand, R. Courtin, D. Gautier, and C. P. McKay (1992), Numerical simulation of the circulation of the atmosphere of Titan, in *Symposium on Titan*, European Space Agency Special Publication ESA SP-338, pp. 101–106, ESA.

Hourdin, F., O. Talagrand, R. Sadourny, R. Courtin, D. Gautier, and C. P. McKay (1995), Numerical simulation of the general circulation of the atmosphere of Titan, *Icarus, 117*, 358–374.

Hourdin, F., O. Talagrand, K. Menou, R. Fournier, J. Dufresnse, D. Gautier, R. Courtin, B. Bezard, and C. P. McKay (1996), Numerical modeling of the circulation of superrotating atmospheres: Venus and Titan, in *Environment Modeling for Space-Based Applications*, European Space Agency Special Publication ESA SP-392, pp. 329–333, ESA.

Joshi, M. M., R. M. Haberle, and R. T. Reynolds (1997), Simulations of the atmospheres of synchronously rotating terrestrial planets orbiting M dwarfs: Conditions for atmospheric collapse and the implications for habitability, *Icarus, 129*, 450–465.

Kasprzak, W. T., A. E. Hedin, H. G. Mayr, and H. B. Niemann (1988), Wavelike perturbations observed in the neutral thermosphere of Venus, *J. Geophys. Res., 93*, 11,237–11,246.

Kasprzak, W. T., H. B. Niemann, A. E. Hedin, and S. W. Bougher (1993), Wave-like perturbations observed at low altitudes by the Pioneer Venus Orbiter Neutral Mass Spectrometer during Orbiter entry, *Geophys. Res. Lett., 20*, 2755–2758.

Keating, G. M., J. Y. Nicholson, and L. R. Lake (1980), Venus upper atmosphere structure, *J. Geophys. Res., 85*, 7941–7956.

Kliore, A. J. (1985), Recent results on the Venus atmosphere from Pioneer Venus radio occultations, *Adv. Space Res., 5*, 41–49.

Kossin, J. P., B. D. McNoldy, and W. H. Schubert (2002), Vortical swirls in hurricane eye clouds, *Month. Wea. Rev., 130*, 3144–3149.

Krasnopolsky, V. A. (1983), Venus spectroscopy in the 3000–8000a region by veneras 9 and 10, in *Venus*, edited by D. M. Hunten, L.

Colin, T. M. Donahue, and V. I. Moroz, pp. 681–765, University of Arizona Press, Tucson.

Kraucunas, I., and D. L. Hartmann (2005), Equatorial superrotation and the factors controlling the zonal-mean zonal winds in the tropical upper troposphere, *J. Atmos. Sci., 62*, 371–389.

Lawrence, B. N. (1997), The effect of parameterized gravity wave drag on simulations of the middle atmosphere during northern winter 1991/1992—General evolution, in *Gravity Wave Processes: Their Parameterization in Global Climate Models, NATO ASI Series*, vol. 1 50, edited by K. Hamilton, pp. 291–307, Springer-Verlag, Berlin.

Lee, C. (2006), Modelling of the atmosphere of Venus, Ph.D. thesis, Oxford University.

Lee, C., S. R. Lewis, and P. L. Read (2006), Superrotation in a Venus general circulation model, *J. Geophys. Res.*, E04S11, doi:10.1029/2006JE002874.

Lellouch, E., J. J. Goldstein, J. Rosenqvist, S. W. Bougher, and G. Paubert (1994), Global circulation, thermal structure, and carbon monoxide distribution in Venus' mesosphere in 1991, *Icarus, 110*, 315–339.

Lellouch, E., T. Clancy, D. Crisp, A. Kliore, D. Titov, and S. W. Bougher (1997), Monitoring of mesospheric structure and dynamics, in *Venus II: Geology, Geophysics, Atmosphere, and Solar Wind Environment*, edited by S. W. Bougher, D. M. Hunten, and R. J. Phillips, pp. 295–324, University of Arizona Press, Tucson.

Leovy, C. B., and N. L. Baker (1987), Zonal winds near Venus' cloud top level: A model study of the interaction between the zonal mean circulation and the semidiurnal tide, *Icarus, 69*, 202–220.

Limaye, S. S. (1985), Venus atmospheric circulation: Observations and implications of the thermal structure, *Adv. Space Res., 5*, 51–62.

Limaye, S. S. (1990), Observed cloud level circulation on Venus: Temporal variations and solar longitude dependence, in *Middle Atmosphere of Venus, Veroffentlichungen des Forschungsbereichs Geound Kosmoswissenschaften*, vol. 18, edited by K. Schäfer and D. Spänkuch, pp. 121–140, Akademie-Verlag, Berlin.

Limaye, S. S. (2007), Venus atmospheric circulation: Known and unknown, *J. Geophys. Res.*, in press.

Limaye, S. S., and V. E. Suomi (1978), Cloud motions on Venus: Global structure and organization, *J. Atmos. Sci., 38*, 1220–1235.

Lindzen, R. S. (1981), Turbulence and stress owing to gravity wave and tidal breakdown, *J. Geophys. Res., 86*, 9707–9714.

Lindzen, R. S. (1990), *Dynamics in Atmospheric Physics*, Cambridge University Press, New York, NY, 310pp.

Mayr, H. G., I. Harris, H. B. Niemann, H. C. Brinton, N. W. Spencer, H. A. Taylor, R. E. Hartle, W. R. Hoegy, and D. M. Hunten (1980), Dynamic properties of the thermosphere inferred from Pioneer Venus mass spectrometer measurements, *J. Geophys. Res., 85*, 7841–7847.

Mayr, H. G., I. Harris, W. T. Kasprzak, M. Dube, and F. Varosi (1988), Gravity waves in the upper atmosphere of Venus, *J. Geophys. Res., 93*, 11,247–11,262.

Mengel, J. G., H. G. Mayr, I. Harris, and D. R. Stevens-Rayburn (1989), Non-linear three-dimensional spectral model of the Venusian thermosphere with superrotation: II. Temperature, composition, and winds, *Planet. Space Sci., 37*, 707–722.

Nakamura, M., T. Imamura, M. Ueno, N. Iwagami, T. Satoh, S. Watanabe, M. Taguchi, Y. Takahashi, M. Suzuki, T. Abe, G. L. Hashimoto, T. Sakanoi, S. Okano, Y. Kasaba, J. Yoshida, M. Yamada, N. Ishii, T. Yamada, K. Uemizu, T. Fukuhara, and K. i. Oyama (2007), PLANET-C: Venus Climate Orbiter misson of Japan, *Planet. Space Sci.*, doi:10.1016/j.pss.2007.01.009, in press.

Newman, M., and C. B. Leovy (1992), Maintenance of strong rotational winds in Venus' middle atmosphere by thermal tides, *Science, 257*, 647–650.

Newman, M., G. Schubert, A. J. Kliore, and I. R. Patel (1984), Zonal winds in the middle atmosphere of Venus from Pioneer Venus radio occultation data, *J. Atmos. Sci., 41*, 1901–1913.

Niemann, H. B., R. E. Hartle, A. E. Hedin, W. T. Kasprzak, N. W. Spencer, D. M. Hunten, and G. R. Carignan (1979), Venus upper atmosphere neutral gas composition: First observations of the diurnal variations, *Science, 205*, 54–56.

Niemann, H. B., W. T. Kasprzak, A. E. Hedin, D. M. Hunten, and N. W. Spencer (1980), Mass spectrometric measurements of the neutral gas composition of the thermosphere and exosphere of Venus, *J. Geophys. Res., 85*, 7817–7827.

Paxton, L. H., D. E. Anderson, and A. I. F. Stewart (1985), The Pioneer Venus Orbiter Ultraviolet Spectrometer experiment : Analysis of H-Lyman-alpha data, *J. Geophys. Res., 59*, 129–132.

Paxton, L. H., D. E. Anderson, and A. I. F. Stewart (1988a), Analysis of the Pioneer Venus Ultraviolet Spectrometer Lyman-alpha data from near the subsolar region, J. *Geophys. Res., 93*, 1776–1772.

Paxton, L. H., D. E. Anderson, and A. I. F. Stewart (1988b), Correction to Analysis of the Pioneer Venus Ultraviolet Spectrometer Lyman-alpha data from near the subsolar region, *J. Geophys. Res., 93*, 11,551.

Pechman, J. B., and A. P. Ingersoll (1984), Thermal tides in the atmosphere of Venus: Comparison of model results with observations, *J. Atmos. Sci., 41*, 3290–3313.

Phillips, T. J., K. AchutaRao, D. Bader, C. Covey, C. M. Doutriaux, M. Fiorino, P. J. Gleckler, K. R. Sperber, and K. E. Taylor (2006), Coupled climate model appraisal: A benchmark for future studies, *Eos Trans. AGU, 87*, 185, 191, 192.

Rannou, R., F. Montmessin, F. Hourdin, and S. Lebonnois (2006), The latitudinal distribution of clouds on Titan, *Science, 311*, 201–205.

Roos-Serote, M., P. Drossart, T. Encrenaz, E. Lellouch, R. W. Carlson, K. H. Baines, and F. W. Taylor (1995), The thermal structure of the middle Venusian atmosphere from the Galileo/NIMS spectra, *Icarus, 114*, 300–309.

Rosenqvist, J., E. Lellouch, T. Encrenaz, and G. Paubert (1995), Global circulation in Venus' mesosphere from IRAM CO observations (1991-1994): A tribute to Jan Rosenqvist, *Bull. Amer. Astron. Soc., 27*, 26.

Rossow, W. B., and G. P. Williams (1979), Large-scale motion in the Venus stratosphere, *J. Atmos. Sci., 36*, 377–389.

Rossow, W. B., S. B. Fels, and P. H. Stone (1980), Comments on 'A three-dimensional model of dynamical processes in the Venus atmosphere', *J. Atmos. Sci., 37*, 250–252.

Rossow, W. B., A. D. Del Genio, and T. Eichler (1990), Cloud-tracked winds from Pioneer Venus OCPP images, *J. Atmos. Sci., 47*, 2053–2084.

Schäfer, K., R. Dubois, R. Haus, K. Dethloff, H. Goering, D. Oertel, H. Becker-Ross, W. Stadthaus, D. Spänkuch, V. L. Moroz, L. V. Zasova, and I. A. Matsygorin (1990), Infrared Fourier-Spectrometer experiment from Venera-15, *Adv. Space Res., 10,* 57–66.

Schloerb, F. P., S. E. Robinson, and W. M. Irvine (1980), Observations of CO in the stratosphere of Venus via its J=0-1 rotational transition, *Icarus, 43,* 121–127.

Schofield, J. T., and F. W. Taylor (1983), Measurements of the mean, solar-fixed temperature and cloud structure of the middle atmosphere of Venus, *Quart. J. R. Meterol. Soc., 109,* 57–80.

Schubert, G. (1983), General circulation and the dynamical state of the Venus atmosphere, in *Venus,* edited by D. Hunten, L. Colin, T. Donahue, and V. Moroz, pp. 681–765, University of Arizona Press, Tucson.

Schubert, G., and J. Whitehead (1969), The moving flame experiment with liquid mercury: Possible implications for the Venus atmosphere, *Science, 163,* 71–72.

Seiff, A. (1983), Thermal structure of the atmosphere of Venus, in *Venus,* edited by D. Hunten, L. Colin, T. Donahue, and V. Moroz, pp. 215–279, University of Arizona Press, Tucson.

Seiff, A., D. B. Kirk, R. E. Young, R. C. Blanchard, J. T. Findlay, G. M. Kelly, and S. C. Sommer (1980), Measurements of thermal structure and thermal contrasts in the atmosphere of Venus and related dynamical observations: Results from the four Pioneer Venus probes, *J. Geophys. Res., 85,* 7903–7933.

Shah, K. P., D. O. Muhleman, and G. L. Berge (1991), Measurement of winds in Venus' upper mesosphere based on Doppler shifts of the 2.6-nm ^{12}CO line, Icarus, 93, 96–121.

Stevens, D. E. (1983), On symmetric stability and instability of zonal mean flows near the equator, *J. Atmos. Sci., 40,* 883–893.

Stewart, A. I. F., J. C. Gerard, D. W. Rusch, and S. W. Bougher (1980), Morphology of the Venus ultraviolet night airglow, *J. Geophys. Res., 85,* 7861–7870.

Stone, P. H. (1975), The dynamics of the atmosphere of Venus, *J. Atmos. Sci., 32,* 1005–1016.

Suarez, M. J., and L. L. Takacs (1995), Documentation of the ARIES/GEOS dynamical core: Version 2, in *NASA Technical Memorandum 104606, Technical Report Series on Global Modeling and Data Assimilation,* vol. 5, p. 45, NASA, Goddard Space Flight Center, Greenbelt, MD.

Suomi, V. E., and S. S. Limaye (1978), Venus: Further evidence of vortex circulation, *Science, 201,* 1009–1011.

Taylor, F. W., R. Beer, M. T. Chahine, D. J. Diner, L. S. Elson, R. D. Haskins, D. J. McCleese, J. V. Martonchik, P. E. Reichley, S. P. Bradley, J. Delderfield, J. T. Schofield, C. B. Farmer, L. Froidevaux, J. Leung, M. T. Coffey, and J. C. Gille (1980), Structure and meteorology of the middle atmosphere of Venus: Infrared remote sensing from the Pioneer Orbiter, *J. Geophys. Res., 85,* 7963–8006.

Taylor, F. W., D. M. Hunten, and L. V. Ksanfomaliti (1983), The thermal balance of the middle and upper atmosphere of Venus, in Venus, edited by D. M. Hunten, L. Colin, T. M. Donahue, and V. I. Moroz, pp. 650–680, University of Arizona Press, Tucson.

Thompson, R. (1970), Venus' general circulation is a merry-go-round, *J. Atmos. Sci., 27,* 1107–1116.

Titov, D. V., E. Lellouch, and F. W. Taylor (2001), Venus Express: Response to ESA's call for ideas for the re-use of the Mars Express platform, *Proposal to European Space Agency,* pp. 1–74.

Titov, D. V., H. Svedhem, D. Koschny, R. Hoofs, S. Barabash, J.-L. Bertaux, P. Drossart, V. Formisano, B. Häusler, O. Korablev, W. J. Markiewicz, D. Nevejans, M. Pätzold, G. Piccioni, T. L. Zhang, D. Merrit, O. Witasse, J. Zender, A. Accomazzo, M. Sweeney, D. Trillard, M. Janvier, and A. Clochet (2006), Venus Express science planning, Planet. *Space Sci., 54,* 1279–1297, doi:10.1016/j.pss.2006.04.017.

Tobiska, W. K., T. Woods, F. Eparvier, R. Viereck, L. Floy, D. Bouwer, G. Rottman, and O. R. White (2000), The SOLAR2000 empirical solar irradiance model and forecast tool, *J. Atmos. Sol. Phys., 62,* 1233–1250.

Tomasko, M. G., L. R. Doose, P. H. Smith, and A. P. Odell (1980), Measurements of the flux of sunlight in the atmosphere of Venus, *J. Geophys. Res., 85,* 8167–8186.

Warner, C. D., and M. E. McIntyre (2001), An ultrasimple spectral parameterization for nonorographic gravity waves, *J. Atmos. Sci., 58,* 1837–1857.

Yamamoto, M., and M. Takahashi (2003a), Superrotation and equatorial waves in a T21 Venus-like AGCM, *Geophys. Res. Lett., 30,* 1449, doi:10.1029/2003GL016,924.

Yamamoto, M., and M. Takahashi (2003b), The fully developed superrotation simulated by a general circulation model of a Venus-like atmosphere, *J. Atmos. Sci., 60,* 561–574.

Yamamoto, M., and M. Takahashi (2004), Dynamics of Venus' superrotation: The eddy momentum transport processes newly found in a GCM, *Geophys. Res. Lett., 31,* L09,701, doi:10.1029/2004GL019,518.

Yamamoto, M., and M. Takahashi (2006a), Superrotation maintained by meridional circulation and waves in a Venus-like AGCM, *J. Atmos. Sci., 63,* 3296–3314.

Yamamoto, M., and M. Takahashi (2006b), Stationary and slowly propagating waves in a Venus-like AGCM: Roles of topography in Venus atmospheric dynamics, *Theor. Appl. Mech. Japan,* in press.

Young, R. E., and J. B. Pollack (1977), A three-dimensional model of dynamical processes in the Venus atmosphere, J. Atmos. Sci., 34, 1315–1351. Young, R. E., and J. B. Pollack (1980), Reply, *J. Atmos. Sci., 37,* 253–254. Young, R. E., A. P. Ingersoll, D. Crisp, L. S. Elson, R. A. Preston, R. L. Walterscheid, G. Schubert, G. S. Golitsyn, J. E. Blamont, V. N. Ivarov, R. S. Sagdeev, V. M. Linkin, and V. V. Kerzhanovich (1987), Implications of the VEGA balloon results for Venus atmospheric dynamics, *Adv. Space Res., 7,* (12)303–(12)305.

Zhang, S., S. W. Bougher, and M. J. Alexander (1996), The impact of gravity waves on the Venus thermosphere and O_2 IR nightglow, *J. Geophys. Res., 101,* 23,195–23,205.

G. Schubert, Department of Earth and Space Sciences and Institute of Geophysics and Planetary Physics, University of California, Los Angeles, CA 90095, USA (schubert@ucla.edu)

Radiation in the Atmosphere of Venus

Dmitry V. Titov[1], Mark A. Bullock[2], David Crisp[3], Nilton O. Renno[4],
Fredric W. Taylor[5], and Ljudmilla V. Zasova[6]

This chapter reviews the observations of the radiative fluxes inside and outside the Venusian atmosphere, along with the available data about the planetary energy balance and the distribution of sources and sinks of radiative energy. We also briefly address the role of the heat budget on the atmospheric temperature structure, global circulation, thermodynamics, climate and evolution. We compare the main features of radiative balance on the terrestrial planets, and provide a general description of the radiative-convective equilibrium models used to study their atmospheres. We describe the physics of the greenhouse effect as it applies to the evolution of the Venusian climate, concluding with a summary of outstanding open issues.

1. INTRODUCTION

Solar and thermal radiation play a dominant role in many chemical and dynamical processes that define the climate of Venus. In some cases the radiative effects are pushed to their extremes in ways that make Venus unique among the Earth-like planets in the Solar System. The thick cloud layer that completely covers Venus and is one of the main climate forcing factors is a product of a photochemical "factory" that forms sulfuric acid aerosols from sulfur dioxide and water vapor in the middle atmosphere. The strongly scattering clouds reflect more than 75% of incoming solar flux back to space, so that Venus absorbs less solar energy than the Earth does, even though it is only 70% as far from the Sun. Meanwhile, the cloud layer and atmospheric gases produce a powerful greenhouse effect that is responsible for maintaining globally-averaged surface temperature as high as 735 K, the highest among solar system planets. Venus is also remarkably different from the other terrestrial planets

in terms of the distribution of energy sources and sinks in the atmosphere. Half of the solar flux received by the planet is absorbed at the cloud tops (~65 km) by CO_2 and the unknown UV absorber. Only 2.6% of the solar flux incident at the top of the atmosphere reaches the surface in the global average. This solar energy deposition pattern drives an unusual global circulation in the form of retrograde zonal superrotation.

This chapter reviews the current knowledge on radiative processes in the Venus atmosphere, including their interaction with atmospheric dynamics and cloud formation, greenhouse effect, as well as their influence on climate stability and evolution. Section 2 gives a synthesis of observations of the radiative fluxes inside and outside the atmosphere. Section 3 summarizes the available data about the planetary energy balance and distribution of sources and sinks of radiative energy. Implications of the radiative heat exchange to the forcing of the global circulation, greenhouse effect, current climate and its evolution are discussed in brief in Section 4. The chapter concludes with a summary of outstanding open issues and the progress expected from the Venus Express mission.

2. RADIATION FIELD IN AND OUTSIDE THE VENUS ATMOSPHERE: A SYNTHESIS OF OBSERVATIONS

Measurements of the radiation field within and outside of the Venus atmosphere during the past several decades have produced great progress in our understanding of condi-

[1] Max Planck Institute for Solar System Research, Germany
[2] Southwest Research Institute, Boulder, Colorado, USA
[3] JPL/NASA, USA
[4] University of Michigan, USA
[5] University of Oxford, England
[6] Space Research Institute /IKI/, Moscow, Russia

Exploring Venus as a Terrestrial Planet
Geophysical Monograph Series 176
Copyright 2007 by the American Geophysical Union.
10.1029/176GM08

tions on Venus and of atmospheric processes on the planet. Remote sensing observations from ground-based telescopes and orbiter instruments describe the spectral dependence of the solar radiation reflected from the planet as well as the thermal radiation emitted to space. *In situ* measurements of scattered solar and thermal radiation from descent probes provide constraints on the vertical distribution of radiation within the atmosphere. This section gives a synthesis of the available observations.

2.1 Venus as Seen From Space

The spectrum of Venus when observed from space has two components: reflected solar radiation and infrared thermal emission. The first component dominates in the UV through the near-infrared range (0.2–4. μm) over the sunlit hemisphere, while the second one prevails at longer infrared wavelengths (4–50 μm). Figure 1 shows the mean spectrum of each of the two components together with the night side emission spectrum.

2.1.1 Reflected solar radiation and albedo. Observations of the reflected solar spectrum of Venus are summarized in Moroz (1983), Moroz et al. (1985), and Crisp and Titov (1997). A major fraction of the incoming solar flux is scattered back to space by thick sulfuric acid clouds that completely cover the planet. Estimates of the Bond albedo vary from 0.8 ± 0.02 (Tomasko et al., 1980b) to 0.76 ± 0.01 (Moroz et al., 1985). This makes Venus the most reflective planet in the Solar System. The spectral dependence of the Venus albedo is presented in Figure 2. The albedo curve shows a broad depression in the UV-blue range. At wavelengths

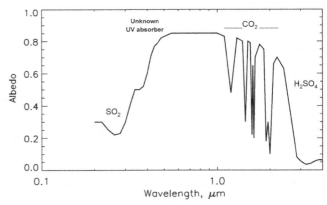

Figure 2. Spectral dependence of the Venus spherical albedo.

below 0.32 μm this absorption can be explained by the presence of SO_2 gas within and above the cloud tops. The absorption at longer wavelengths (0.32 μm–0.5 μm) is attributed to the presence of an unknown absorber in the upper cloud (58–65 km). Variability of the upper cloud structure and abundance of the absorbing species produces the well-known UV markings on the Venus disc with albedo variations of up to 30%. At near infrared wavelengths between 1 and 2 μm, sharp spectral features associated with absorption by CO_2 and H_2O within and above the cloud tops are clearly seen in the Venus spectrum. At wavelengths longer than 2.5 μm, the Venus albedo quickly drops to a few percent due to strong absorption by sulfuric acid aerosols (Fig. 2). Above 4 μm thermal radiation emitted by the cloud tops becomes comparable to the reflected solar component and its intensity rapidly increases with wavelength (Fig. 1).

2.1.2 Thermal emission from the cloud tops. Even though Venus has a very high surface temperature, from space it appears quite cold because the sulfuric acid clouds are opaque at wavelengths throughout the infrared and the temperature of the cloud tops varies between 220K and 260K. Venus thermal radiation has been measured both from the ground and from space. The most comprehensive data sets were delivered by the Pioneer Venus Orbiter Infrared Radiometer (Taylor et al., 1980), the Venera 15 Fourier spectrometer (Oertel et al., 1987), and the NIMS/Galileo infrared mapping spectrometer (Carlson and Taylor, 1993). Figure 3 shows examples of the thermal infrared spectra of Venus. The spectrum of the planet at thermal wavelengths (5 – 50 μm) is close to that of a blackbody at the cloud top temperatures with spectral features mainly belonging to CO_2, H_2O, SO_2, and other gases that absorb at levels within and above the clouds as well as the broad signature of sulfuric acid aerosols. The fundamental absorption bands of CO_2 at 4.3 μm, 4.8 μm, and 15 μm produce the strongest spectral features. These bands have been

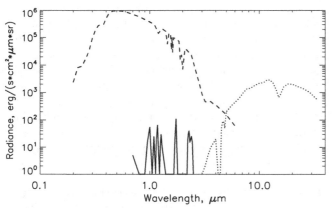

Figure 1. General view of the Venus spectrum at a resolving power, $\lambda/\Delta\lambda$, of 200 as seen from space composed of the reflected solar component (dashed line), the thermal emission from the cloud tops and mesosphere (dots), and the night side emission escaping from the lower atmosphere (solid line).

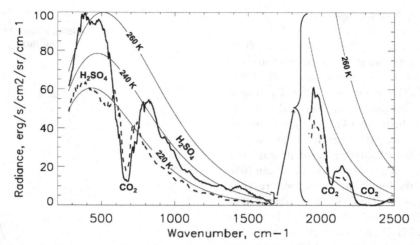

Figure 3. Examples of the Venus thermal infrared spectra measured by the FTS/Venera 15 and NIMS/Galileo experiments in the equatorial (solid) and middle (dashes) latitudes. The NIMS spectra (1900–2500 cm⁻¹) are multiplied by a factor of 100. Thin solid lines show the blackbody spectra for the temperatures of 260 K, 240 K, and 220 K.

used to retrieve the temperature and aerosol altitude profiles in the mesosphere (Roos-Serote et al., 1995; Zasova et al., 2007). The thermal structure of the Venus mesosphere varies strongly with latitude, with polar regions up to 20 K warmer than equatorial latitudes at most pressure levels between the cloud tops and 100 km. The thermal structure varies to a lesser extent with local time. Thermal infrared spectra also show a wealth of weaker gas absorption features belonging to the hot and isotopic bands of CO_2 as well as SO_2, H_2O, and CO gases above the clouds and in the upper cloud.

2.1.3 Thermal emission from the lower atmosphere. An efficient atmospheric greenhouse mechanism maintains Venus surface temperatures as high as 735 K. At this temperature, the surface emits substantial amounts of energy even in the near infrared range (1–5 µm). This emission was detected for the first time by the spectrophotometers onboard the Venera landers as an increase in the measured spectra at wavelengths above 0.8 µm. However, the importance of this emission was fully realized only after Allen and Crawford (1984) discovered that thermal radiation from the lower atmosphere leaks to space through partially transparent atmospheric "windows" – spectral gaps between strong CO_2 and H_2O absorption bands in the near infrared range (0.8–2.4 µm). At these wavelengths the H_2SO_4 clouds are transluscent: the aerosols are almost purely scattering and have little variability with wavelength. The altitude of the origin of the outgoing thermal radiation is wavelength dependent and varies from the very surface at 1 µm to ~35 km at 2.3 µm. These emissions are 4 orders of magnitude weaker than the reflected solar component (Fig. 1) so they can be observed only on the night side. Figure 4 shows examples

of the Venus night side spectra measured by NIMS/Gallileo (Carlson and Taylor, 1993) and high resolution spectra of the same spectral "windows" from the ground-based observations (Crisp, 1989; Crisp et al., 1991a,b; Bezard et al., 1990; Meadows and Crisp, 1996; Taylor et al., 1997). Images of the Venus night side taken in the spectral windows (Plate 1) revealed up to factor of 10 in brightness variations that corresponded to variations in cloud opacity between 20 and 40. The discovery of the night side thermal emission provided a powerful remote sensing tool for studying the composition of the lower atmosphere, to map the surface, and to monitor cloud opacity and atmospheric dynamics—tasks achievable only by descent probes before.

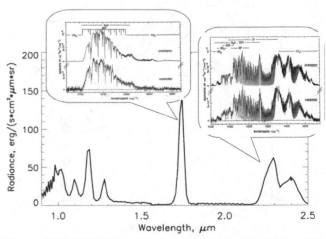

Figure 4. NIMS/Galileo near infrared spectrum of the Venus night side. Ground based high resolution spectra of the emission at 1.7 µm and 2.3 µm (from Bezard et al., 1990) are shown in the insets.

2.2 Radiation Field Inside The Atmosphere

2.2.1 Scattered solar radiation. The solar radiation field inside the atmosphere was measured by spectrophotometers on several Venera descent probes (Ekonomov et al., 1983; Moroz et al., 1983). These observations provided a unique data set for characterizing the angular and spectral distribution of solar scattered light between 0.4 and 1.2 μm from ~65 km down to the surface. Figure 5 shows examples of the Venera-13 spectra. The intensity of solar radiation propagating downwards gradually decreases with altitude. Less than 10% of the radiation that hits the top of the atmosphere reaches the surface and only about 2.5% (~17 W/m^2) is absorbed at the ground (Crisp and Titov, 1997). Increasing absorption by near-infrared CO_2 and H_2O bands as well as absorption at the blue end of the spectrum is also evident as the probe descends. Analysis of these data supported the conclusion from ground-based near-infrared observations (De Bergh et al., 1995) that the H_2O mixing ratio is nearly constant (30 ± 10 ppm) at altitudes between the cloud base and ~16 km but shows a probable increase up to 50–70 ppm in the lower scale height (Ignatiev et al., 1997).

Vertical profiles of the solar downward and upward fluxes and their divergence provide valuable information about

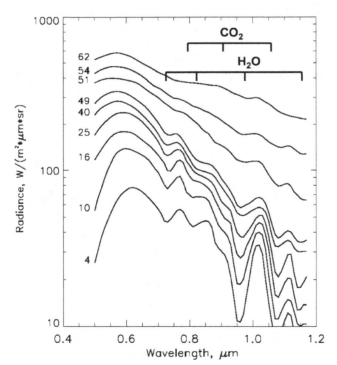

Figure 5. Spectra of the downward scattered solar radiation measured by the Venera 13 descent probe. Lines are labelled with probe altitude, in km.

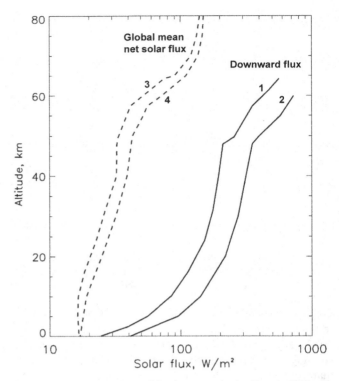

Figure 6. Vertical profiles of the downward solar flux (solid lines) and global mean net solar flux (dashed lines) in the Venus atmosphere. The downward fluxes were measured by the Pioneer-Venus LSFR experiment (curve 1) (Tomasko et al., 1980a, table 8) and spectrophotometers onboard the Venera 11, 13, 14 landers (curve 2) (Moroz et al. 1985, table 6–7). The calculated global mean net solar fluxes are from the Pioneer-Venus (curve 3) and Venus International Reference Atmosphere (curve 4) (Moroz et al., 1985).

the structure of the atmosphere and solar heating rates. The solar flux radiometer (LSFR) aboard the Pioneer Venus Large Probe measured solar fluxes during descent in a series of channels covering a broad range from 0.4 through 1.8 μm (Tomasko et al., 1980a). The Venera-13 and 14 descent probes also carried a photometric experiment to measure the distribution of UV radiation (Ekonomov et al., 1984). Figure 6 shows the vertical profiles of downward solar fluxes measured by the descent probes. Their altitude derivative is proportional to the extinction coefficient of the scattering medium. These measurements show that UV and blue radiation are strongly absorbed in the upper cloud layer above 57 km. The abrupt change in the broadband radiance at ~48 km clearly marks the location of the cloud base.

To find the global mean net solar flux, numerical models must be used to extend the spectral and spatial coverage provide by the experimental results. Figure 6 shows the results of two models (Tomasko et al., 1980a; Moroz et al., 1985). The deposition of solar energy occurs at altitudes of positive net flux divergence. The measurements and subsequent radiative transfer modeling reveal two regions where deposition of the solar energy is significant. The first one is located in the upper cloud layer above ~57 km and is possibly caused by the unknown UV absorber and by absorption beyond 3 μm by the H_2SO_4 cloud particles. The second region is located in the lower atmosphere (40 – 20 km) where solar energy is absorbed by CO_2 and H_2O in the near-infrared and possibly elemental sulfur in the UV-blue range. Surprisingly, almost no absorption of sunlight occurs between 57 and 48 km where the bulk of the cloud layer is located.

2.2.2 Thermal fluxes in the atmosphere. The vertical distribution of thermal fluxes depends on the profiles of temperature and opacity sources in the atmosphere. The vertical net thermal flux divergence defines the thermal cooling rate which, in combination with net solar heating, determines the radiative energy balance in the atmosphere. Experimental characterization of the thermal fluxes was the goal of three net flux radiometers (SNFR) on the Pioneer Venus Small Probes (Suomi et al., 1980) and an infrared radiometer (LIR) on the PV Large Probe (Boese et al., 1979). The original data were affected by a significant error whose source was discovered in laboratory testing after the mission was completed. Revercomb et al. (1985) corrected the measurements and analyzed their implications for the cloud structure, water vapor distribution in the lower atmosphere, and radiative cooling. Figure 7 shows the corrected net fluxes at four descent sites and compares them to model calculations for different water abundances below the clouds.

The measured thermal fluxes indicate the presence of an additional either gaseous or particulate source of opacity above 60 km. Below the clouds, the measurements suggest a strong increase of the net flux and its divergence from equator to pole. Because the temperature structure below the clouds varies little with latitude, this trend implies a strong latitude variability of infrared opacity sources. Since sulfur dioxide has fairly low absorption, Revercomb et al. (1985) concluded that the net flux measurements imply a significant latitude trend in the water mixing ratio below the clouds. This conclusion, however, contradicts the results of a recent re-analysis of spectrophotometry on the Venera descent probes (Ignatiev et al., 1997) and observations in the near IR windows (Crisp et al., 1991b; Drossart et al., 1993; de Bergh et al., 2006) which imply an H_2O mixing ratio of ~30 ppm without significant latitude variability below the clouds. However, the near IR observations show that there is significant spatial variability in the cloud optical depth, with persistently low optical depths at latitudes between 40 and 60 degrees (Crisp et al., 1991b). The low cloud optical depths at latitudes where the Pioneer Venus North Probe

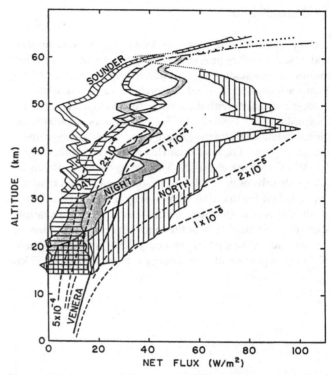

Figure 7. Net thermal fluxes in the Venus atmosphere derived from the SNFR and LIR measurements on the Pioneer Venus descent probes (from Revercomb et al., 1985). Dashed lines show the modeled profiles of the net thermal fluxes for different water vapor mixing ratio (numbers at the curves).

entered the atmosphere may explain the large thermal net flux divergences near the cloud base inferred from the net flux radiometer results (Crisp and Titov, 1997).

The vertical divergence of the measured net thermal fluxes that defines the cooling rate also shows a significant latitude gradient. Comparisons of the thermal flux divergence in Figure 7 with the divergence of solar flux (heating rate) in Figure 6 suggest strong net radiative cooling of the lower atmosphere at high latitudes. To maintain the observed thermal structure in the presence of this high latitude cooling, the net atmospheric transport must be descending there.

3. RADIATIVE ENERGY BALANCE

The measurements described in Section 2 were used to develop models of the distribution of opacity sources and temperatures in the atmosphere. These models were then used to calculate broadband radiative fluxes in and outside the atmosphere, and to assess the distribution of radiative energy sources and sinks. This section summarizes the results of this radiative balance modeling.

3.1 Global Budget

Observations by the Pioneer Venus and Venera orbiters and descent probes provided a substantial amount of information about scattering and absorbing properties of the Venus atmosphere. Several comprehensive radiative transfer models consistent with the data were developed at that time (Tomasko et al., 1980a,b; 1985). The authors calculated the global balance of radiative energy and solar heating rates in the atmosphere. The total solar flux at the Venus orbit is 2622 $\pm$ 6 W/m^2 (Moroz et al., 1985). Due to its high albedo the planet absorbs only 157 $\pm$ 6 W/m^2 on average, less than that deposited on Earth (~240 W/m^2), despite the fact that Venus is 30% closer to the Sun. Both models and observations show that less than 10% of the total solar energy incident on Venus reaches the surface, and only 2.5% is absorbed there. The largest portion of solar energy is absorbed above 57 km by the unknown UV absorber at the cloud tops. This is in contrast with the Earth, where 74% of the solar energy is absorbed directly at the ground (Arking, 1996).

The outgoing thermal radiation has been characterized by the Pioneer Venus Orbiter Infrared Radiometer and the Venera 15 Fourier transform spectrometer (Taylor et al., 1980; Oertel et al., 1987). These observations were analyzed to retrieve the temperature and aerosol structure of the Venus mesosphere and to calculate the outgoing thermal flux (Schofield and Taylor, 1982; Zasova et al., 2007). The effective globally averaged effective temperature as measured by Pioneer Venus is ~230 K which corresponds to an outgoing thermal flux of ~160 W/m^2. This value is slightly different from the mean solar flux deposited on the planet but given the uncertainties in both values this discrepancy cannot be interpreted as an indication of global energy imbalance.

Available observations and models quantify the planetary radiation budget (Tomasko et al., 1980b; Schofield and Taylor, 1982). Figure 8 compares the latitude distribution of solar radiation received by the planet and thermal radiation emitted to space for Venus and Earth. Both planets receive solar energy mainly at low latitudes with the incident flux strongly decreasing toward the poles. Earth has similar trend in latitude distribution of the outgoing thermal radiation. On Venus, however, the outgoing thermal flux is almost constant with latitude indicating that atmospheric dynamics are more efficient in transporting energy from equator to pole.

3.2 Distribution of Sources and Sinks

The vertical divergence of the net solar and thermal fluxes yields the radiative heating and cooling rates in the atmosphere. These sources and sinks of radiative energy force the atmospheric dynamics. Radiative transfer models of the Venus atmosphere have progressed together with the accumulation of observational data (Pollack et al., 1980; Tomasko, 1983). Tomasko et al. (1985) used an atmospheric radiative transfer model derived from the measurements obtained by the Pioneer Venus and Venera descent probes to calculate solar fluxes and heating rates (Figure 9).

Subsequently, remote sensing measurements by the Pioneer Venus and Venera 15 orbiters significantly improved our knowledge of the temperature and aerosol structure of the mesosphere (Schofield and Taylor, 1982; Zasova et al., 2007). This inspired extensive numerical modeling of the radiative balance in the Venus mesosphere (Crisp 1986, 1989; Haus and Goering, 1990; Titov, 1995), summarized by Crisp and Titov (1997). These studies confirmed that the radiative forcing is very sensitive to atmospheric parameters such as temperature structure, aerosol composition and distribution,

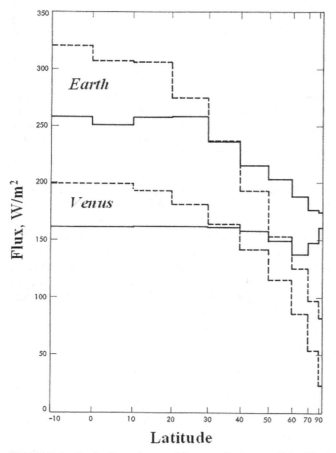

Figure 8. Latitude dependence of the zonally averaged incident solar (dashes) and outgoing thermal (solid lines) fluxes (from Schofield and Taylor, 1982). Similar curves for the Earth are also shown for comparison.

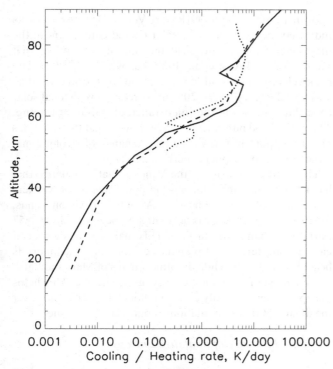

Figure 9. Vertical profiles of globally averaged solar heating (solid line - Tomasko et al., 1985; dashed line - Crisp and Titov, 1997) and thermal cooling rates (dots - Crisp and Titov, 1997).

and abundance of trace gases, all of which remain poorly constrained by available observations.

Figure 9 shows the vertical profiles of globally averaged cooling and heating rates in the Venus atmosphere. Above the clouds the radiative energy exchange occurs mainly in the CO_2 bands. Both cooling and heating rates gradually decrease from ~50 K/day near the top of the mesosphere (90 km) to few K/day at the cloud tops (70km). Absorption by the unknown UV absorber and the H_2SO_4 cloud particles above 57 km in the upper cloud creates a heating rate peak of up to 8K/day in the global average. About 10% of the solar energy incident on the top of the atmosphere or half of the solar energy absorbed by the planet is deposited in this altitude range. Radiative cooling cannot compensate for this heating at low latitudes, creating a region of net heating in the upper cloud. Below the clouds (< 48 km), both cooling and heating rates gradually decrease from 0.1–1 K/day at the cloud base to ~0.001 K/day at the surface. Figure 10 shows the altitude-latitude field of net radiative heating calculated for the mesospheric temperature structure retrieved from the Pioneer Venus OIR remote sensing data (Crisp, 1989). The observed thermal structure can be maintained in the presence of the net solar heating at low latitudes and net thermal cooling near the poles if the mesospheric circulation is characterized by rising motion at

low latitudes, poleward flow near the mesopause (~100 km) and subsidence over at high latitudes (Crisp, 1986).

Measurements of scattered solar radiation below the clouds (Moroz et al., 1983; Ekonomov et al., 1984) indicate that the atmosphere at these levels is only slightly heated by absorption of sunlight in weak near-IR CO_2 and H_2O bands and by additional UV absorption. Thermal infrared fluxes and cooling rates in the lower atmosphere are very poorly constrained by the observations and the models. Although Revercomb et al. (1985) managed to correct the Pioneer Venus thermal flux measurements, the error bars are still quite large such that these results should still be used with caution in thermal balance studies. Spatial variability of the cloud opacity, which can be as large as a factor of 2 (20 to 40), can also cause significant variations of thermal radiative forcing at levels within and below the clouds affecting the local temperatures and heat fluxes (Crisp and Titov, 1997). For example, the temperatures measured by the Vega-1 balloon were systematically by 6.5 K higher than those observed by Vega-2 at similar levels within the middle cloud region at equatorial latitudes (Linkin et al., 1986; Crisp et al., 1990). The amplitude of this zonal temperature contrast was surprising because it is almost as large as the pole to equator gradient at these levels. This phenomenon can be explained if Vega-1 balloon flew in a denser cloud that was heated more strongly as it absorbed upwelling thermal flux from the deep atmosphere (Crisp and Titov, 1997).

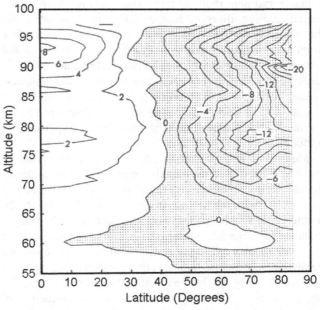

Figure 10. The altitude-latitude field of the net radiative heating (K/day) in the Venus mesosphere calculated for the PV OIR thermal structure.

4. ROLE OF RADIATION ON THE CLIMATE AND EVOLUTION OF VENUS

Radiation plays an important role on the physical and chemical processes on Venus due to high density of its atmosphere, great amount of radiatively active species, and peculiarities of the energy deposition pattern. This section gives a brief summary of these implications and compares Venus to the other terrestrial planets.

4.1 Radiative Forcing of the Atmospheric Circulation

Remote sensing by the Pioneer Venus and Venera-15 orbiters revealed significant latitudinal variations in the mesospheric temperature and cloud structure (Taylor et al., 1980; Lellouch et al., 1997; Zasova et al., 2007). The vertical structure of temperature and aerosols in the upper cloud layer (58–70 km) also vary with latitude. In particular, the aerosol scale height at the cloud tops decreased from equator to the "cold collar" at 60–70 N, with simultaneous development of a strong temperature inversion in this region (Plate 2).

Estimates of the net radiative forcing have been combined with measurements of the anomalous thermal structure to infer several interesting features of the mesospheric dynamics (Crisp, 1986, 1989; Haus and Goering, 1990; Titov, 1995; Crisp and Titov, 1997). First, radiative transfer models confirmed that the mesospheric temperature contrasts between equator and pole indicate that these regions are in strong radiative disequilibrium. Radiative energy transport would have quickly destroyed this feature; its persistence therefore implies significant meridional heat transport by the global circulation. A meridional circulation could produce adiabatic cooling in its rising branch at low latitudes and compressional heating in a descending polar branch. Crisp (1989) found that subsidence velocities of ~1cm/s were needed to produce adiabatic heating large enough to compensate for radiative cooling.

Measurements of the mesospheric temperature structure have also been used to infer the zonal circulation at theses levels (Lellouch et al., 1997; Zasova et al., 2007) and to correlate it with the net radiative heating pattern (Plate 2). If the zonal winds at these levels remain in cyclostrophic balance (e.g. the latitudinal pressure gradient is balanced by centrifugal force in the zonal circulation), the observed temperature structure indicates that the superrotating cloud top zonal winds should decay with altitude. It also suggests the presence of a strong jet stream with wind velocities of 120–140 m/s near the cloud tops at mid-latitudes (50–60 degrees), the transition region between net heating and cooling (Plate 2).

Both Pioneer Venus OIR and Venera-15 observations indicated the existence of a solar-locked component in the mesospheric temperature field, the diurnal and semi-diurnal thermal tides (Taylor et al., 1983, Zasova et al., 2007). The peak velocity of the mid-latitude, cloud-top thermal wind jet was found to vary by 10–20% in correlation with local solar time thus indicating the importance of radiative forcing. Theoretical and numerical studies showed that thermal tides play an important role in the maintenance of strong zonal winds in the mesosphere (Forbes, 2004).

The third peculiarity of the Venus radiative balance that has implications for atmospheric dynamics is the vertical distribution of radiative heating. As noted above, on Venus, most of the solar energy is deposited in the upper cloud (>57 km) rather than at the surface. This makes Venus a special case among terrestrial planets because its atmosphere is heated from the top while the atmospheres of Mars and Earth receive most of their solar energy at the surface. This helps to explain their radically different global dynamics, however the details of these circulations remain an open issue.

4.2 Global Balance of Radiative Entropy

Radiation carries not only energy but also entropy. Its flux is proportional to that of the radiative flux divided by the temperature at which energy is deposited or emitted. Since heating by solar radiation and cooling by thermal emission occur at different altitudes and temperatures, planets have non-zero (usually negative) balance of radiative entropy. This flux is balanced by the entropy production in the irreversible (dissipative) processes such as viscous and turbulent dissipation, phase transitions, and precipitation. The radiative entropy flux can be estimated from the radiative energy balance and is a measure of cumulative effect of all dissipative processes on a planet.

The radiative energy balance in the Venusian atmosphere is described in section 3. Figure 11 shows a sketch of the global mean budgets of energy and entropy on the planet. At the cloud level Venus receives ~ 400 mW/m^2/K of entropy from solar radiation and looses ~ 600 mW/m^2/K from thermal emission. A relatively small amount of energy and entropy is delivered to the lower atmosphere and the surface by the solar flux (Figure 11). The net global mean budget of radiative entropy on Venus is about -100 mW/m^2/K.

Comparison of the entropy budget on Venus with that on Earth can shed light on the differences in irreversible processes on both planets. Goody (2000) assessed their effect on Earth. The total mean flux of radiative entropy on Earth is about -60 mW/m^2/K. Viscous and turbulent dissipation, water phase transitions and precipitation were found to be the main sources of entropy on Earth (Goody 2000, Renno

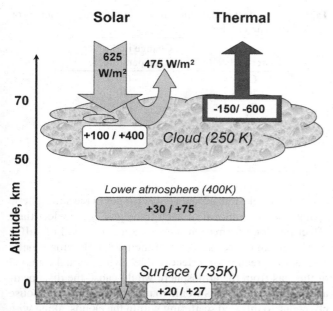

Figure 11. Globally mean budgets of energy E and entropy (S=E/T) on Venus. The figures separated by slashes are energy in W/m² and entropy in mW/m²/K.

2001). The Venus dry climate suggests negligible role of the dissipative mechanisms associated with transport, phase transitions of water and precipitation. Observations suggest rather weak turbulence within the cloud layer with eddy diffusion coefficient of 10^4–10^5 cm²/s (Kerzhanovich and Marov, 1983) and corresponding entropy production not exceeding 1 mW/m²/K. Thus, the entropy sources and sinks on Earth are quite different from those on Venus. However, the role of large scale dynamics, chemical reactions, and cloud processes on the Venusian entropy budget still needs to be studied.

Another difference between the entropy balances on Earth and Venus is the vertical distribution of the radiative sinks and sources of entropy. Earth receives solar energy and entropy mainly at the surface and emits them to space mostly from the atmosphere (Renno and Ingersoll, 1996; Renno, 1997; Goody, 2000). On Venus the sources of radiative entropy at the surface and in the lower atmosphere represent only 20% of the total balance (Figure 11). The largest sinks and sources of radiative entropy on Venus are both located in the upper cloud deck.

General principles of dynamical system when applied to the Venus climate can lead to interesting, although so far qualitative, conclusions. Nonlinear dynamics shows that the emergence of higher levels of order in dissipative systems that exchange energy and entropy with environment is typical of both hydrodynamics (e.g., Rayleigh-Benard convection) and chemistry (e.g., Zhabotinsky reaction) (Prigogine,

1980). Strong external forcing can push such systems to a stationary state that is far from thermodynamic equilibrium and which is characterized by the maximum entropy production (MEP). The MEP principle was applied to planetary atmospheres and successfully predicted such general parameters of the Earth climate as surface temperature, cloud coverage, meridional energy transport (Paltridge, 1975; Grassl, 1990) and equator-to-pole temperature contrasts (Lorenz et al., 2001). Venus gives another example of such systems. The general circulation is in steady but strongly non-equilibrium state. The super-rotating Venusian atmosphere (Gierasch et al., 1997) has rather high level of organization and, thus, low entropy maintained by energy and entropy exchange with the Sun and space. However, a detailed analysis of the Venus climate based on these principles is still awaited.

4.3 Greenhouse Effect

The planets receive solar energy in the ultraviolet, visible, and near-infrared spectral ranges, in which their atmospheres are relatively transparent. The surface and the lower atmosphere get rid of the absorbed energy by emitting radiation at infrared wavelengths. Strong absorption bands of gases and aerosols trap the heat in the lower atmosphere that results in raising the surface temperature. This process is referred to as the atmospheric *greenhouse effect*. The larger the infrared opacity, the higher surface temperature is required to balance the incoming solar flux. The difference between the actual surface temperature and the effective temperature of the planet without an atmosphere is a measure of the greenhouse effect. Its value reaches ~500 K on Venus, which is by far the largest on the terrestrial planets.

Bullock and Grinspoon (2001) used a one-dimensional, two-stream radiative model to calculate the globally averaged temperature structure of the Venus atmosphere, study its sensitivity to various model parameters, and possible evolution with time. Gaseous opacities were obtained from the correlated-k coefficients (Goody et al., 1989) for the nine radiatively active gases (Fig. 12) with line parameters taken from the HITRAN (Rothman et al., 1998) and HITEMP (Wattson and Rothman, 1992) spectral databases. CO_2 (Moskalenko et al., 1979) and H_2O continuum opacities (Liou, 1992) were included, as were Rayleigh scattering by CO_2 and N_2 (van de Hulst, 1981). Absorption and scattering in the clouds was calculated assuming spherical particles in the globally-averaged cloud model based on the Pioneer Venus measurements (Knollenberg and Hunten, 1980).

The lower atmosphere is opaque at infrared wavelengths, CO_2 being the principal source of absorption at wavelengths near 2.0, 2.7, 4.3, 4.8, 5.2, and 15 microns. Water vapor, SO_2, CO, and the H_2SO_4 clouds provide the primary sources of opacity between these bands thus strongly affecting the energy balance of the atmosphere (Plate 3). The most significant windows between CO_2 bands are between 1–1.2 μm, where only H_2O absorbs, at 1.7 μm where HCl and CO are also detectable, and between 2.2 and 2.5 μm, where absorption by CO and SO_2 may be seen. Because absorption by these species occurs between bands of the major absorber, CO_2, changes in their abundance have a large, disproportionate influence on the Venus greenhouse effect. For instance, the water vapor mixing ratio on average is only 30 ppm, but it contributes about 70 K to the greenhouse effect due to absorption bands at 1.4, 1.9, and 2.5 μm. Therefore natural variations of the atmospheric water abundance have a significant effect on the surface temperature. To a lesser extent than water, perturbations to atmospheric sulfur gas abundance can be expected to alter the efficiency of the Venus greenhouse effect.

By subtracting out one atmospheric absorber at a time and re-calculating the thermal fluxes and the equilibrium state of the atmosphere we obtain the contribution of each constituent to the greenhouse effect (Table 1).

Contributions of the atmospheric components are not additive because of strong overlap between the absorption bands. Thus, the cumulative contribution of all opacity sources in the table exceeds the observed greenhouse effect. These numerical

Table 1. The effect of the removal of infrared opacity sources on the surface temperature

Source Deleted	Change in Surface Temperature, K
CO_2	-420
Clouds	-140
H_2O	-70
OCS	-12
CO	-3
SO_2	-3
HCl	-2

experiments take into account only changes in the atmospheric opacity due to removal of each trace gas, but neglect their effect on cloud formation. For instance, removal of sulfur dioxide as an optical agent from the model only marginally affects the greenhouse effect (Table), while actual removal of this gas from the atmosphere would cause the disappearance of the clouds. This would in turn reduce the greenhouse effect due to infrared scattering within the clouds, along with a decrease in the planetary albedo. The implications of this kind of change for the surface temperature, with competing albedo and greenhouse forcing, is not obvious without detailed calculations. The surface temperature sensitivity of a fully coupled radiative/convective and cloud microphysical model to the variations of water vapor and sulfur dioxide abundance is discussed in detail by Bullock and Grinspoon (2001).

The greenhouse effect also acts on Mars and Earth, although it results in less dramatic changes in surface temperature. On both planets it is mainly due to the presence of H_2O and CO_2 in the atmosphere and reaches few degrees on Mars and 30–40K on Earth. Although the greenhouse effect on our planet is rather moderate it is obviously responsible for maintaining surface temperature above the freezing point and climate conditions comfortable for life. Without the greenhouse effect the mean temperature would fall to ~ -20C, ocean would freeze, and life would become impossible. Since the conditions on the terrestrial planets and especially the amount of CO_2 and H_2O in the atmosphere could have varied over the geological history the greenhouse effect could have played more important role in the past (see section 4.5).

4.4 Radiative-Convective Equilibria in Earth-Like Atmospheres.

Complex and poorly understood feedbacks makes planetary climate modeling a difficult endeavor. One-dimensional (1-D) radiative-convective equilibrium (RCE) models are useful for studying the atmospheric heat budget, vertical structure, and radiation-climate feedbacks. These models calculate the vertical temperature profile that results from the balance of radiative, convective, and diffusive heat fluxes. The atmospheric

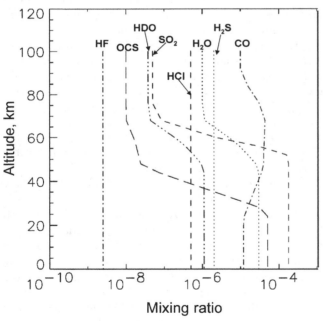

Figure 12. A model of globally averaged chemical abundances of the radiatively active gases in the Venus atmosphere based on the Pioneer Venus and the ground-based observations. N_2 is also present, as the second most abundant gas in the Venus atmosphere, with constant mixing ratio of 0.0365 (Oyama et al. 1980).

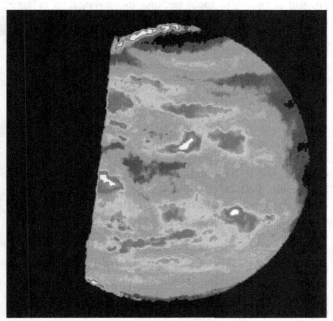

Plate 1. NIMS/Galileo image of the Venus night side taken in the 2.3 μm "window" (Carlson et al., 1993). False colours mark variations of brightness produced by spatial inhomogeneity of the total cloud opacity.

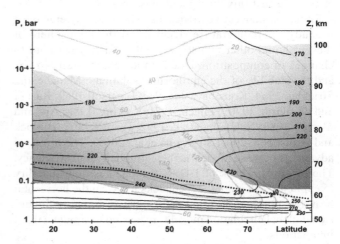

Plate 2. Latitude-altitude fields of physical parameters in the mesosphere of Venus. Black and green curves, respectively, are the isolines of atmospheric temperature (in K) and thermal wind field (in m/s) derived from the Venera-15 remote sounding experiment (Lellouch et al., 1997). Dotted line shows the cloud top altitude (Zasova et al., 2007). Pink and blue areas mark the regions with net radiative heating and cooling whose maximum/minimum values reach ±10 K/day (Crisp and Titov, 1997).

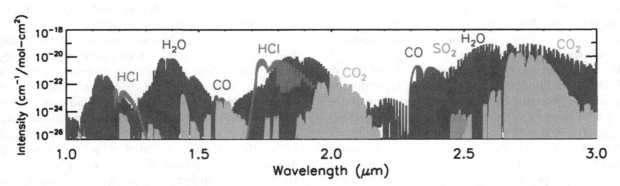

Plate 3. Spectral line intensities as a function of wavelength in microns, for CO_2 (green), H_2O (blue), CO (red), SO_2 (orange) and HCl (purple). Line intensities are from the HITRAN 2004 data base (Rothman et al., 2005) and do not reflect their weighting due to mixing ratios in the Venus atmosphere.

composition is usually assumed to be constant, except that the amount of condensates such as water vapor are assumed to depend on temperature. In general, radiative-convective models perform detailed calculations of the radiative and convective heat flux throughout the atmosphere.

Radiative-convective models have been widely employed to study the temperature structure of the atmosphere of the Earth and other planets, as well as their sensitivity to changes in composition, clouds, and solar radiation flux, and the stability of the equilibrium solutions (Manabe and Strickler, 1964; Komabayashi, 1967; Ingersoll 1969; Rasool and de Bergh, 1970; Pollack, 1971; Kasting, 1988; Abe and Matsui, 1988; Nakajima et al., 1992; Renno, 1997; Bullock and Grinspoon, 2001; Sugiyama et al., 2005). The principal advantage of these models is their simplicity and efficiency. The major drawback of 1-D RCE models is their inability to explicitly calculate the feedbacks between the horizontal heat transport and the temperature structure. Three-dimensional global circulation models are needed to simulate these interactions.

Significant efforts have been made in understanding the behavior and stability of the climate of Earth-like planets with respect to variations in incoming solar radiation. In particular, most studies have focused on the role of water vapor, a strong greenhouse gas whose atmospheric abundance is strongly dependent on temperature. Although these studies are for hypothetical planets with infinite supply of water vapor, they shed light on the climate evolution and the history of water on Earth, and on early Venus because it most likely had significant water reservoirs in the past (Donahue et al., 1997).

Simpson (1927) pointed out that atmospheres dominated by water vapor should have limited ability to emit radiation to space. If the solar forcing (i.e. solar radiation flux deposited at the surface) exceeds a critical value the atmosphere cannot balance the incoming energy and equilibrium is not possible. Komabayashi (1967), Ingersoll (1969) and Rasool and de Bergh (1970) investigated the role of solar flux on the radiative-convective equilibrium of Earth-like atmospheres with large water reservoirs. Komabayashi (1967) and Ingersoll (1969) used grey models to calculate radiative transfer and an analytic analysis of the thermodynamics of water to study the conditions for radiative-convective equilibrium. The presence of a reservoir of condensable infrared absorber like water substance makes the opacity of the atmosphere dependent on surface temperature. This produces a positive feedback in the system; increases in the surface temperature causes increases in the amount of atmospheric water vapor that further heat the surface. Komabayashi (1967) and Ingersoll (1969) independently discovered that in such atmosphere equilibrium between phases is impossible when the solar flux exceeds a critical

value of about 385 W/m^2 (the Komabayashi-Ingersoll limit). Below this value, the condensable atmospheric component can be in equilibrium with large surface reservoirs such as an ocean. Such an atmosphere can radiate only a limited amount of energy to space because its total opacity is large and only energy emitted by the relatively cold tropopause can escape to space. If the solar flux exceeds the critical value, the heat that cannot be radiated away is used to evaporate the surface reservoir and increase its concentration in the atmosphere. This state of disequilibrium is known as "the runaway greenhouse."

Although the grey radiative equilibrium models described above demonstrate the principal behavior of an atmosphere with positive water vapor feedback, they do not take into account the spectral properties of water vapor, as well as the transport of heat and water substance by atmospheric convection. Pollack (1971) developed a non-gray radiative equilibrium model and studied the influence of clouds on its solutions. Later Kasting et al. (1984) and Kasting (1988) used a non-grey radiative-convective model with moist convective adjustment to calculate the temperature and water vapor profiles as a function of the solar heat flux for an Earth-like atmosphere. Neglecting cloudiness, they estimated that the critical solar flux required to initiate the runaway greenhouse of the type envisioned by Komabayashi and Ingersoll is about 1.45 the present flux value on Earth. This corresponds to the solar flux at the orbit of Venus about 4 billion years ago, when the Sun was 30% fainter than today (Sagan and Mullen, 1972). However, Kasting et al. (1984) and Kasting (1988) suggested that a slightly subcritical Venus, with hot oceans and a wet stratosphere existed before this time. They used the term "moist greenhouse" to distinguish this steamy regime from "the runaway greenhouse".

Renno et al. (1994a, b) and Renno (1997) studied the stability of moist atmospheres with a non-gray radiative-convective model and an explicit hydrological cycle for the first time. They showed that adequate representation of the vertical transport of infrared-active volatiles such as water vapor is crucial for accurate climate simulations. Recently Sugiyama et al. (2005) examined how relative humidity affects the thermal emission of radiation to space and the existence of multiple equilibria. According to Renno (1997) the radiative-convective equilibrium conditions of moist atmospheres can be expressed in terms of four critical net solar forcing values (Figure 13). For net solar flux smaller than $F_1 \cong 275$ W/m^2 the equilibrium is unique. This case corresponds to a cold and optically thin atmosphere with small amounts of water vapor and direct emission of radiation to space by the surface. For net forcing between F_1 and $F_2 \cong 290$ W/m^2 two equilibria can occur. Sugiyama et al. (2005) attributed this bifurcation to the feedback between relative humidity and temperature. Either multiple equilibria or runaway greenhouse can occur when the net solar radiation flux is between F_2 and F_3. Moreover, a runaway greenhouse can be caused by a finite amplitude instability when the solar forcing exceeds F_2, depending on the amplitude and duration of the perturbation. Finally, a runaway greenhouse always occurs when the net solar radiation flux is above F_4 $\cong 310$ W/m^2. These critical values are much smaller than the classical value for a runaway greenhouse predicted by Komabayashi (1967) and Ingersoll (1969).

It follows from the above that there are two distinct physical mechanisms that limit the amount of radiation emitted by hypothetical planets with Earth-like atmospheres with a condensable greenhouse active agent such as water vapor. The first mechanism arises in the upper atmosphere when the temperature structure is determined by radiative equilibrium. It is referred to as the Komabayashi-Ingersoll or stratospheric limit (Renno, 1997; Sugiyama et al., 2005). The second mechanism arises in the troposphere where convection is responsible for maintaining the temperature structure (Simpson,1927; Kasting, 1988; Nakajima et al., 1992). It is referred to as the tropospheric limit. Its value is much smaller than the stratospheric limit (Figure 13). Nakajima et al. (1992) returned to a simple gray model and provided a clear interpretation of both limits and studied the conditions in which the transition between them occurs. When the atmosphere is optically thick ($\tau \gg 1$) in the thermal infrared, the emitted flux is defined by the atmospheric temperature profile at the level $\tau \sim 1$ and does not depend on surface temperature. The temperature lapse rate of an atmosphere in radiative equilibrium is larger than that of an atmosphere with moist convection. This difference in temperature lapse rate and consequently in the distribution of the temperature-dependent absorber causes the difference between the two limits. Runaway greenhouse occurs when the net solar radiation flux is above the tropospheric limit, the lower of the two. A bifurcation takes place and leads to two linearly stable climate equilibria when the net solar radiation flux is between these two other limits (Renno, 1997). This is not a surprise because existence of multiple equilibria is typical of non-linear dynamical systems.

Studies of Earth-like planets with water vapor as the major condensable greenhouse gas are relevant to the evolution of the climate of Earth and early Venus. The global mean solar flux reaching the top of Venus' atmosphere (625 W/m^2) significantly exceeds both the tropospheric and stratospheric limits and suggests that the runaway greenhouse might have played an important role of its evolution. Rasool and de Bergh (1970) argued that any Venusian surface water reservoir would have boiled early in its history. The results of Nakajima et al. (1992), Renno (1997) and Sugiyama et

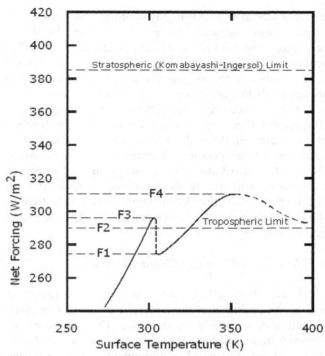

Figure 13. Dependence of the surface temperature on the solar net forcing for the moist atmosphere (after Renno (1997)). Multiple equilibria (bifurcations) occur in certain range of solar flux. Solid lines show stable and dashed lines unstable solutions.

al. (2005) also suggest that both multiple climate equilibria and a runaway greenhouse could have occurred on early Venus.

The two radiative-convective equilibria solutions found by Renno (1997) (Figure 13) can be interpreted in the framework of a convective heat engine driven by exchange of radiative heat between an energy source (the Sun) and an energy sink (space). The first solution corresponds to an optically thin atmosphere. In this regime, the climate system is in a state of weak dissipation. Renno (1997) argues that this solution corresponds to the weakest natural heat engine possible within the constraints of moist convection and a given value of the solar forcing. Solar radiation is mainly absorbed at the surface, while the bulk of the emission of radiation to space originates on the warm surface and lower troposphere. The entropy flux is minimum because energy is absorbed and emitted at relatively similar temperatures. The convective layer is shallow and the thermodynamic efficiency is small. This regime is stable to finite amplitude perturbations when increase in the atmospheric opacity leads to increase in emission of radiation to space.

The second equilibrium corresponds to an optically thick atmosphere. The convective layer is relatively deep and the thermodynamic efficiency is large. Renno (1997) argues that this second regime is highly nonlinear and is in a state of maximum dissipation. Moreover, he argues that it corresponds to the most powerful natural heat engine possible within the constraints of moist convection for a given solar forcing value. In this regime, the atmosphere's opacity to infrared radiation is the largest. The bulk of the emission of radiation to space originates in the cold upper troposphere. The entropy flux is maximum because energy is absorbed at warm temperatures in the deep atmosphere and emitted at relatively cold temperatures from the cloud tops. Knowing the spectral distribution of the energy emitted by Venus can give us insights into its climate equilibrium. Indeed, it can indicate whether Venus could have reached either the stratospheric or the tropospheric limit for a runaway greenhouse.

4.5 Greenhouse Effect and Climate Evolution

The greenhouse effect seems to have played an important role in the evolution of the atmospheres on all terrestrial planets and, in particular, on Venus. Theories about the planets formation suggest that the neighboring planets Earth, Mars and Venus received similar amounts of water during their formation (Ip and Fernandez, 1988). However, the atmosphere of Venus currently has about 100,000 times less water than the oceans and atmosphere of the Earth and is strongly enriched in deuterium. The measurements on Venus give D/H = 150 ± 30 of the terrestrial value (Donahue et al., 1997; de Bergh et al., 2006). This suggests that Venus lost most of its water at some time in the past. The efficiency of escape depends strongly on the amount of hydrogen and deuterium in the upper atmosphere. The important result of the radiative-convective models of the early Venus atmosphere is that the greenhouse mechanism efficiently forces water into the stratosphere, where solar UV photons readily dissociate H_2O molecules thus creating favorable conditions for hydrogen escape. The hydrodynamic escape is rapid enough to eliminate a large ocean in less than 600 million years (Kasting, 1988).

Current Venus climate is a system with complex feedback between radiative effects, cloud formations processes, and surface-atmosphere interactions with water vapor and sulfur dioxide playing the key role in this balance. Bullock and Grinspoon (2001) used their RCE model to study sensitivity of the Venus climate to perturbations in these species due to global resurfacing event that assumed to have happened several hundreds millions years ago (McKinnon et al. 1997). The Venus climate appears to be stable within broad range of H_2O mixing ratios. However, at about 50 times the current abundance the enhanced water greenhouse begins to warm the atmosphere, evaporating and thinning the clouds from below. This reduces the albedo, further warming the atmosphere, and further evaporating the clouds. This positive feedback destroys the massive H_2SO_4/H_2O clouds and replaces them with thin, high water-rich clouds that makes the surface temperature increase to about 920 K. Further increases in atmospheric H_2O content do not increase surface temperatures substantially, mostly because they does not enhance the greenhouse effect as much as before, and also because the high water clouds become thicker and more reflective, offsetting the greenhouse effect. Thus, the Venusian climate reaches equilibrium with surface temperatures limited to about 920 K, over a wide range of atmospheric water abundance. If either H_2O or SO_2 abundance fall below 0.5 ppm the clouds also disappear but the surface temperature stabilizes at ~700 K.

5. OPEN ISSUES AND PERSPECTIVES

Observations, numerical modeling and theoretical studies have all revealed the extremely important role that radiation plays in various processes on Venus, both now and early in its history. The large opacity of the atmosphere and the presence of great amounts of radiatively active gases and aerosols gives Venus its unique place among the terrestrial planets. The greenhouse mechanism has clearly been very effective in forming the current and early climate on Venus, while the peculiar distribution of the radiative energy sinks and sources drives the remarkable super-rotation of the entire atmosphere. Chemical interactions between different gaseous and aerosol species and the importance of non-linear feedbacks make the Venus climate a very complex system.

Previous investigations of radiation in the Venus atmosphere provided a general understanding of the distribution of fluxes, sources and sinks of radiative energy, and of the radiative forcing of the atmospheric dynamics and climate. At the same time they left a great number of unsolved problems. One of the most important open issues in this field is the variability of atmospheric properties such as the abundance of radiatively active gases, cloud microphysical and optical properties and total opacity, and the influence of these on the energy balance. The second problem concerns the radiative forcing of the atmospheric global circulation. How does the distribution of the sources and sinks of radiative energy drive the atmospheric dynamics? The thermodynamics of the Venus atmosphere is the third issue needing to be clarified and quantified. New studies of the fluxes and balance of energy and entropy, the role of dissipative processes and efficiency of the Venus heat engine will help us to understand the Venus climate-forming mechanisms. The forth open question concerns the role of radiation in the evolution of the Venus atmosphere, the early greenhouse effect and the loss of water from the planet, as well as recent climate perturbations during global resurfacing. Resolving these open issues in Venus physics would result in significant progress in comparative planetology and climatology of the terrestrial planets in general and in the study of the Earth's climate evolution in particular.

Future work will require a combination of new observations, laboratory studies, numerical modeling, and theoretical investigations. Remote sensing of the Venus mesosphere will provide the global temperature and aerosol structure, distribution and variations of the radiatively active gases, and characterize the aerosol properties and total cloud opacity with complete latitude and local time coverage. Observations of the night side in the near-IR transparency "windows" from orbit will provide access to the composition and cloud properties in the middle and lower atmosphere—the regions that had been so far reachable by descent probes only. Measurements of the reflected solar and outgoing thermal fluxes from orbit are required to quantify the details of energy and entropy budgets.

While the remote sensing observations from orbit provide global coverage, the details of the atmospheric properties and especially their vertical distribution require in-situ measurements. The first priority here is the characterization of the aerosol population including number density, particle size distribution and optical properties as well as their chemical origin. Special attention should be paid to the identification of the unknown UV absorber due to its important role in the radiative balance. Accurate measurements of the vertical profiles of the radiatively active gases and their spatial variations are important for development of the models of the Venus greenhouse. Vertical profiles of the radiative heat fluxes are poorly known and new in-situ observations are required. And finally, precise measurements of the vertical profiles of atmospheric temperatures in the deep atmosphere would constrain models of the Venus greenhouse and to determine the regions of convective stability.

The Venus lower atmosphere is an example of an extreme environment. The spectral properties of the atmospheric gases at such conditions must be measured in laboratories to support the analysis of remote and in-situ measurements and state-of-the-art radiative transfer models to study the radiative energy balance and forcing of the global circulation and climate and their sensitivity to the variation of the atmospheric optical properties.

The Venus Express spacecraft that began orbital operations on April 11, 2006 (Svedhem et al., 2007) carries a powerful suite of spectral and imaging instruments that aims at a global study of the structure, composition and dynamics of the Venus atmosphere and plasma environment. A versatile program of orbital observations includes global monitoring and close-up imaging of atmospheric phenomena, solar, stellar and Earth radio occultation to study the structure and composition of the atmosphere, and in-situ measurements of neutral atoms, plasma, and the magnetic field (Titov et al., 2006b). The mission is planned for the period of 2 Venus sols (500 days) with a possible extension for two more sols. More details about how the Venus Express mission contributes to Venus research can be found in Titov et al. (2006a) and Taylor et al. (2007). The remote sensing investigations of Venus will be continued by the Japanese Planet-C mission to be launched in 2010 (Imamura et al., 2006). This mission will focus on Venus meteorology and will carry out long-term monitoring of the dynamics of the atmosphere from orbit by a suite of five cameras exploiting different spectral ranges and windows to get access to different levels in the atmosphere.

Although it is anticipated that Venus Express and Planet-C will provide a global survey of the Venus atmosphere and circumplanetary plasma, these missions have the limitations inherent to remote sensing methods. The next logical step is in-situ studies by descent probes and balloons that will investigate the composition and microphysical and optical properties of the clouds, as well as measure the fluxes of solar and thermal radiation within the atmosphere. Although no such investigation is so far approved by the world's space agencies, the science goals, payloads and possible space platforms are well identified and are well within current technological capabilities (Crisp et al., 2002; Chassefiere et al., 2002; Baines et al., 2007).

REFERENCES

Abe, Y. and T. Matsui, Evolution of an impact-generated H_2O-CO_2 atmosphere and formation of a hot proto-ocean on Earth, *J. Atmos. Sci. 45*, 3081–3101, 1988.

Arking., A., Absorption of solar energy in the atmosphere: Discrepancy between model and observations, *Science 273*, 779–782, 1996.

Allen, D.A. and J.W. Crawford, Cloud structure on the dark side of Venus, *Nature 307*, pp.222–224, 1984.

Baines, K.H., S. Atreya, R.W. Carlson, D. Crisp, D. Grinspoon, C.T. Russel, G. Schubert, K. Zahnle, Experiencing Venus: Clues to the origin, evolution, and chemistry of terrestrial planets via in-situ exploration of our sister world, in *Exploring Venus as a Terrestrial Planet*, edited by L.W. Esposito, American Geophysical Union, Washington, DC, 2007 (this book).

Bézard, B., C. de Bergh, D. Crisp, and J.-P. Maillard, The deep atmosphere of Venus revealed by high-resolution nightside spectra, *Nature 345*, 508–511, 1990.

Boese, R.W., J.B. Pollack, and P.M. Silvaggio, First results from the large probe infrared radiometer experiment, *Science 203*, 797–800, 1979.

Bullock, M.A. and D.H. Grinspoon, The recent evolution of climate on Venus, *Icarus 150*, 19–37, 2001.

Carlson, R.W. and F.W. Taylor, The Galileo encounter with Venus: results of the near-infrared mapping spectrometer, *Planet. Space Sci. 41*, 475–476, 1993.

Carlson, R.W., L.W. Kamp, K.H. Baines, J.B. Pollack, D.H. Grinspoon, T. Encrenaz, P. Drossart, E. Lellouch, and B. Bezard, Variations in Venus cloud particle properties: A new view of Venus's cloud morphology as observed by the Galileo near-infrared mapping spectrometer, *Planet. Space Sci. 41*, 477–486, 1993.

Chassefière, E., J. J. Berthelier, J.-L. Bertaux, E. Quemèrais, J.-P. Pommereau, P. Rannou, F. Raulin, P. Coll, D. Coscia, A. Jambon, P. Sarda, J.-C. Sabroux, G. Vitter, A. L. Pichon, B. Landeau, P. Lognonné, Y. Cohen, S. Vergniole, G. Hulot, M. Mandéa, J.-F. Pineau, B. Bézard, H.-U. Keller, D. Titov, D. Breuer, K. Szego, C. Ferencz, M. Roos-Serote, O. Korablev, V. Linkin, R. Rodrigo, F. W. Taylor, and A.-M. Harri, The Lavoisier mission: A system of descent probe and balloon flotilla for geochemical investigation of deep atmosphere and surface of Venus, *Adv. Space Res. 29*, 255–264, 2002.

Crisp, D., Radiative forcing of the Venus mesosphere: I. Solar fluxes and heating rates, *Icarus 67*, 484–514, 1986.

Crisp, D., Radiative forcing of the Venus mesosphere: II. Thermal fluxes, cooling rates, and radiative equilibrium temperatures, *Icarus 77*, 391–413, 1989.

Crisp, D., A.P. Ingersoll, C.E. Hildebrand, and R.A. Preston, VEGA balloon meteorological experiments. *Adv. Space Res. 10*, 109–124, 1990.

Crisp, D., D.A. Allen, D.H. Grinspoon, and J.B. Pollack, The dark side of Venus: near-infrared images and spectra from the Anglo-Australian Observatory, *Science 253*, 1263–1266, 1991a.

Crisp, D., S. McMuldorch, S.K. Stephens, W.M. Sinton, B. Ragent, K.-W. Hodapp, R.G. Probst, L.R. Doyle, D.A. Allen, and J. Elias, Ground-based near-infrared imaging observations of Venus during the Galileo encounter, *Science 253*, 1538–1541, 1991b.

Crisp, D. and D.V. Titov, The thermal balance of the Venus atmosphere, *in Venus-II* edited by S.W. Bougher, D.M. Hunten, and R.J. Phillips, pp. 353–384, The University of Arizona Press, Tucson, Arizona, 1997.

Crisp, D., M.A. Allen, V.G. Anicich, R.E. Arvidson, S.K. Atreya, K.H. Baines, W.B. Banerdt, G.L. Bjoraker, S.W. Bougher, B.A. Campbell, R.W. Carlson, G.Chin, A. Chutjian, R.T. Clancy, B.C. Clark, T.E. Cravens, A.D. DelGenio, L.W. Esposito, B. Fegley, M. Flasar, J.L. Fox, P.J. Gierasch, R.M. Goody, D.H. Grinspoon, S.L. Gulkis, V.L. Hansen, R.R. Herrick, D.L. Huestis, D.M. Hunten, M.A. Janssen, J. Jenkins, C.L. Johnson, G.M. Keating, A.J. Kliore, S.S. Limaye, J.G. Luhmann, J.I. Lunine, P. Mahaffy, P.J. McGovern, V.S. Meadows, F.P. Mills, H.B. Niemann, T.C. Owen, K.I. Oyama, R.O. Pepin, J.J. Plaut, D.C. Reuter, M.I. Richardson, C.T. Russel, R.S. Saunders, J.T. Schofield, G. Schubert, D.A. Senske, M.K. Shepard, T.G. Slanger, S.E. Smrekar, D.J. Stevenson, D.V. Titov, E.A. Ustinov, R.E. Young, Y.L. Yung, Divergent evolution among Earth-like planets: The case for Venus exploration, The future of Solar System exploration, 2003–2013, Community contributions to the NRC Solar System Exploration Survey, ASP Conference Series, vol. 272, 5–34, 2002.

De Bergh, C., B. Bezard, D. Crisp, J.P. Maillard, T. Owen, J.B. Pollack, and D.H. Grinspoon, Water in the deep atmosphere of Venus from high-resolution spectra of the night side, *Adv. Space Res. 15*, 79–88, 1995.

De Bergh, C., V. Moroz, F.W. Taylor, D. Crisp, B. Bézard, L.V. Zasova. The composition of the atmosphere of Venus below 100 km altitude: an overview, *Planet. Space Sci. 54*, 1389–1397, 2006.

Donahue, T.M., D.H. Grinspoon,, R.E. Hartle, and R.R. Hodges, Ion/neutral escape of hydrogen and deuterium: evolution of water. *in Venus-II* edited by S.W. Bougher, D.M. Hunten, and R.J. Phillips, pp. 385–414, The University of Arizona Press, Tucson, Arizona, 1997.

Drossart, P., Bézard, B., Th. Encrenaz, E. Lellouch, M. Roos, F.W. Taylor, A.D. Collard, S.B. Calcutt, J. Pollack, D.H. Grinspoon, R.W. Carlson, K.H. Baines, and L.W. Kamp, Search for spatial variations of the H_2O abundance in the lower atmosphere of Venus from NIMS-Galileo, *Planet. Space Sci. 41*, 495–504, 1993.

Ekonomov, A.P., Yu.M. Golovin, V.I. Moroz, and B.E. Moshkin, Solar scattered radiation measurements by Venus probes, *in Venus*, edited by D.M. Hunten, L. Colin, T.M. Donahue, and V.I. Moroz, pp. 632–649, The University of Arizona Press, Tucson, Arizona,1983.

Ekonomov, A.P., V.I. Moroz, B.E. Moshkin, V.I. Gnedykh, Yu.M. Golovin, and A.V. Grigoriev, Scattered UV solar radiation within the clouds of Venus, *Nature 307*, 345–347, 1984.

Forbes, J.M., Tides in the middle and upper atmospheres of Mars and Venus, *Adv. Space Res. 33*, 125–131, 2004.

Gierasch, P., R.M. Goody, R.E. Young, D. Crisp, C. Edwards, R. Kahn, D. McCleeses, D. Rider, A. Del Genio, R. Greely, A. Hou, C.B. Leovy, and M. Newman, The general circulation of the Venus atmosphere: an assessment, *in Venus-II* edited by S.W. Bougher, D.M. Hunten, and R.J. Phillips, pp. 459–500, The University of Arizona Press, Tucson, Arizona, 1997.

Goody, R.M., R.A. West, L. Chen, and D. Crisp, The correlated-k method for radiation calculations in nonhomogeneous atmospheres, *Journal of Quantitative Spectroscopy and Radiative Transfer 42*, 539–550, 1989.

Goody, R. M., Sources and sinks of climate entropy. *Q.J.R. Meteorol. Soc. 126*, 1953–1970, 2000.

Goody, R.M., On the mechanical efficiency of deep, tropical convection. *J. Atmos. Sci. 60*, 2827–2832, 2003.

Grassl, H., The climate at maximum entropy production by meridional and atmospheric heat fluxes, *Q.J. Roy. Met. Soc. 107*, 153–166, 1990.

Haus, R. and H. Goering, Radiative energy balance in the Venus mesosphere, *Icarus 84*, 62–82, 1990.

Ignatiev, N.I., V.I. Moroz, B.E. Moshkin, A.P. Ekonomov, V.I. Gnedykh, A.V. Grigoriev, and I.V. Khatuntsev, Water vapor the lower atmosphere of Venus: a new analysis of optical spectra measured by entry probes, *Planet. Space Sci. 45*, 427–438, 1997.

Ingersoll, A.P., The runaway greenhouse: A history of water on Venus, *J. Atm. Sci. 26*, 1191–1198, 1969.

Ip, W.-H., and J.A. Fernandez, Exchange of condensed matter among the outer and terrestrial proto-planets and the effect on surface impact and atmospheric accretion, *Icarus 74*, 47–61, 1988.

Kasting, J.F., J.B. Pollack, and T.P. Ackerman, Response of Earth's atmosphere to increases in solar flux and implications for loss of water from Venus, *Icarus 57*, 335–355, 1984.

Kasting, J.F., Runaway and moist greenhouse atmospheres and the evolution of Earth and Venus, *Icarus 74*, 472–494, 1988.

Kerzhanovich, V.V. and M. Ya. Marov, The atmospheric dynamics of Venus according to Doppler measurements by the Venera entry probes, *in Venus* edited by D.M. Hunten, L. Colin, T.M. Donahue, and V.I. Moroz, pp. 766–778, The University of Arizona Press, Tucson, Arizona, 1983.

Knollenberg, R.G., and D.M. Hunten, The microphysics of the clouds of Venus: Results of the Pioneer Venus particle size spectrometer experiment, *Journal of Geophysical Research 85*, 8039–8058, 1980.

Komabayashi, M., Discrete equilibrium temperatures of a hypothetical planet with the atmosphere and the hydrosphere of one component-two pahse system under constant solar radiation, *J. Meteor. Soc. Japan 45*, 137–139, 1967.

Lellouch, E., T. Clancy, D. Crisp, A.J. Kliore, D. Titov, S.W. Bougher, Monitoring of mesospheric structure and dynamics, *in Venus-II* edited by S.W. Bougher, D.M. Hunten, and R.J. Phillips, pp. 295–324, The University of Arizona Press, Tucson, Arizona, 1997.

Linkin, V.M., V.V. Kerzhanovich, A.N. Lipatov, A.A. Shurupov, A. Sieff, B. Ragent, R.E. Young, A.P. Ingersoll, D. Crisp, L.S. Elson, R.A. Preston, J.E. Blamont, Thermal structure of the Venus atmosphere in the middle cloud layer, *Science 231*, 1420–1422, 1986.

Liou, K.N., *Radiation and Cloud Processes in the Atmosphere*, 487 pp., Oxford University Press, New York, 1992.

Lorenz, R.D., J.I. Lunine, P.G. Withers, and C.P. McCay, Titan, Mars, and Earth: Entropy production by latitudinal heat transport, *Geophys. Res. Lett. 28*, 415–418, 2001.

Manabe, S., and R. F. Strickler, Thermal equilibrium of the atmosphere with a convective adjustment. *J. Atmos. Sci. 21*, 361–385, 1964.

McKinnon, W.B., K.J. Zahnle, B.A. Ivanov, and H.J. Melosh, Cratering on Venus: Models and observations, in *Venus II*, edited by S.W. Bougher, D.M. Hunten, and R.J. Phillips, pp. 969–1014, University of Arizona Press, Tucson, 1997.

Meadows, V.S. and D. Crisp, Ground-based near-infrared observations of the Venus night side: The thermal structure and water abundance near the surface, *J. Geophys. Res. 101*, 4595–4622, 1996.

Moroz, V.I., Stellar magnitude and albedo data of Venus, *in Venus*, edited by D.M. Hunten, L. Colin, T.M. Donahue, and V.I. Moroz, pp.27–35, The University of Arizona Press, Tucson, Arizona, 1983.

Moroz, V.I., A.P. Ekonomov, Yu.M. Golovin, B.E. Moshkin, and N.F. San'ko, Solar radiation scattered in the Venus atmosphere: The Venera 11, 12 data, *Icarus 53*, 509–537, 1983.

Moroz, V.I., A.P. Ekonomov, B.E. Moshkin, H.E. Revercomb, L.A. Sromovsky, J.T. Schofield, D. Spaenkuch, F.W. Taylor, and M.G. Tomasko, Solar and thermal radiation in the Venus atmosphere, *Adv. Space Res. 5*, 197–232, 1985.

Moskalenko, N.I., Y.A. Il'in, S.N. Parzhin, and L.V. Rodinovon, Pressure-induced IR radiation absorption in atmospheres, *Bulletin of the Academy of Science USSR, Atmospheric and Oceanic Physics 15*, 632–637, 1979.

Nakajima, S., Y.–Y. Hayashi, and Y. Abe, A study on the "Runaway Greenhouse Effect" with a one-dimensional radiative-convective model. *J. Atmos. Sci. 49*, 2256–2266, 1992.

Nakamura, M. T. Imamura, M. Ueno, N. Iwagami, T. Satoh, S. Watanabe, M. Taguchi, Y. Takahashi, M. Suzuki, T. Abe, G.L. Hashimoto, T. Sakanoi, S. Okano, Y. Kasaba, J. Yoshida, M. Yamada, N. Ishii, T. Yamada, K. Uemizu, T. Fukuhara and K.-I. Oyama. Planet-C: Venus Climate Orbiter mission of Japan, *Planet. Space Sci. 55*, p. 1831, 2007.

Oertel, D., V.I. Moroz, D. Spaenkuch, V.M. Linkin, H. Jahn, V.V. Kerzhanovich, H. Becker-Ross, I.A. Matsygorin, K. Stadthaus, A.N. Lipatov, J. Nopirakowsky, A.A. Shurupov, W. Dohler, L.V. Zasova, K. Schaefer, E.A. Ustinov, J. Guldner, and R. Dubois, Infrared spectrometry from Venera -15 and Venera-16, *Adv. Space Res. 5*, 25, 1987.

Oyama, V.I., G.C. Carle, F. Woeller, J.B. Pollack, R.T. Reynolds, and R.A. Craig, Pioneer Venus gas chromatography of the lower atmosphere of Venus, *J. Geophys. Res. 85*, 7891–7902, 1980.

Paltridge, G.W. Global dynamics and climate – a system of minimum entropy exchange, *Q.J.R.Meteorol. Soc. 101*, 475–484, 1975.

Pollack, J. B., A nongray calculation of the runaway greenhouse: Implications for Venus' past and present. *Icarus 14*, 295–306, 1971.

Pollack, J.B., O.B. Toon, and R. Boese, Greenhouse models of Venus' high surface temperature, as constrained by Pioneer Venus measurements, *J. Geophys. Res. 85*, 8223–8231, 1980.

Prigogine, I., *From being to becoming: Time and complexity in the physical sciences*, Freeman and company, 1980.

Rasool, S.I. and C. de Bergh, The runaway greenhouse and the accumulation of CO_2 in the Venus atmosphere. *Nature 226*, 1037–1039, 1970.

Renno, N.O., K.A. Emanuel and P.H. Stone, A Radiative-Convective Model with an Explicit Hydrologic Cycle: 1.Formulation and Sensitivity to Model Parameters. *J. Geophys. Res. 99*, 14,429–14,441, 1994a.

Renno, N.O., P.H. Stone, and K.A. Emanuel, A Radiative-Convective Model with an Explicit Hydrological Cycle: 2.Sensitivity to Large Changes in Solar Forcing. *J. Geophys. Res. 99*, 17001–17020, 1994b.

Renno, N. O. and A. P. Ingersoll, Natural convection as a heat engine: A theory for CAPE. *J.Atmos. Sci. 53*, 572–585, 1996.

Renno, N.O., Multiple-equilibria in radiative-convective atmospheres. *Tellus 49A*, 423–438, 1997.

Renno, N. O., Comments on "Frictional dissipation in a precipitating atmosphere." *J. Atmos. Sci. 58*, 1173–1177, 2001.

Revercomb, H.E., L.A. Sromovsky, V.E. Suomi, and R.W. Boese, Net thermal radiation in the atmosphere of Venus, *Icarus 61*, 521–538, 1985.

Roos-Serote, M., P. Drossart, Th. Encrenaz, E. Lellouch, R.W. Carlson, K.H. Baines, F.W. Taylor, and S.B. Calcutt, The thermal structure and dynamics of the atmosphere of Venus between 70 and 90 km from the Galileo-NIMS spectra, *Icarus 114*, 300–309, 1995.

Rothman, L.S., C.P. Rinsland, A. Goldman, S.T. Massie, D.P. Edwards, J.M. Flaud, A. Perrin, C. Camy-Peyret, V. Dana, J.Y. Mandin, J. Schroeder, A. McCann, R.R. Gamache, R.B. Watson, K. Yoshino, K.V. Chance, K.W. Jucks, L.R. Brown, V. Nemtchinov, and R. Varanasi, The HITRAN molecular spectroscopic database and HAWKS (HITRAN Atmospheric Workstation): 1996 edition, *J. Quant. Spectr.Rad. Transfer 60*, 665–710, 1998.

Rothman, D.H., D. Jacquemart, A. Barbe, D.C. Benner, M. Birk, L.R. Brown, M.R. Carleer, C. Chackerian, K.V. Chance, L.H. Coudert, V. Dana, V.M. Devi, J.M. Flaud, R.R. Gamache, A. Goldman, J.-M. Hartmann, K.W. Jucks, A.G. Maki, J.Y. Mandin, S.T. Massie, J. Orphal, A. Perrin, C.P. Rinsland, M.A.H. Smith, J. Tennyson, R.N. Tolchenov, R.A. Toth, J.V. Auwera, P. Varanasi, and G. Wagner, The HITRAN 2004 molecular spectroscopic database, *J. Quant. Spectr. Rad. Transfer 96*, 139–204, 2005.

Sagan, C. and G. Mullen, Earth and Mars: evolution of atmosphere and surface temperatures, *Science 177*, 52–56, 1972.

Schofield, J.T. and F.W. Taylor, Net global thermal emission from the Venusian upper atmosphere, *Icarus 52*, 245–262, 1982.

Simpson, G. C., Some studies in terrestrial radiation. *Mem. Roy. Meteor. Soc. 16*, 69–95, 1927.

Sugiyama, M., P. H. Stone, and K. A. Emanuel, The role of relative humidity in radiative-convective equilibrium. *J. Atmos. Sci. 62*, 2001–2011, 2005.

Suomi, V.E., L.A. Sromovsky, and H.E. Revercomb, Net radiation in the atmosphere of Venus: measurements and interpretation, *J. of Geophys. Res. 85*, nA13, 8200–8218, 1980.

Svedhem, H., D.V. Titov, D. McCoy, J.-P. Lebreton, S. Barabash, J.-L. Bertaux, P. Drossart, V. Formisano, B. Häusler, O. Korablev, W.J. Markiewicz, D. Nevejans, M. Pätzold, G. Piccioni, T.L. Zhang, F.W. Taylor, E. Lellouch, D. Koschny, O. Witasse, M. Warhaut, A. Accommazzo, J. Rodriguez-Canabal, J. Fabrega, T. Schirmann, A. Clochet and M. Coradini, Venus Express – the first European mission to Venus. *Planet. Space Sci. 55*, 1636–1652, 2007.

Taylor, F.W., R. Beer, M.T. Chahine, D.J. Diner, L.S. Elson, R.D. Haskins, D.J. McCleese, J.V. Martonchik, P.E. Reichley, S.P. Bradley, J. Delderfield, J.T. Schofield, C.B. Farmer, L. Froidevaux, J. Leung, M.T. Coffey, and J.C. Gille, Structure and meteorology of the middle atmosphere of Venus: infrared remote sensing from the Pioneer obiter, *J. of Geophys. Res. 85*, nA13, 7963–8006, 1980.

Taylor, F.W., D.M. Hunten, and L.V. Ksanfomaliti, The thermal balance of the middle and upper atmosphere of Venus, *in Venus* edited by D.M Hunten, L. Colin, T.M. Donahue, and V.I. Moroz, pp. 650–680, The University of Arizona Press, Tucson, Arizona, 1983.

Taylor, F.W., D. Crisp, and B. Bézard, Near-infrared sounding of the lower atmosphere of Venus, *in Venus-II* edited by S.W. Bougher, D.M. Hunten, and R.J. Phillips, pp. 325–351, The University of Arizona Press, Tucson, Arizona,1997.

Taylor, F.W., H. Svedhem, and D.V. Titov, Venus Express and terrestrial planet climatology, in *Exploring Venus as a Terrestrial Planet*, edited by L.W. Esposito, American Geophysical Union, Washington, DC, 2007 (this book).

Titov, D.V., Radiative balance in the mesosphere of Venus from the Venera-15 infrared spectrometer results, *Adv. Space Res. 15*, 73–77, 1995.

Titov, D.V., H. Svedhem, and F.W. Taylor, The atmosphere of Venus: current knowledge and future investigations, *in Solar System Update* edited by Ph. Blondel and J.W. Mason, Springer-Praxis, pp.87–110, 2006a.

Titov D.V., H. Svedhem, D. Koschny, R. Hoofs, S. Barabash, J.-L. Bertaux, P. Drossart, V. Formisano, B. Häusler, O. Korablev, W.J. Markiewicz, D. Nevejans, M. Pätzold, G. Piccioni, T.L. Zhang, D. Merritt, O. Witasse, J. Zender, A. Accommazzo, M Sweeney, D. Trillard, M. Janvier, A. Clochet. Venus Express science planning, *Planet. Space Sci. 54*, 1279–1297, 2006b.

Tomasko, M.G., The thermal balance of the lower atmosphere of Venus, in *Venus*, edited by D.M. Hunten, L. Colin, T.M. Donahue, and V.I. Moroz, pp. 604–631, University of Arizona Press, Tucson, 1983.

Tomasko, M.G., L.R. Doose, P.H. Smith, and A.P. Odell, Measurements of the flux of sunlight in the atmosphere of Venus, *J. of Geophys. Res. 85*, nA13, 8167–8186, 1980a.

Tomasko, M.G., P.H. Smith, V.E. Suomi, L.A. Sromovsky, H.E. Revercomb, F.W.Taylor, D.J. Martonchik, A. Sieff, R. Boese, J.B. Pollack, A.P. Ingersoll, G. Schubert, and C.C. Covey, The thermal balance of Venus in light of the Pioneer Venus mission, *J. of Geophys. Res. 85*, nA13, 8187–8199, 1980b.

Tomasko, M.G., L.R. Doose, and P.H. Smith, The absorption of solar energy and the heating rate in the atmosphere of Venus, Adv. Space *Res. 5*, 71–79, 1985.

van de Hulst, H.C., *Light Scattering by Small Particles*, Dover, New York, 1981.

Wattson, R.B. and L.S. Rothman, Direct numerical diagonalization-Wave of the future, *J. Quant. Spectr. Rad. Trans. 48(5-6)*, 763–780, 1992.

Zasova, L.V., N.I. Ignatiev, I.A. Khatountsev, and V.M. Linkin, Structure of the Venus atmosphere, *Planet. Space Sci. 55*, 1712–1728, 2007.

Dmitry V. Titov, Max Planck Institute for Solar System Research, Max-Planck-Str 2, 37191 Katlenburg-Lindau, Germany, email: titov@mps.mpg.de

Venus Upper Atmosphere and Plasma Environment: Critical Issues for Future Exploration

C. T. Russell,[1,2] J. G. Luhmann,[3] T. E. Cravens,[4] A. F. Nagy,[5] and R. J. Strangeway[1]

This chapter briefly summarizes our state of knowledge about the upper atmosphere and plasma environment of Venus. This is followed by a discussion of some of the outstanding remaining issues in the field beginning with the Venus Express epoch and continuing to Planet C and beyond. We compare with other planets, especially Mars, and emphasize open issues. The goal is to highlight key outstanding problems rather than to be encyclopedic, and to acknowledge opposing views in order to determine where progress needs to be made.

1. INTRODUCTION

Early in the space age Venus was our most thoroughly explored planetary neighbor. While its surface was inhospitable, its proximity to both the Earth and the Sun made it the most frequent target of flyby missions, atmospheric probes and orbiters. Mariner 2 ,5 and 10, a dozen Venera probes, orbiters and landers together with the Pioneer Venus orbiter and probes successfully returned a plethora of data. These were followed by a pair of balloons deployed by Vega 1 and 2 and the Venera and Magellan radar measurements. Then the Venus program became silent and no further missions were targeted for that destination, except for a couple of outer planet missions seeking gravitational boosts on their way to Jupiter and Saturn. Only in April 2006 when Venus

[1]Institute of Geophysics and Planetary Physics and Department of Earth and Space Sciences, University of California, Los Angeles, California, USA.

[2]Dept. of Earth and Space Sciences, University of California, Los Angeles, California, USA.

[3]Space Sciences Laboratory, University of California, Berkeley, California, USA.

[4]Dept. of Physics & Astronomy, University of Kansas, Lawrence, Kansas, USA.

[5]Dept. of Atmos., Oceans and Space Sci., University of Michigan, Ann Arbor, Michigan, USA.

Exploring Venus as a Terrestrial Planet
Geophysical Monograph Series 176
Copyright 2007 by the American Geophysical Union.
10.1029/176GM09

Express was safely inserted into orbit, did Venus exploration spring back to life.

It was clear from the initial flyby of Mariner 2 that the intrinsic magnetic field of Venus was weak, but the determination of how weak remained for the long-duration Pioneer Venus Orbiter (PVO) mission. Unlike the other seven planets studied to date, an intrinsic planetary magnetic field plays no role in Venus' interaction with the solar wind. Venus is the prototypical body with a purely atmospheric/ionospheric interaction with the solar wind. Pioneer Venus arrived at solar maximum when the ionospheric pressure was high and made measurements at low altitude for only two years. Then gravitational perturbations raised the altitude of periapsis out of the Venus ionosphere, while at the same time the ionosphere weakened with the decreasing phase of the solar activity cycle and the associated EUV flux. After another 12 years had passed, for a brief period before PVO entered the atmosphere for the last time as its orbit decayed, it again measured the upper atmosphere and ionosphere at solar maximum. Thus much of our detailed understanding of the interaction is based on solar maximum in-situ measurements when the ionosphere is highly electrically conducting and extends to moderately high altitudes. This situation leads to the virtually complete exclusion of the solar wind from the ionosphere by the magnetic barrier established by electromagnetic induction in the electrically conducting ionosphere.

Our basic picture of the solar wind interaction with the atmosphere and ionosphere of Venus from the results of the PVO mission is illustrated in Figure 1. The magnetic field of the Sun and hence of the solar wind vectorially sums to

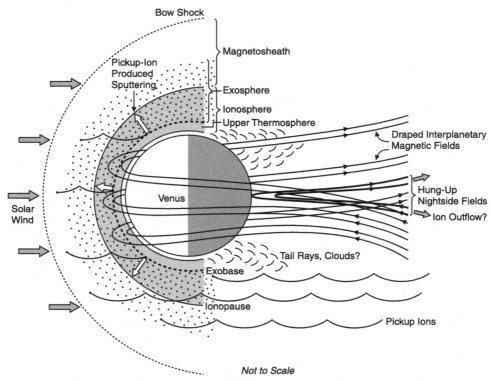

Figure 1. Schematic diagram of the solar wind interaction with Venus including the oxygen escape processes.

zero. It is outward as much as inward, upward as much as downward, and eastward as much as westward. At Venus the interplanetary magnetic field sums almost identically to zero every solar rotation (about 28 Earth days seen from Venus) but in actuality sums very close to zero in much shorter time periods. These time scales are too short for the field to diffuse into the conducting ionosphere or equivalently for the shielding currents flowing on the ionopause to resistively decay. The interplanetary magnetic field and the solar wind plasma pile up outside the ionospheric obstacle and deflect the solar wind around the planet. Because the solar wind is supersonic, a bow shock forms on the upstream side of this obstacle to its flow. The bow shock is the outermost indication that the Venus ionosphere is capable of balancing most, if not all, of the incident solar wind pressure by means of induced currents, in analogy with magnetospheric obstacles to the solar wind flow. However, the similarities between the two solar wind interactions stop at the obstacle boundary.

The Earth's magnetic field shields the Earth's atmosphere by deflecting the solar wind flow far above the Earth's atmosphere. In contrast the exosphere of Venus penetrates the solar wind because dissociative recombination of O_2^+ in the ionosphere creates a hot O corona that extends to 4000 km altitude. Figure 2 illustrates the structure of the exosphere extending far above the exobase at about 250 km altitude.

The zeroth order picture of the Venus ionosphere as a virtually impenetrable conducting obstacle is modified slightly when the solar wind dynamic pressure becomes enhanced. The boundary of the solar wind interaction, the ionopause shown in Figure 2, moves closer to Venus where the ionosphere is collisional, the currents are resistive, and the magnetic field can diffuse into the ionosphere. The solar wind can apply tangential stress to the ionosphere and scavenge plasma out of the ionosphere at the boundary between the two. The upper atmosphere also finds itself exposed to flowing solar wind. Any newly created ions in the region above the upper boundary of the ionosphere will experience the interplanetary electric field and be convected downstream with the solar wind flow. As noted above, this interaction is sensitive to the phase of the solar activity cycle, so the arrival of Venus Express at the heretofore unexplored solar minimum is very welcome.

While the exploration of Venus has lain in abeyance for almost 15 years, needed progress has been made in other areas, especially in understanding the structure and dynamics of the solar wind and its dependence on the solar activity cycle. We had glimmerings of this toward the end of the Pioneer Venus mission but the more complete modern understanding of interplanetary phenomena has given us much deeper appreciation of the variability of the solar wind interaction with Venus, especially at solar maximum. As we discuss below, we need to examine the solar wind interaction with Venus most carefully at these times. For just as many geologic features were carved in the Earth's crust by the extreme events rather than slow erosion at more normal times, so too we believe much of the atmospheric loss occurs during these infrequent but strong interactions.

Extreme solar wind conditions lead to magnetized ionospheres and magnetized ionospheres lead to magnetic fields below the ionosphere [*Luhmann*, 1991]. Under such conditions we may be able to sound the electrical conductivity of the interior of Venus, but, although the means to explore this possibility exist via balloons, or aerostats, we have yet to develop plans for such exploration.

Also in the interim, the detection of lightning from telescopes on Earth [Hansell et al., 1995] should have removed the last objection to the interpretation of purported lightning-associated radio waves, plasma waves and optical signals from orbit, (and from the surface). The importance of lightning depends on its intensity and occurrence rate. Since we have effectively established that lightning occurs, we need to determine how much is present, its intensity and location, as well as its implications.

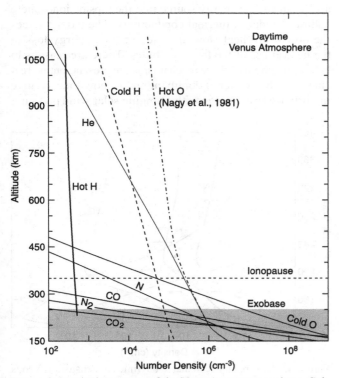

Figure 2. Vertical structure of the Venus upper atmosphere. Collisional region, below exobase, is shaded.

In the Sections that follow we examine first our current understanding of the ionosphere and what questions remain after PVO. The important potential consequences of the ionospheric properties and the solar wind interaction for the evolution of the planet are then considered the possible loss of a significant amount of water from the planet over the course of its history wherein the solar wind interaction plays a critical role. [*Donahue and Russell*, 1997] A key element in this discussion is solar variability, both the EUV flux and the eruptive disturbances in the active solar corona that can lead to enhanced atmospheric loss. We also examine briefly what we expect to learn with Venus Express. Then we examine what more we can learn about the solid body of the planet with magnetic measurements and finally we examine the controversy concerning the existence of lightning in the atmosphere of Venus.

2. THE IONOSPHERE OF VENUS: WHAT IS KNOWN AND WHAT STILL REMAINS TO BE LEARNED AT THIS TIME?

We know more about the upper atmosphere and ionosphere system of Venus than any other solar system body besides the Earth. This is mainly the result of the extensive and extended observations by the Pioneer Venus Orbiter during its 14-year lifetime. However, there remain outstanding unanswered questions of considerable importance in spite of PVO's accomplishments, and in many cases because PVO observations allow us to ask more detailed and specific ones. Here we briefly summarize what we know about the ionosphere and what issues still need to be addressed.

The major source of ionization at Venus is solar extreme ultraviolet (EUV) radiation. The photoionization rate peaks near an altitude of about 140 km above the surface of the planet. The main neutral constituent at this altitude is CO_2, with about 15% of atomic oxygen also present. Although the major initial ion is CO_2^+ the presence of atomic oxygen leads to chemical reactions making O_2^+ as the major ion at these altitudes. The main ion chemistry is shown in Figure 3, and Figure 4 shows measured and calculated ion densities. The transition from O_2^+ to O^+ as the major ion takes place near 200 km. The electron density peak is a photochemical peak whereas the peak O^+ density is analogous to the terrestrial F_2 peak.

The chemical lifetime in the Venus daytime ionosphere becomes comparable to the transport lifetime near 200 km, which is also the approximate altitude of the exobase. Measurements of the ion velocities [*Miller and Whitten*, 1991] have indicated that the horizontal velocity increases with altitude and with solar zenith angle, reaching a few km/sec at the terminator and becoming supersonic on the

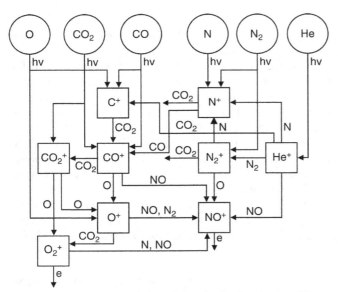

Figure 3. Block diagram of the ionospheric chemistry at Venus (from Nagy et al., 1983).

nightside. These velocities are driven mainly by day to night pressure gradients.

There is generally a sharp break in the topside ionospheric density at an altitude where thermal plasma pressure is approximately equal to the magnetic pressure. This sharp gradient in the ionospheric thermal plasma density is called the ionopause. The pressure transition is a form of tangential discontinuity in magnetohydrodynamic terminology. The sum of these two pressures is equal to the dynamic pressure of the unperturbed solar wind outside the bow shock. In the ionosphere there is a pressure gradient balancing the weight of the ionosphere above that point and the solar wind pressure. When the solar wind pressure increases, the height of the ionopause must decrease, but it levels off near 300 km when the pressure exceeds about 4×10^{-8} dynes cm^{-2} as shown in Figure 5a and collisions become important. Also, the mean ionopause height rises from about 350 km at the subsolar location to about 900 km at a solar zenith angle of 90°, as indicated in Figure 5b.

The effective night on Venus lasts about 58 Earth days, during which time the ionosphere could be expected to disappear because no new photoions and electrons are created to replace the ones lost by recombination. Therefore, initially it was a surprise when Mariner 5 found a significant nightside ionosphere at Venus [*Kliore et al.*, 1967]. Subsequently wide ranging measurements confirmed the existence of a nightside ionosphere. It is now well accepted that day to night plasma flows, along with direct electron impact ionization on the nightside are responsible for the observed densities; the relative importance of these two source mechanisms varies

with solar cycle, solar wind pressure and the given species under consideration. The nightside ionosphere varies greatly with both time and location. Order of magnitude changes have been seen by the instruments carried on PVO along a single path and/or between consecutive orbits. Terminology evolved that talked about disappearing ionospheres, ionospheric holes, tail rays, plasma clouds etc in order to attempt a classification of the different changing conditions. Figure 6 shows electron densities measured along a single orbit (PVO orbit 530), clearly showing two "ionospheric holes". Strong radial magnetic fields were found to be present in these holes, which allows easy escape of the thermal plasma into the tail region, [*Hartle and Grebowski*, 1990], providing a plausible mechanism for these observed sharp density drops. Of course the question remains why these magnetic fields are at certain locations and not others, or in other words what is the basic mechanism/reason for the appearance of these fields/holes.

The observed ion and electron temperatures at Venus are significantly higher than the neutral gas temperature and cannot be accounted for in terms of simple EUV heating and classical thermal conduction, processes controlling the behavior of the mid-latitude terrestrial ionosphere [*Cravens et al.*, 1980]. Two proposed mechanisms which lead to calculated temperatures consistent with the observed ones are: 1) an ad-hoc energy source into the topside ionosphere and/or 2) reduced thermal conductivity. The latter reduces the downward heat flow and the eventual energy loss to the neutral gas at the lower altitudes. There are reasons to believe that both of these mechanisms are present, but no real clues which if either is dominant and/or how their relative importance may change with changing solar wind pressure,

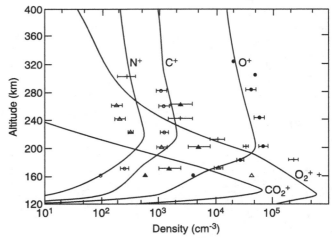

Figure 4. Measured and calculated daytime ion densities at Venus (from Nagy et al., 1980). Solid triangles denote N$^+$; open circles C$^+$ solid dots O$^+$, plus signs, O$_2^+$ and open triangles CO$_2^+$.

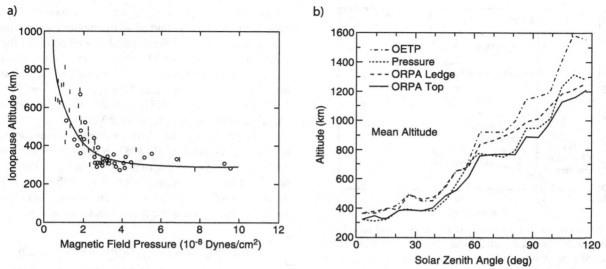

Figure 5. a) Measured variation of the ionopause altitude as a function of the magnetic pressure (from Brace and Kliore, 1991); Letters I and O indicate inbound and outbound passes respectively. b) Measured variation of the ionopause altitude as a function of the solar zenith angle (from Brace and Kliore, 1991).

solar EUV flux or other unanticipated factors. Observed small fluctuations in the magnetic field make the suggested decrease in the thermal conductivity not unreasonable, and such reduced thermal conductivity does lead to temperatures consistent with the observed ones, as shown in Figure 7 [*Cravens et al.*, 1980].

The plasma wave instrument carried by PVO did indicate the presence of wave activity at the top of the ionosphere. Estimates of the energy input into the ionosphere from this wave activity were of the order of 10^{10} eV cm^{-2}, about the right value to again lead to calculated temperatures con-

sistent with the observations, as indicated in Figure 8. The situation is similar at Mars and the same suggestions also lead to calculations in line with the observed temperatures at that planet, A couple of more recent attempts have been made, using the PVO data base, to try to elucidate the relative contribution of these two proposed explanations for the observed temperatures, but they did not lead to any definitive conclusions [*Dobe, et al.*, 1993; *Nagy et al.*, 1997]. Gan et al. (1990) took a somewhat different approach to the energetics which was applicable to the case of the magnetized Venus ionosphere, in which "organized" large-scale

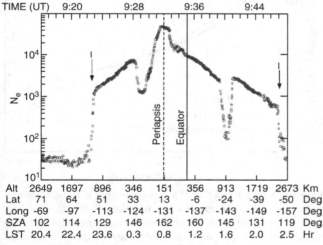

Figure 6. Electron densities measured during one orbit (530) by the Pioneer Venus Orbiter (from Brace et al., 1982).

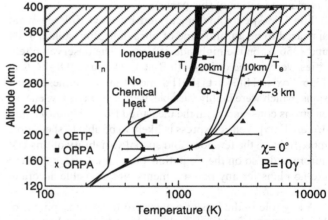

Figure 7. Measured (solid squares and triangles) calculated (solid lines) electron and ion temperatures for zero topside heat input; the influence of the fluctuating magnetic field is indicated by the various effective mean free-paths (from Cravens et al., 1980).

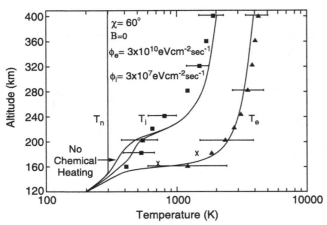

Figure 8. Measured (solid squares and triangles) and calculated (solid lines) electron and ion temperatures for zero magnetic field; the assumed heat input at the boundary is indicated in the figure (from Cravens et al., 1980).

magnetic fields are present. They solved the transport equation for superthermal electron (photoelectrons and incident solar wind electrons) fluxes along magnetic field lines and also solved the electron energy equation for T_e along the field lines. They found that the rather high temperatures for this type of ionosphere could be explained by shocked solar wind electron fluxes entering the ionosphere. However, a clear understanding of the mechanism(s) controlling the energetics of the ionospheres of Venus and Mars is still unresolved as discussed in some detail by an earlier review paper [*Nagy and Cravens*, 1997]. The energetics are important in part because they determine the height of the pressure balance ionopause, which in turn affects the efficacy of various solar wind related atmospheric escape processes (also discussed in the next section of this paper).

The UV spectrometer carried by PVO did see what appeared to be auroral type of ultraviolet emission at 1304 and 1356 A, on the nightside of Venus. The observed intensities were on the order 4–10 R at 1304 A and 0.8–1.6 R at 1356. Fox and Stewart [1991] demonstrated that these emissions, which were highly variable, were caused by low energy electrons consistent with the measured fluxes [*Knudsen and Miller,* 1985]. Venus Express is likely to be able to shed some more light on the temporal and spatial variability of this UV aurora and also on the excitation source(s). Currently there are no plans for any measurements on the visible aurora at Venus.

A key role in the loss of oxygen from the atmospheres of Venus (as well as Mars) is played by exospheric hot atoms (e.g. the hot oxygen corona). The sources of hot atoms include: (1) charge exchange of hot ionospheric O^+ ions with other neutrals, (2) sputtering of exospheric oxygen by

hot H atoms (also produced by charge exchange of hot ionospheric H^+ ions), (3) sputtering by incident solar wind ions and atmospheric ions accelerated or picked up in the solar wind, (4) hot O produced by the dissociative recombination of ionospheric O_2^+ ions. The last mechanism is the dominant hot O source at Venus and Mars [cf. *Nagy and Cravens*, 1988; *Fox and Hac*, 1997]. Figure 9 shows measured and calculated hot oxygen profiles for Venus.

As the previous discussion explained [also see ionospheric review papers Brace et al., 1983; Nagy et al., 1983; Fox and Kliore, 1997], O_2^+ is the major ion species at Venus and is produced by the reaction of CO_2^+ ions with neutral O or by the reaction of O^+ ions with CO_2. The major loss mechanism for O_2^+ is the dissociative recombination reaction:

$$
\begin{aligned}
O_2^+\left(X\,^2\Pi_g\right)+e \rightarrow O\left(^3P\right)+O\left(^3P\right)+[6.95\,eV] & \quad \{0.22\} \\
O\left(^3P\right)+O\left(^1D\right)+[4.99\,eV] & \quad \{0.42\} \\
O\left(^1D\right)+O\left(^1D\right)+[3.02\,eV] & \quad \{0.31\} \quad (1) \\
O\left(^3P\right)+O\left(^1S\right)+[2.77\,eV] & \quad \{<0.01\} \\
O\left(^1D\right)+O\left(^1S\right)+[0.80\,eV] & \quad \{0.05\}
\end{aligned}
$$

The excess energy for a given branch is shown inside the [] brackets and the measured branching ratios are indicated by the numbers in the {} brackets [*Schunk and Nagy*, 2000]. The rate coefficient decreases with increasing temperature [*Fox and Hac*, 1997]. As indicated the reaction exothermicity is 6.95 eV, for ground state atoms, which implies a maximum the speed of about 6.5×10^5 cm/s. For Venus, even this speed is

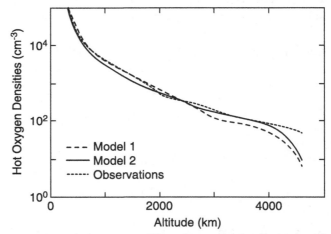

Figure 9. Vertical structure of Venus hot oxygen density. Measured and calculated hot O densities as a function of altitude. The short dashed curve indicates the observed values, the solid and long dashed curves indicate the calculated values based on two different branching ratios of the dissociative recombination reaction indicated by equation (1)) (Adapted from Nagy and Cravens, 1988)

less than the escape speed and this mechanism only produces atoms in ballistic trajectories. However, such non-escaping trajectories lead to a robust, far-reaching oxygen exosphere at Venus. A comparison with Mars is useful. At this planet, with its lower escape speed, about half of the O_2^+ recombinations taking place above the exobase produce O atoms that can directly escape. However, the Martian ionospheric densities are lower than at Venus and the hot densities are correspondingly less.

Yet loss of O from the atmosphere occurs in a number of different ways. Direct escape of O from dissociative recombination of O_2^+ is not effective but photoionization (by solar photons) or solar wind electron impact ionization of the resulting exospheric oxygen population creates pick-up O^+ ions, which are often able to escape by being "carried off" by the solar wind. For Venus, this is an important loss process. Sputtering of this exospheric population by energetic solar wind protons is another escape mechanism of potential importance. Oxygen can also be sputtered away by picked-up oxygen ions that impact the atmosphere instead of being carried down the wake. In the next section we focus on the issue of oxygen escape and its potential importance for Venus' evolutionary path.

3. AN OCEAN'S-WORTH OF OXYGEN ESCAPE?

The question of whether post-accretion and post-impact era atmosphere escape has significantly affected Venus' atmosphere and perhaps climate history is very analogous to that for Mars. It has been speculated that Venus may have had a liquid water ocean and a milder climate in the first 1–2 billion years of its history, which it lost during a runaway-greenhouse catastrophe (e.g. Kumar et al., 1983; Donahue and Hartle, 1992; Chassefiere, 1997; Kulikov et al., 2006). But the surface evidence for that hypothetical early ocean is largely hidden from us by the clouds and lack of samples that might contain hydrated minerals and other signs of crustal alteration by water. On the other hand, the atmosphere may provide us with the needed evidence for this early ocean if we can but properly interpret it.

3.1. Current Loss Rates

Current-day atmosphere escape rates on this weakly magnetized planet, though poorly constrained by observations, are estimated to be modest. These include thermal (Jeans) escape and photochemically induced nonthermal escape of neutral atoms, both sensitive to the solar EUV flux (e.g. McElroy et al., 1982; Kumar et al., 1982; Nagy and Cravens, 1988), and solar wind interaction-related mechanisms (e.g. Brace et al., 1995; Luhmann and Bauer, 1992). The latter

involve pickup or acceleration and direct escape of ionized species from the collisionless or weakly collisional regions of the upper atmosphere. Planetary ions are picked up when the atmosphere is penetrated by the interplanetary magnetic field. In addition, there is a related loss of neutrals from the region around the exobase due to collisions with some of these energized ions (sputtering, ion drag acceleration). There is also a potential for Earth-like polar wind escape of ions along draped interplanetary field lines with sufficient vertical components to support the creation of an ambipolar (charge separation) electric field (e.g. Hartle and Grebowsky, 1990), or enough divergence to enable mirror force-driven loss of ions heated at lower altitudes in nightside auroral processes. Other losses due to fluid-like shear instabilities at the boundary of the solar wind and ionospheric plasmas have also been suggested based on irregular structures routinely observed near the ionopause on PVO [e.g. see Brace et al., 1982, 1987, 1990, 1995], but these ionospheric "clouds" and "tail rays" have not been confirmed as actually escaping. Figure 1, presented earlier, summarizes the various oxygen escape processes envisioned acting today, which have presumably operated at some level throughout Venus' history.

Two major differences between the Martian and the Venus cases are the huge contrast in the (current) amount of atmosphere (~10 mb vs ~90 bar), and the planets' different proximity to the Sun (~0.7 AU vs ~1.5 AU). Mars also has lower surface gravity related to its smaller size, with an escape velocity for a particle near the surface of ~5 km/s compared to Venus' ~11 km/s. In common with Mars, the atmosphere of Venus (see Figure 8) is primarily CO_2, with a predominantly O_2^+ ionosphere, and contains very little water vapor relative to Earth. Both atmospheres lose water by photodissociation near the exobase at ~150–250 km altitude. Some of the hydrogen easily escapes by Jeans escape, charge exchange with solar wind protons, or energization by collisions with neighboring hot particles in the exosphere. [Donahue, 1999] However, it is the fate of the left-over oxygen that may hold the key to an evolutionary scenario involving a Venus ocean.

Fegley [2004] points out that if Venus once had an Earth-like inventory of water, there is a major oxygen budget problem. The runaway greenhouse phase must have turned the ocean into a massive water vapor atmosphere, making it easy for the photodissociation process occurring in the upper layers to lead to the escape of the hydrogen component [e.g. see the discussions in Chassefiere, 1997; Lammer et al, 2006]. Unless the surface took up the oxygen left behind by oxidation of the rocks and soil, Venus would have been left with much more atmospheric oxygen than is present today. Yet the surface had to have overturned vigorously or over a long time to have digested the left-over oxygen by oxidation to great depth—and there is no sign of such vigorous (or long-duration) overturning in the available images from the Magellan radar. While Venus appears to have been resurfaced at some point in the last half billion years, it is generally considered that an ocean would have disappeared before that event. Could the great bulk of the oxygen have instead escaped to space? The intriguing aspect of this hypothesis is that we observe oxygen escaping today. If we can understand how to extrapolate similar escape back in time, we could make an educated guess as to its possibly critical role in Venus' history.

As discussed in the preceding section on the ionosphere, photochemistry in the CO_2 atmospheres of Venus and Mars is responsible for their primarily O_2^+ ionospheres. This leads to the regular occurrence of a process called dissociative recombination in which pairs of suprathermal or hot O atoms are produced when O_2^+ ions come into contact with a photoelectron. This process has been described by many authors including McElroy et al. [1982], Nagy and Cravens [1988] and Fox and Hac [1997]. The process produces an extended hot O corona that forms part of the Venus exosphere, but negligible directly escaping O, in contrast to Mars (e.g. see Kim et al., 1998). For Venus oxygen, the primary route of present-day escape is via the solar wind acceleration of O^+ ions born in the hot O corona and the upper thermosphere whose vertical profile is shown in Figure 2. As mentioned above, this acceleration of planetary ions is a comet-like process involving the solar wind convection electric field, E = -VXB (where V is the local plasma velocity and B the magnetic field) that effectively penetrates the atmosphere together with the draped interplanetary magnetic fields. The "picked-up" O^+ ions born in the upper atmosphere can be either swept into the wake, to be lost to space if their accelerated velocity exceeds the escape velocity, or they can impact the exobase region where they collide with other particles and deposit their energy in the upper atmosphere. Some of the returned ions are expected to sputter additional atmospheric neutral particles from the exobase, although this response has not been observationally verified [*Luhmann and Kozyra*, 1991]. Most of these neutrals simply contribute to the trapped suprathermal component of the exosphere, but a few have enough energy to directly escape. However, even those that are added to the nonthermal corona of Venus cycle back into the pickup ion source population in a feedback loop.

The PVO plasma analyzer results suggested O^+ escape rates of ~6 x 10^{24} s^{-1} [*McComas et al.*, 1986; *Moore et al.*, 1990]. If these processes, and the photochemical losses described above, operated at their current rates for ~3.5 billion years, the estimated loss of oxygen is ~10^{42} O atoms. One can calculate the O content of an Earth-like ocean of water that would need to be lost to explain Venus' current dry state as ~10^{46} O atoms. At the current rate of escape,

the loss falls short by a factor of ~10^4. Even if the initial amount of water was less than Earth's inventory by a factor of ~0.2, as suggested by PVO measurements of the D/H ratio in Venus' atmosphere [*Donahue*, 1999], it would still take several thousand times the current loss rate. But one needs to ask whether the current rate is appropriate, and under what circumstances it could have been much greater.

3.2. Solar History and Solar Variability

One of the greatest challenges to understanding planetary atmosphere evolution is the unknown history of the Sun. Solar history is constrained mainly by observations of stellar counterparts at various ages, and from inferences derived from the irradiation history of lunar samples and meteorites. Modern observations of early Sun-like stars confirm the expectation that they are more rapidly rotating than the current Sun, and hence more active, an interpretation consistent with observations in spectral lines such as CaK that are associated with solar activity [e.g. *Ayres*, 1995]. The stars suggest that the Sun may in fact have been dimmer overall, with reduced luminosity compared to today, but brighter at the short wavelengths critical to upper atmospheric heating and ionization. Moreover, if solar maximum is any indication, the higher intensities may be averages over a series of episodic increases at much greater levels than the typically stated increases of 2–4 times at EUV wavelengths of interest would suggest.

It has been previously pointed out that a higher solar EUV flux would produce more than a proportional increase in ion production and loss through its multiple roles in both enhancing upper atmosphere densities and increasing ion production. Preliminary modeling results based on these assumptions exist for Mars for the assumption of modest (<10X) past solar EUV flux increases [*Luhmann et al.*, 1992; *Lammer et al.*, 2006]. The measurements necessary to study the relationships between solar EUV fluxes and ion pickup and related sputtering are not currently available for either Mars or Venus (although the Venus Langmuir Probe provided an EUV proxy), although they are fully within our technical capabilities. But such observations must also include the response to brief extreme excursions in EUV fluxes of orders of magnitude due to solar events such as flares. Although flares occur only a fraction of the time, they are ubiquitous during solar maximum. They also provide the possibility of investigating the loss-process consequences of EUV enhancements of orders of magnitude, which Sun-like stars indicate were a common occurrence in the early solar system.

A second less widely discussed but potentially important element of the early-Sun behavior may have been increased frequency of the coronal eruptions known as coronal mass ejections or CMEs. CMEs are the cause of almost every major high solar wind dynamic pressure episode experienced on the PVO mission (e.g. Lindsay et al., 1994,1995). An example of these CME effects on the solar wind plasma and magnetic field is shown in Figures 10a and b, together with an interpretative illustration. These interplanetary disturbances from CMEs that routinely occur around solar maximum may have been more frequent and/or powerful in the first few billion years of the Sun's life. Their impact on planetary atmospheres like that of Venus has been examined only indirectly. However, PVO observations showed that they drive the solar wind interaction effects more deeply into the Venus atmosphere (e.g. Dryer et al., 1982). High solar wind pressure phenomena include magnetized ionospheres and "disappearing" nightside ionospheres, possible indications of significantly enhanced atmosphere escape [e.g. *Cravens et al.*, 1982; *Brace et al.*, 1995]. An especially auspicious combination for enhancing escape is a combination of the flare EUV effects on ion production and the penetration of the solar wind interaction effects together. While in isolated events the flare EUV enhancements precede the arrival of the interplanetary disturbances, temporally separating each of their effects by a few days, at solar maximum flaring can occur from repeating or independent sources during the passage of an ejected coronal cloud. However, the relevant pre-Venus Express instruments on PVO and Venera could not observe the consequences of these disturbed conditions on the ion escape—either due to instrumental or orbital or solar cycle sampling limitations.

Further episodic high solar wind dynamic pressure events and interplanetary field increases are caused by the passage of solar wind stream interaction regions [e.g. *Taylor et al.*, 1985; *Lindsay et al.* 1994; *Jian et al.*, 2006]. These produce spiral-shaped pressure ridges that may survive more than one solar rotation, giving the appearance of a repeating pattern of ~day-long high dynamic pressure intervals several times a month. Figures 11a and b show plasma and field data obtained on PVO during the passage of a stream interaction region near Venus, together with an illustration of the overall geometry of such structures in the heliosphere. While the stream interaction-associated increases rarely exceed those caused by CMEs, they are still a known second cause of ionospheric magnetization at Venus at solar maximum.

A simplified calculation suggests the potential importance of an experimental investigation. Suppose that the normal escape rate due to ion pickup and its sputtering consequences is conservatively estimated from the global integration of the neutral atmosphere density above the exobase (instead of the ionopause) and the rate of photoionization (neglecting the normally smaller contributions from solar wind electron impact and charge exchange with solar wind protons). If we ignore the solar cycle changes in the upper atmosphere (e.g. Fox and Sung, 2001), and assume that under high dynamic pressure conditions, every ion produced above the exobase is either removed by pickup or has a sputtering yield of one, then the loss rates for solar minimum and maximum based on the average photoion production rates are roughly 2×10^{26} s^{-1} and 5×10^{26} s^{-1}, respectively. If these loss rates further do not change with time, the cumulative loss over 3.5 billion years is then $\sim 10^{44}$, much greater than estimated above from the available PVO measurements. As pointed out by Brannon and Fox [1994] and Brace et al. [1995], the ultimate regulators of escape to space are the processes that define the exobase composition and particle energetics. Brace et al. [1995] found that the atmosphere encountered on PVO can supply more than the necessary flux of oxygen ions to the exobase to feed these estimated greater loss rates. However, to lose the Earth-like ocean's oxygen content would require at least 100 times more loss.

The nature of solar EUV flux evolution may easily account for this factor. If we propose that the most significantly enhanced EUV flux prevailed in the first ~1.5 billion years of the period of present-day loss process domination, the O loss rate would have to be ~500 times greater than estimated here. In a simplified scenario in which solar EUV intensity, I, was about 65 times greater than present, simplified photochemistry suggests the ionosphere density was about sqrt(I) or 8 times greater. If the ionosphere enhancement of 8 times leads to a hot O corona enhancement of ~8 times, then the O^{+} production rate in the corona (assuming it is optically thin) is ~65 x 8 or 520 times the PVO production. But is a factor of 65 in EUV flux over 1.5 billion years of solar history starting ~ 3.5 billion years ago a reasonable assumption? Ayres [1997] gives a flux range of only ~5–10 for the period 3.5–2.0 billion years ago based on observations of Sun-like stars.

Our understanding of the EUV fluxes of solar type stars is rapidly increasing thanks to missions such as FUSE and Chandra, as well as improved ground-based observations. Higher time and spatial resolution measurements are suggesting that these stars exhibit sporadic activity outbursts, rather than (or in addition to) continuously enhanced EUV and X-ray outputs. Ayres (1997) also notes that because early Sun-like stars may lack organized magnetic cycles, they can produce higher time-averaged photoionization rates than stars with well-behaved activity cycles like the Sun's. Consider the effect on the average EUV intensity of episodic events such as flares instead of a steady enhanced flux. The integration of episodic events occupying a small fraction of the time may make little contribution to the overall average but dominate the escape. For example, assume, as in the paragraph above, that the production rate of the main ionosphere goes as the square root of the EUV intensity, and that the resulting O coronal density produced from the ionosphere has a similar EUV dependence. If in addition we conservatively assume the only source of the pickup O^+ production is the photoionization of the hot O exosphere, the production of ions within the reservoir above the exobase goes as the EUV intensity to the 1.5 power. As noted above, a 65-fold enhancement in EUV flux over 3.5 billion years would be needed to create the necessary O^+ reservoir for escape. However, if we instead assume a 1000-fold EUV enhancement over present-day fluxes (not unusual for flares) ~10% of the time during the first 1.5 billion years, we can get the same total escape.

This effect can be illustrated by simply integrating the escape flux assuming proportionality to $I^{1.5}$, for the two hypothetical EUV intensity time histories in Figure 12. In one case a steady EUV intensity enhancement of 16 is maintained, and in the other case an enhancement of 240 is assumed to last 1/15 of the overall time period shown. While the average EUV fluxes for these histories is nearly the same, the nonlinearly-related ion production rate is much greater for the case with the brief large enhancement. Note that CMEs or solar wind stream interactions occurring during the same period , either related or unrelated to the flaring region(s), must be sufficiently frequent to maintain (for much of the time) the upper ionospheric magnetization needed for ion pickup down to exobase levels through their associated solar wind pressure enhancements. Charge exchange and solar wind electron impact-related additional enhancements of ionization also accompany these periods of enhanced solar wind pressure due to the density and sometimes temperature increases in the compressed or shocked solar wind plasma. Thus ion production in the ion escape reservoir above the exobase at active times is greater than for undisturbed times even in the absence of ongoing flare EUV enhancements.

Of course, in addition to uncertainties about the solar history, there are other caveats that must be considered in any historical extrapolation model. We note that the above speculations do not account for possible atmospheric chemistry differences in the past, and neglect likely significant contributions of accompanying thermospheric density changes – which may in fact serve to increase the O reservoir for pickup ion production in a higher solar EUV era. It has been suggested that mass loading of the solar wind plasma by the production of pickup ions will limit the escape rate by diminishing the resulting pickup ion velocities. However, as long as the average ambient plasma flow allows pickup velocities of >11 km/s to be attained, Venus gravity will not be sufficient to retain a particle with an appropriate outward trajectory. For ion pickup, the background flow velocity (in the plasma carrying the penetrated interplanetary magnetic field) can be even less because the ion can gain up to twice the background flow speed. Thus ~5.5 km/s would suffice if the picked up particle then left the influence of Venus' gravitational potential well (perhaps as a neutral from a charge exchange reaction in the ionosphere or exosphere). Note that 5 km/s flows are the order of the observed bulk thermal ion velocity in the thermosphere at the terminator observed on PVO [*Knudsen and Miller*, 1985] and attributed to antisolar ionospheric thermal plasma pressure gradient-driven flow. The key is that the ion needs to be produced in the essentially collisionless or weakly collisional region of the atmosphere near or above the exobase, and that region has to be magnetized, at least in places, with a component perpendicular to a locally flowing plasma (either solar wind or ionospheric) moving at least 5.5 km/s in the antisolar direction. Any EUV-related enhancements of either neutral atmosphere or ion production in this region increases escape, at least in principle. While ion acceleration by convection electric fields can also occur in magnetized collisional regions of the deeper atmosphere, any energy gained by the ion will be shared with neighboring particles—and probably diluted to below that needed to escape (e.g. an ion drag process instead of pickup). This latter scenario may be prevalent at times of current solar minimum activity. Although the ionosphere at solar minimum is expected to be magnetized most of the time, the low EUV fluxes will limit the escape rates.

3.3. Prospects for Progress

Venera first detected the escaping flux of planetary ions in the Venus wake [e.g. *Vaisberg et al.*, 1995], with PVO later improving on those observations [*Mihalov and Barnes*, 1982; *McComas et al.*, 1986; *Moore et al.*, 1990; *Kasprzak et al.*, 1991; *Mihalov et al.*, 1995; *Luhmann et al.*, 2006]. Global loss rates could not be measured due to the limitations of both sets of instruments and of the spacecraft orbital sampling. Nevertheless estimates supported by these observations greatly exceed photochemical loss rates derived from fitting altitude profiles of the hot O corona obtained with the PVO UV Spectrometer measurements [*Nagy et al.*, 1981], and from calculations based on the observed ionospheric O_2^+ source of the hot O [*Nagy and Cravens*, 1988]. Ion mass spectrometer and supporting measurements on Venus Express have the potential to much better constrain the O^+ (and other species) loss rates, as well as the associated atmospheric sputtering. But it will not be sufficient to simply measure these fluxes at any time, as they can give a false impression of the influence of these processes. On time scales ranging from the solar flare time scale of a few minutes, to the passage of an interplanetary coronal mass ejection or solar wind stream interaction region taking hours to days, to the monthly to decadal cycles of solar activity, one can expect to find significant variations that must be characterized to extrapolate related atmospheric oxygen losses into the past.

The Aspera-4 experiment on Venus Express [*Barabash et al.*, 2006] coupled with the magnetometer [*Zhang et al.*, 2006] is capable of measuring the local solar wind plasma and interplanetary field around apoapsis with sufficient detail to identify and characterize a passing CME or stream interaction region-associated disturbance (though such apoapsis observations are not routinely obtained at the time of writing). Interpretations of the observations from these investigations must consider that a flare on the Sun creates an impulse of energetic photons and associated atmospheric ionization that occurs within minutes of the solar event and then as rapidly fades. It may also be followed in a few tens of minutes by additionally ionizing solar energetic electrons and ions. But a CME disturbance is a relatively slow ejection of coronal material that plows through the ambient ~300–800 km/s solar wind at speeds of several hundred km/s up to 2500 km/s. This disturbance reaches Venus hours to several days after its eruption in the solar corona. It is typically observed in-situ as a sequence that begins with an interplanetary shock and associated increases in solar wind plasma density, velocity, and temperature that last a fraction of a day. Figure 10 shows an example of the interplanetary conditions accompanying a CME disturbance observed on PVO, together with an illustration of the interpretation. The initial period of dynamic pressure enhancement is followed by a longer period of often enhanced interplanetary magnetic field magnitude during which the field sometimes exhibits a slow rotation including unusual out-of-ecliptic orientations. The overall duration of the passing event is several days. At solar maximum activity levels, Earth experiences a few CME disturbances per month.

Many more CMEs occur on the Sun, but their interplanetary consequences have finite extent and may miss Venus' location in the solar system. Nevertheless, PVO observations showed that during the period around solar maximum, the chance of encountering a CME-related disturbance interacting with Venus is comparable [*Lindsay et al.*, 1994; *Mulligan et al.*, 1998; *Jian et al.*, 2006]. Similarly, Figure 11 illustrates the solar wind stream structure and its appearance in PVO observations. An aim of Venus Express observations should be to observe the solar wind interaction response to these events both at periapsis and in the wake region, including confirmation of any ionospheric enhancement or magnetization, and to measure the pickup ion response. The longer time series of solar wind behavior in Figure 13, from a year (1980) just prior to the solar maximum experienced on PVO, indicates there are numerous opportunities to catch the enhanced solar wind pressure periods and characterize their observable effects on oxygen escape.

ASPERA-4 and the Venus Express magnetometer investigators can analyze the atmospheric and ionospheric consequences of solar and interplanetary events in ways not possible on PVO. In particular, ASPERA-IMS can measure the pickup ion fluxes during these periods to see if they significantly increase as well as their energy spectrum, angular distribution, and composition. Together with the magnetometer observations, these measurements can establish whether the combination of a magnetized upper Venus atmosphere and high EUV flux during solar wind disturbance passage is a recipe for enhanced ion pickup and escape. The ASPERA-ELS can determine whether increased photoionization and/or electron impact ionization are occurring at the same time, while the magnitude of any EUV increase can be quantified from its

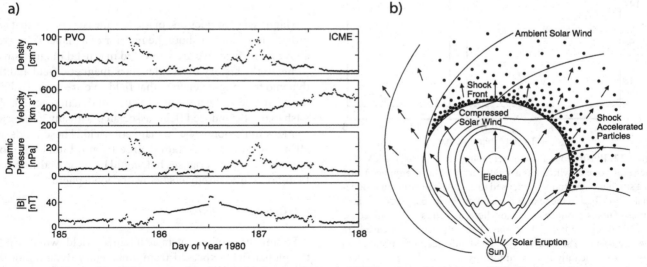

Figure 10. a) Time series of solar wind behavior during an ICME passage, b) Cartoon of ICME propagation.

a)   b)

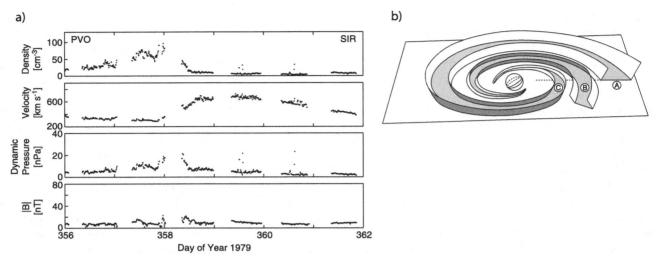

Figure 11. a) Time series of the variation of the solar wind and interplanetary magnetic field during a stream interaction, b) Cartoon of stream interaction.

photoelectron measurements. Evidence of additional charge exchange ion production can be inferred from the ASPERA-NPD observations. The ~250 km prime mission periapsis is, however, marginal for observing the magnetized upper ionosphere, although adjusted periapsis phases have been discussed for the extended mission. The new measurements can provide an observational basis for better sorting out effects

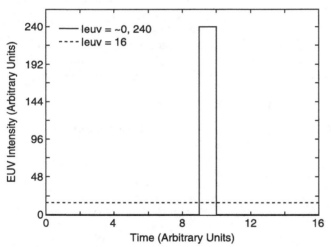

Figure 12. Illustration of two different hypothetical solar EUV flux time series, one of which has a small period of large enhanced flux and a smaller background flux, used to calculate pickup ion production as described in the text. It is assumed the ionosphere produced by these fluxes produces the exosphere, which is then ionized by the EUV to create the pickup ions. The enhanced flux case here plays the role of a solar activity-related enhancement. Production rates are significantly increased for this case, as shown in the legend box, even though the average EUV flux is nearly the same.

and estimating the potential role of escape to space of oxygen from Venus over time, and whether or not it helps solve the problem of the fate of a Venus ocean. An extended Venus Express mission that both samples the plasma and fields to lower altitudes and the solar wind at apoapsis, and includes a higher solar activity period than the mission's first few years allow, is required for definitive observations of potentially history-defining episodes of Venus oxygen loss. Finally we note that at Solar minimum the Venus ionosphere may become magnetized at lower dynamic pressures than it solar maximum. Whether this magnetization acts as a shield or leads to further loss of atmosphere remains to be seen.

4. PROBING THE VENUS INTERIOR FROM SPACE

Magnetism provides one of the few measurements that can provide information about the interior of a planet from well above its surface. We are interested in crustal magnetism as a clue to the history of the planetary magnetic field and the dynamo that might generate that field. We are curious about the existence and nature of any current dynamo action. We wish to use the induced fields associated with any large scale changes in the ionospheric magnetic field. These fields are all measurable to a certain degree from orbiters with low altitude ~150 km periapes but would be improved with balloon-borne measurements below the ionosphere.

4.1. Crustal Magnetism

Examination of the intrinsic magnetic field, whether it be the global field expected from a planetary dynamo, or the more localized field associated with magnetized crust was

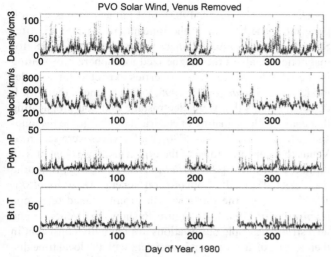

Figure 13. Illustration of the variability of solar wind plasma parameters of particular importance to the removal of atmosphere from Venus, here as measured on PVO with the Ames Plasma Analyzer. Both the dynamic pressure and the interplanetary magnetic field magnitude contribute to the magnetization of the Venus ionosphere during solar maximum, when ion production rates are highest. The peaks in these two quantities are due to solar wind stream interaction regions and ICMEs. The largest peaks are from ICMEs.

most extensively carried out with the Pioneer Venus Orbiter. The upper limit to the dipole field was 8.4×10^{10} Tm^3 or 10^{-5} of the terrestrial moment [*Phillips and Russell*, 1987]. The same study also searched for geographically correlated surface magnetism and found no evidence for such fields. We do not expect remanent magnetic fields on Venus because

the high temperature at the surface, even though they do not rule out some long-lived crustal magnetism, requires that the thickness of crust that has temperatures below the Curie point is small. Thus the resulting magnetic field would be very small. While the discovery of such magnetic fields would be very important and therefore should be explored if at all possible, the expectation of such a discovery is slim. Yet there are reasonably achievable measurements that can be carried out relevant to these open questions.

4.2. Sounding the Interior Conductivity

The interplanetary disturbance that has become known as the interplanetary coronal mass ejection (ICME), mentioned earlier, is possibly the most significant agent for the removal of Venus oxygen, but it also provides a means of sounding the electrical conductivity of the interior of Venus. When the solar wind dynamic pressure is high as it is during an ICME, the altitude of the solar wind – ionosphere interface (the ionopause) drops, and the strong field of the magnetic barrier is pushed into a region of finite electrical resistivity in which the currents that flow to exclude the magnetic field from the interior of the plasma are too weak to exclude it and the ionosphere becomes magnetized. Luhmann [1991] has modeled the lower boundary of the ionosphere to infer how strong the field below the ionosphere becomes based on PVO measurements of magnetized ionospheres. As illustrated in Figure 14 asymptotic field strengths of many 10's of nT are found. These, if monitored at different (atmospheric) sites around the planet with balloons say, could lead to an understanding of the size of the electrically conducting core

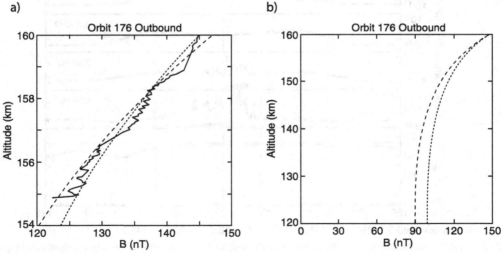

Figure 14. a) Altitude profile of the magnetic field magnitude observed on the outbound leg of PVO orbit 176 (solid curve). The dashed and dotted curves show diffusion models (for different values of c) that approximate the data, b) Curves in a) extrapolated to below the ionosphere.

of Venus. If the ionosphere allows step-function increases in the field strength to occur, the response of the induction currents to these sudden increases could provide information on the radial structure of the interior electrical conductivity. While it may be some time before a surface mission is realized, magnetometers at balloon altitudes and on probes can provide a first look at the magnetic fields below the ~150 km minimum periapsis of PVO. In the meantime the ASPERA-4 ELS experiment of Venus Express can provide some insights into the depths of penetration of the draped interplanetary fields from the photoelectrons that stream out of the ionosphere along them. The intensities of these photoelectron beams from Venus should be related to the altitudes to which the fields "probe" the ionospheric production layer. If Venus Express extended mission persists into the period where nightside ionospheric holes are present, ELS can also explore their depth and their connectivity to the interplanetary field.

5. LIGHTNING

A topic that unnecessarily became controversial during the Venera and Pioneer Venus epochs was the detection of lightning on Venus. Much of the evidence of lightning has been reviewed by Russell [1991], and Grebowsky et.al [1997] who summarized the Venera and earlier Pioneer Venus data. Lightning was reported based on electromagnetic radiation in the atmosphere from both the Venera 11–14 landers [*Ksanfomality et al.*, 1983], and Pioneer Venus when it penetrated the bottom of the ionosphere (Figure 15, after Strangeway et al., 1993). Pioneer Venus also detected electromagnetic signals in the magnetized ionosphere propagating in the whistler mode with properties expected for whistler mode waves [*Strangeway*, 1995, and references therein]. It detected signals that were electrostatic and non-propagating that could be the signature of electrical discharges to the ionosphere [*Singh and Russell*, 1986]. Radio waves were seen near Venus that appeared to be of the type usually associated with lightning on Earth [*Gurnett et al.*, 1991]. Finally, lightning was detected optically from orbit on Venera 9 [*Krasnopol'sky*, 1983] and from the Earth with a ground based telescope [*Hansell et al.*, 1995]. As Figure 16 shows the Venera 9 and terrestrial telescopic observations are mutually consistent in their spatial distribution and consistent with the local time distribution from the plasma wave data on PVO [*Russell*, 1991].

There have also been negative searches for lightning on Venus. Gurnett et al. [2001] argued that radio frequency bursts detected by Cassini during the Venus encounters were too weak or infrequent to be caused by lightning with terrestrial cloud-to-ground lightning properties. Gurnett et al. did allow for intracloud or cloud-to-ionosphere lightning. Since the cloud deck at Venus is around 50 km, intracloud and cloud-to-ionosphere lightning is much more likely than cloud-to-ground lightning.

One of the limitations of the Pioneer Venus observations was that the wave instrument only detected electric fields.

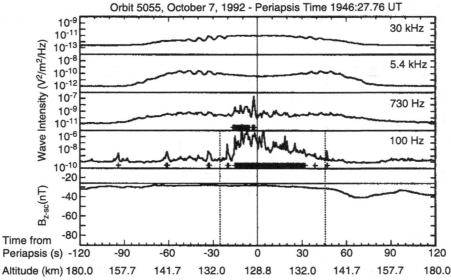

Figure 15. Electric field data acquired on the last orbit of Pioneer Venus. Four minutes of data are shown, with periapsis in the center of the plot. The spacecraft apparently dipped below the ionospheric density peak, and very strong electric field bursts were observed, stronger than any other electric field signals detected throughout the Pioneer Venus mission. These bursts are consistent with an atmospheric source, such as lightning. (After Strangeway et al., 1993).

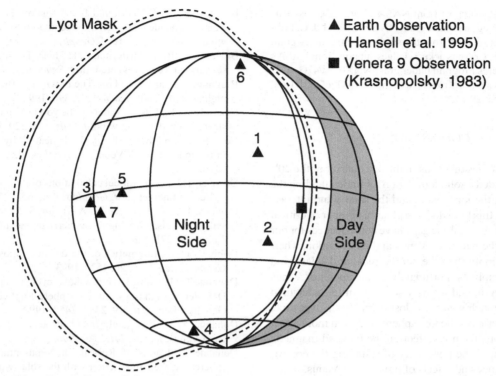

Figure 16. Optical signatures of Venus lightning. Triangles show the signals detected from the Earth [*Hansell* et al., 1995]. The square shows the location of the Venera observations [*Krasnopolsky*, 1983]. The optical observations are consistent with the distribution of bursts observed by the Pioneer Venus wave electric field instrument.

While there are strong arguments for suggesting that the 100 Hz waves detected in and above the nightside ionosphere of Venus are indeed whistler-mode waves [*Strangeway*, 1995], most notably that the burst rate for 100 Hz waves maximizes when the magnetic field is vertical, it has been argued that such waves would excessively heat the ionosphere [*Cole and Hoegy*, 1996]. The arguments of Cole and Hoegy are essentially based on steady state, and given the transient nature of the waves this may be invalid. Nevertheless, some heating should occur. How much heating depends on the energy flux of the waves. *Russell et al.* [1989] estimate electromagnetic energy fluxes of the order 10^{-7} Wm^{-2}. *Strangeway* [2000] performed a calculation that included Joule dissipation and found that the bottomside of the ionosphere could be heated to several eV, assuming an energy flux a factor of 30 higher than that reported by Russell et al. [1989]. This amount of heating, while transitory, will almost certainly have consequences.

That lightning induced waves may heat the ionosphere is important as such heating may cause enhanced ionization and ionospheric escape. By the same token lightning of sufficiently high rate and energy will have consequences for the atmosphere. Recently Krasnopolsky [2006] reported observations of absorption features due to NO in the Venus atmosphere. He argued that lightning is the only source for NO in the lower Venus atmosphere, and further he estimated flash rates comparable to terrestrial rates (6 $km^{-2}y^{-1}$), assuming typical terrestrial flash energies of 10^9 J.

While Krasnopolsky [2006] argues for lightning flash rates and energies comparable to terrestrial lightning to explain the NO absorption features, the other data cited above are insufficient to fully constrain the properties of Venus lightning. Does this lightning affect the chemistry of the Venus atmosphere as suggested by Krasnopolsky [2006]? Are the lightning rate and current densities sufficient to be of danger to balloons and aerostats that we might desire to explore the Venus atmosphere? Can we better constrain the energy and flash rate of Venus lightning? Venus Express will help to answer these questions by providing observations of electromagnetic radiation that complements the Pioneer Venus electric field observations, as well as optical measurements of flash rates.

Acknowledgements. The author wishes to thank both the Pioneer Venus project personnel and the scientific team who made this mission such a great success and provided the observations and interpretations that have enabled us to write this article. C. T.

Russell acknowledges support from NASA's Venus Express Participating Scientist Program under grant NNG05GC87G. J. G. Luhmann acknowledges support from the same program under grant NNG06GC6IG. T. E. Cravens acknowledges support from NASA's Planetary Atmospheres Program under grant NNX07AF47G. A. F. Nagy acknowledges support from NASA grant NNG06GF31G and NSF grant ATM0455729

6. CONCLUSIONS

Our extensive studies of Venus in the last quarter of the 20th century have provided a solid foundation for understanding the upper atmosphere, the ionosphere and the solar wind interaction. Nevertheless, this knowledge still has a number of critical gaps. As described above, these gaps have consequences for our understanding of the nature of Venus and its evolution. Their resolution affects more than the aeronomical and solar wind interaction subdisciplines. Fortunately a new epoch of Venus exploration has begun and we may soon begin to understand whether and how much water was lost from Venus over time, more about the physics of the ionosphere, and even more about the interior of Venus from its effect on the induced magnetic field. Finally we have the possibility of resolving the controversy on the existence and effects of lightning at Venus.

REFERENCES

Ayres, T. R., Evolution of the solar ionizing flux, *J. Geophys. Res.*, *102*, 1641, 1997.

Barabash, S. et al., R. Lundin, H. Andersson, K. Brinkfeldt, A. Grigoriev, H. Gunell, M. Holmströmm. Yamauchi, K. Asamura, P. Bochsler, P. Wurz, R. Cerulli-Irelli, A. Mura, A. Milillo, M. Maggi, S. Orsini, A. J. Coates, D. R. Linder, D. O. Kataria, C. C. Curtis, K. C. Hsieh, B. R. Sandel, R. A. Frahm, J. R. Sharber, J. D. Winningham, M. Grande,E. Kallio, H. Koskinen, P. Riihelä, W. Schmidt, T. Säles, J. U. Kozyra, N. Krupp, S. Livi, J. Woch, J. G., Luhmann, S. Mckenna-Lawlor, E. C. Roelof, D. J. Williams, J.-A. Sauvaud, A. Fedorov, and J.-J. Thocaven , The Analyzer of Space Plasmas and Energetic Atoms (ASPERA-3)for the Mars Express Mission, Aspera-4 paper, Space Sci. Rev., in press, 2006.

Brace, L. H., and A. J. Kliore, The structure of the Venus ionosphere, *Space Sci. Rev.*, *55*, 81, 1991.

Brace, L. H., R. F. Theis, and W. R. Hoegy, Plasma clouds above the ionopause of Venus and their implications, *Planet. Space Sci.*, *30*, 29, 1982.

Brace, L. H., H. A. Taylor, Jr., T. I. Gombosi, A. J. Kliore, W. C. Knudsen, and A. F. Nagy, The ionosphere of Venus: Observations and their interpretation p.779 in *Venus*, eds. D. M. Hunten, L. Colin, T. M. Donahue, and V. I. Moroz, Univ. of Arizona Press, Tucson, 1983.

Brace, L. H., W. T. Kasprzak, H. A. Taylor, R. F. Theis, C. T. Russell, et al., The ionotail of Venus: Its configuration and evidence for ionospheric escape, *J. Geophys. Res.*, *92*, 15, 1987.

Brace, L. H., R. F. Theis, and J. D. Mihalov, The response of the Venus nightside ionosphere and ionotail to solar EUV and solar wind dynamic pressure, *J. Geophys. Res.*, *95*, 4075, 1990.

Brace, L. H., R. E. Hartle, and R. F. Theis, The nightward ion flow scenario at Venus revisited, *Adv. Space Res.*, *16*, 99, 1995.

Brannon, J. J., and J. L. Fox, The downward flux of O+ over the nightside of Venus, *Icarus*, *112*, 396, 1994.

Chassefiere, E., Loss of water on the young Venus: The effect of a strong primitive solar wind, *Icarus*, *126*, 229, 1997.

Cole, K. D., and W. R. Hoegy, Joule heating by ac electric fields in the ionosphere of Venus, *J. Geophys. Res.*, *101*, 2269–2278, 1996.

Cravens, T. E., et al., Model calculations of the dayside ionosphere of Venus: Energetics, *J. Geophys. Res.*, *85,* 7778, 1980.

Cravens, T. E., L. H. Brace, H. A. Taylor, S. J. Quenon, C. T. Russell, et al., Disappearing ionospheres on the nightside of Venus, *Icarus*, *51*, 271, 1982.

Dobe, Z. et al., Energetics of the dayside ionosphere of Venus, *Geophys. Res. Lett.*, *20*, 1523, 1993.

Donahue,T. M. and R. E. Hartle, Solar cycle variations in H(+) and D(+) densities in the Venus ionosphere-Implications for escape, Geophys. Res. Lett., 19, 2449–2454, 1992.

Donahue, T. M., New analysis of hydrogen and deuterium escape from Venus, *Icarus*, *141*, 226, 1999.

Donahue, T. M. and C. T. Russell, The Venus atmosphere and ionosphere and their interaction with the solar wind: an overview. Venus II Geology, Geophysics, Atmosphere, and Solar Wind Environment. S. W. Bougher, D. M. Hunten and R. J. Phillips. Tucson, University of Arizona Press: 3–31, 1997.

Dryer, M., H. Perez-de-Tejada, H.A. Taylor Jr., D.S. Intriligator, J.D. Mihalov, B. Rompolt, Compression of the Venusian ionosphere on May 10, 1979 by the interplanetary shock generated by the solar eruption of May 8, 1979, J. Geophys. Res. 87, 9035–9044, 1982.

Fegley, B., Atmospheric evolution on Venus, in The Encyclopedia of Paleoclimatology and Ancient Environments, ed. V. Gornitz, Kluwer Academic Publishers, 2004.

Fox, J. L. and A. Hac, Spectrum of hot O at the exobase of the terrestrial planets, *J. Geophys. Res.*, *102*, 24005, 1997.

Fox, J. L. and A. J. Kliore, Ionosphere: Solar cycle variations, p.161, in *Venus II: Geology, Geophysics, Atmosphere, and Solar Wind Environment*, eds. S. W. Bougher, D. M. Hunten, and R. J. Phillips, Univ. of Arizona Press, Tucson, 1997.

Fox, J. L. and A. I. F. Stewart, The Venus ultraviolet aurora: A soft electron source, *J. Geophys. Res.*, *96*, 9821, 1991.

Fox, J. L. and K. Y. Sung, Solar activity variations of the Venus thermosphere/ionosphere, *J. Geophys. Res.*, *106*, 21,305, 2001.

Gan, L., T. E. Cravens, and M. Horanyi, Electrons in the ionopause boundary layer of Venus, *J. Geohys. Res.*, *95*, 19023, 1990.

Grebowsky, J. M. ,R. J. Strangeway and D. M. Hunten, Evidence for Venus lightning, p.125 in *Venus II: Gelogy, Geophysics, Atmosphere and Solar Wind Environment*, eds S. W. Bougher, D. M. Hunten and R. J. Phillips, Univ. of Arizona Press, 1997.

Gurnett, D. A., W. S. Kurth, A. Roux, et al., Lightning and plasma wave observations from the Galileo flyby of Venus, *Science*, *253*, 1522–1525, 1991.

Gurnett, D. A., P. Zarka, R. Manning, et al., Non-detection at Venus of high-frequency radio signals characteristic of terrestrial lightning, *Nature*, *409*, 313–315, 2001.

Hansell, S. A., W. K. Wells, and D. M. Hunten, Optical-detection of lightning on Venus, *Icarus*, *117*, 345–351, 1995.

Hartle, R. E., and J. M. Grebowsky, Upward ion flow in ionospheric holes on Venus, *J. Geophys. Res.*, *95*, 31, 1990.

Jian, L. C. T. Russell, and J. G. Luhmann, Interplanetary Coronal Mass Ejections: Solar Cycle Variation at 0.72 AU and 1AU *Adv. Space Res.*, in press, 2007

Kasprzak, W. T., J. M. Grebowsky, H. B. Niemann, and L .H. Brace, Suprathermal >36 eV ions observed in the near-tail region of Venus by the Pioneer Venus Orbiter neutral mass spectrometer, *J. Geophys. Res.*, *96*, 11,175, 1991.

Kim, J., A. F. Nagy, J. L. Fox, and T. E. Cravens, Solar cycle variability of hot oxygen atoms at Mars, *J. Geophys. Res.*, *103*, 29339, 1998.

Knudsen, W. C., and K. L. Miller, Pioneer Venus superthermal electron flux measurements in the Venus umbra, *J. Geophys. Res.*, *90*, 2695, 1985..

Krasnopol'sky, V. A., Venus spectroscopy in the 3000–8000 Å region by Veneras 9 and 10, in Venus, edited by D. M. Hunten at al., pp.459–483, Univ. of Arizona Press, Tucson, 1983.

Krasnopol'sky, V. A., A sensitive search for nitric oxide in the lower atmospheres of Venus and Mars: Detection on Venus and upper limit for Mars, *Icarus*, *182* (1), 80–91, 2006.

Ksanfomality, L. V., F. L. Scarf, and W. W. L. Taylor, The electrical activity of Venus, in *Venus*, edited by D. M. Hunten at al., pp. 565–603, Univ. of Arizona Press, Tucson, 1983.

Kulikov, Y. N., H. Lammer, H. Lichtenegger, N. Terada, C. Kolb, D. Langmayr, R. Lundin, E. Guinan, S. Barabash and H. Biernat, Atmospheric and water loss from early Venus. *Planet. Space Sci.*, UK, 54, 1425–44, 2006.

Kulikov, Y. N., et al., Atmospheric and water loss from early Venus, *Planet. Space Sci.*, 54, 1457–1471, 2006.

Kumar, S. D.M. Hunten and J.B. Pollack, Nonthermal escape of hydrogen and deuterium from Venus and implications for loss of water, Icarus, 55, 369–389, 1983.

Lammer, H., Y. N. Kulikov, and H. I. M. Lichtenegger, Thermospheric x-ray and EUV heatingby the young Sun on early Venus and Mars, *Space Sci. Rev.*, *122*, 189, 2006.

Lindsay, G. M., C. T. Russell, J. G. Luhmann, P. Gazis, On the sources of interplanetary shocks at 0.72 AU, *J. Geophys. Res.*, *99*, 11, 1994.

Lindsay, G. M., C. T. Russell, and J. G. Luhmann, Coronal mass ejection and stream interaction region characteristics and their potential geoeffectiveness, *J. Geophys. Res.*, *100*, 16,999, 1995.

Luhmann, J. G., Induced magnetic fields at the surface of Venus inferred from Pioneer Venus Orbiter near-periapsis measurements, *J. Geophys. Res.*, *96*, 18,831–18,840, 1991.

Luhmann, J.G. and S.J. Bauer, Solar wind effects on atmosphere evolution at Venus and Mars, p. 417 in in *Venus and Mars: Atmo-spheres, Ionospheres and Solar Wind Interactions*, Geophysical Monograph 66, eds. J.G.Luhmann, M. Tatrallyay, R.O. Pepin, American Geophysical Union, 1992

Luhmann, J. G., and T. E. Cravens, Magnetic fields in the ionosphere of Venus, *Space Sci. Rev.*, *55*, 201, 1991.

Luhmann, J. G., and J. U. Kozyra, Dayside pickup oxygen ion precipitation at Venus and Mars: Spatial distributions, energy deposition and consequences, *J. Geophys. Res.*, *96*, 5457, 1991.

Luhmann, J. G., R. E. Johnson, and M. H. G. Zhang, Evolutionary impact of sputtering of the Martian atmosphere by O^+ pickup ions, *Geophys. Res. Lett.*, *19*, 2151, 1992.

Luhmann, J. G., T. L. Zhang, S. M. Petrinec, C. T. Russell, P. Gazis, and A. Barnes, Solar cycle 21 effects on the interplanetary magnetic filed and related parameters at 0.7 AU and 1.0 AU, *J. Geophys. Res.*, *98*, 5559, 1993.

Luhmann, J. G., W. T. Kasprzak, and J. M. Grebowsky, On removing molecular ions from Venus, *J. Geophys. Res.*, *100*, 14,515, 1995.

Luhmann, J. G., S. A. Ledvina, J. G. Lyon, and C. T. Russell, Venus O^+ pickup ions: Collected PVO results and expectations for Venus Express, *Planet. and Space Sci.*,54, 1457–1471, 2006.

McComas, D. J., H. E. Spence, C. T. Russell, and M. A. Saunders, The average magnetic field draping and consistent plasma properties of the Venus magnetotail, *J. Geophys. Res.*, *91*, 7939, 1986.

McElroy, M. B., M. J. Prather, and J. M. Rodriguez, Loss of oxygen from Venus, *Geophys. Res. Lett.*, *9*, 649, 1982.

Mihalov, J. D., and A. Barnes, The distant interplanetary wake of Venus: Plasma observations from Pioneer Venus, *J. Geophys. Res.*, *87*, 9045, 1982.

Mihalov, J. D., C. T. Russell, W. T. Kasprzak, and W. C. Knudsen, Observations of ionospheric escape on Venus' nightside, *J. Geophys. Res.*, *100*, 19579, 1995.

Miller, K. L., and R. C. Whitten, Ion dynamics in the Venus ionosphere, *Space Sci. Rev.*, *55*, 81, 1991.

Moore, K. R., D. J. McComas, C. T. Russell, and J. D. Mihalov, A statistical study of ions and magnetic fields in the Venus magnetotail, *J. Geophys. Res.*, *95*, 12,005, 1990.

Mulligan, T., C. T. Russell, and J. G. Luhmann, Solar cycle evolution of the structure of magnetic clouds in the inner heliosphere, *Geophys. Res. Lett.* 25, 2959, 1998.

Nagy, A. F., and T. E. Cravens, Hot oxygen atoms in the upper atmosphere of Venus and Mars, *Geophys. Res. Lett.*, *15*, 433, 1988.

Nagy, A. F., and T. E. Cravens, Ionosphere: Energetics, in Venus II, edited by S. W. Bougher, D. M. Hunten and R. J. Phillips, The U. of Arizona Press, Tucson, 1997.

Nagy, A. F., et al., Model calculations of the dayside ionosphere of Venus: Ionic composition, *J. Geophys. Res.*, *85*, 7795, 1980.

Nagy, A. F., T. E. Cravens, J. H. Yee, and A. I. F. Stewart, Hot oxygen atoms in the upper atmosphere of Venus, *Geophys. Res. Lett.*, *8*, 629, 1981.

Nagy, A. F., T. E. Cravens and T. I. Gombosi, Basic theory and model calculations of the Venus ionosphere, p.841, in *Venus*, eds. D. M. Hunten, L. Colin, T. M. Donahue, and V. I. Moroz, Univ. of Arizona Press, 1983.

Nagy, A. F., et al., The magnetic field control of the dayside ion temperature in the ionosphere of Venus, *J. Geophys. Res., 102,* 435, 1997.

Ong, M., J. G. Luhmann, C. T. Russell, R. J. Strangeway, and L. H. Brace, Venus ionospheric clouds: Relationship to the magnetosheath field geometry, *J. Geophys. Res., 96,* 11,133, 1991.

Phillips , J. L., and C. T. Russell, Revised upper limit on the internal magnetic moment of Venus, *Adv. Space Res., 7*(12), 291–294, 1987.

Russell, C. T., Venus lightning, *Space Sci. Rev., 55,* 317–356, 1991.

Russell, C. T., M. von Dornum, and R. J. Strangeway, VLF bursts in the night ionosphere of Venus: estimates of the Poynting flux, *Geophys. Res. Lett., 16,* 579–582, 1989.

Schunk, R. W. and A. F. Nagy, *Ionospheres,* Cambridge University Press, 2000.

Shinagawa, H., A two-dimensional model of the Venus ionosphere 2. Magnetized ionosphere, *J. Geophys. Res., 101,* 26921, 1996.

Singh, R. N., and C. T. Russell, Further evidence for lightning on Venus, *Geophys. Res. Lett., 13,* 1051–1054, 1986.

Sonett, C. P., M. Giampapa, and M. S. Matthews, eds., The Sun in Time, University of Arizona Press, 1991.

Strangeway, R. J., Plasma wave evidence for lightning on Venus, *J. Atmos. Terr. Physics, 57,* 537–556, 1995.

Strangeway, R. J., Whistler-mode propagation in the collisional ionosphere of Venus, *Adv. Space Res., 26*(10), 1613–1618, 2000.

Strangeway, R. J., C. T. Russell, and C. M. Ho, Observation of intense wave bursts at very low altitudes within the Venus nightside ionosphere, *Geophys. Res. Lett., 20,* 2771–2774, 1993.

Tanaka, T., Effects of decreasing ionspheric pressure on the solar wind interaction with non-magnetized planets, *Earth Planets Space, 50,* 259, 1998.

Taylor, H. A., P. A. Cloutier, M. Dryer, S. T. Suess, A. Barnes, and R. S. Wolf, Response of Earth and Venus ionospheres to corotating solar wind stream of 3 July 1979, *Earth, Moon and Planets, 32,* 275, 1985.

Vaisberg, O., A. Federov, F. Dunjushkin, A. Kozhukhovsky, V. Smirnov, L. Avanov, C. T. Russell, and J. G. Luhmann, Ion populations in the tail of Venus, *Adv. Space Res., 16*(4) 105, 1995.

Zhang, T. L., W. Baumjohann, M. Delva, H-U Austen, A. Balogh, C. T. Russell, et al., Magnetic field investigation of the Venus plasma environment: Expected new results, *Planet. Space Sci.,54,* 1336–1343, 2006.

Venus Express and Terrestrial Planet Climatology

Fredric W. Taylor

University of Oxford, England

Håkan Svedhem

European Space Agency, Netherlands

Dmitri M. Titov

Max Planck Institute for Solar System Research, Germany

After a delay of more than a decade, the exploration of Venus has resumed through the European Venus Express mission, now in orbit around the planet. The mission payload, its implementation in an elliptical polar orbit, and the science operations planned, all focus on outstanding problems associated with the atmosphere and climate of Venus. Many of these problems, such as understanding the extreme surface warming produced by the carbon dioxide-driven greenhouse effect, and the role of sulfate aerosols in the atmosphere, have resonances with climate-change issues on the Earth and Mars. As data on all three terrestrial planets accumulates, and models of the energy balance and general circulation of their atmospheres improve, it becomes increasingly possible to define and elucidate their behavior in a common, comparative framework. Venus Express seeks to contribute to progress in this area.

INTRODUCTION AND SCIENTIFIC BACKGROUND

Venus Express is Europe's first mission to one of the best explored and least understood of our planetary neighbors. The spacecraft, its planned operations, and its scientific goals have all been described in some detail elsewhere (Svedhem et al., 2007; Titov et al., 2006; Taylor, 2006) and here only a brief synopsis of these factors is given. Our goal in the present article is to place the scientific objectives of Venus Express in the overall context of Venus exploration, past and future, with emphasis not only on the expected progress, but also on the many issues Venus Express is not designed to address and which will require further investi-

Exploring Venus as a Terrestrial Planet
Geophysical Monograph Series 176
Copyright 2007 by the American Geophysical Union.
10.1029/176GM10

gation. We also discuss how any approach to understanding Venus relates to similar issues with Earth and Mars, now that a preliminary exploration of the terrestrial planet family has provided the framework for this, not least by highlighting the many first-order questions that remain.

The scientific focus of the Venus Express mission is on the planet's atmosphere, in particular the key processes that together determine the climate, defined as the current mean state. From its elliptical, polar orbit, Venus Express aims to provide a remote sensing investigation of the global atmosphere and the plasma environment around Venus, and to address some important aspects of the surface physics. With the data, several poorly understood aspects of Venus climate physics can be addressed, including the following list:

Atmospheric structure and dynamics: Observations of global temperature contrasts and the general circulation of the atmosphere; the inference of their coupling with cloud density and minor constituent abundance variations.

Composition and chemistry: Measurements of the variability in the distributions of CO, H_2O, and sulfur-bearing gases, and model studies of their role in cloud formation and the greenhouse effect.

Clouds and their radiative properties: Obtain detailed data on variations in cloud profile and opacity, including time-resolved, long term data that allows the study of clouds as dynamical tracers.

Sources and sinks of atmospheric gases: Clarify certain surface-atmosphere interactions, including finding evidence for active volcanism and its extent; quantify the main exospheric escape processes; and model the long-term effects on climate change.

With Venus Express it is hoped to establish a new picture of the climate on Venus, based on the results expected from the entire suite of instruments on board. This includes four remote sensing experiments for measurements of atmospheric properties, motions, and surface interactions, and a magnetospheric package for new findings about the loss rates of atmospheric gases to the solar wind. Through the use of analysis tools that include radiative transfer models, general circulation models, and climate evolution models, a better understanding, not only of conditions on our planetary neighbor, but also of why they appear to diverge so much from conditions on the Earth, is expected to emerge. An improved understanding of the present-day climate of Venus, with a physical representation of the atmosphere and an assessment of time-dependent effects, will permit better-informed speculation about the evolution and possible future development of the temperature, composition and general circulation of the atmosphere.

THE VENUS EXPRESS MISSION

The spacecraft (figure 1) was launched on November 9, 2005 by a Russian Souz-Fregat launcher from the Baykonur cosmodrome in Kazakhstan. On April 11, 2006 it reached the planet, there to be maneuvered over the next several weeks into in a polar orbit with pericentre and apocentre altitudes of 250 and 66,000 km and a revolution period of 24 hours. The remote sensing instruments collect data both in the vicinity of the planet when over the North polar regions, and from a distance of some ten planetary radii when over the South pole, thus combining global context and detailed close-up views. Observations are planned for 500 days of nominal mission, with a possible extension for another 500 days.

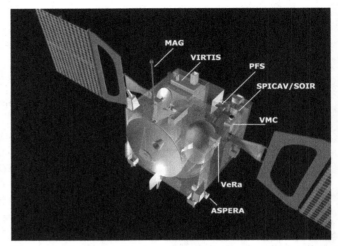

Figure 1. The Venus Express spacecraft, showing the locations of the scientific instruments. The acronyms are explained in the text. (ESA)

The spacecraft re-uses the Mars Express bus and carries seven scientific instruments, five of which were inherited from the Mars Express and Rosetta projects and two of which are new. The payload has as its core a suite of spectro-imaging remote sensing instruments, VIRTIS, PFS, SPICAV/SOIR, and VMC which we now briefly describe. Full descriptions can be found in the publications listed at the top of the References section below.

The Visible-Infrared Thermal Imaging Spectrometer VIRTIS maps Venus in the spectral range from 0.27–5.2 μm with moderate spectral resolution ($\lambda/\Delta\lambda\sim200$) and also provides high-resolution ($\lambda/\Delta\lambda\sim1200$) spectra for the 1.8–5 μm range. It has an instantaneous field of view of 0.25 mrad that corresponds to a spatial resolution ranging from hundreds of meters at pericentre to ~15 km at apocentre. The instrument uses cooled detectors to obtain the high sensitivity necessary for sounding the composition of the lower atmosphere, total cloud opacity, and surface thermal mapping by measuring weak night-side emissions in the transparency "windows", while its mapping capabilities address atmospheric dynamics by tracking cloud features and compositional variations in the UV and IR ranges. The temperature and aerosol structure of the mesosphere between 60 and 90 km is deduced from measurements of thermal infrared emission in the 4–5 μm range.

The Planetary Fourier Spectrometer (PFS) is an infrared spectrometer covering the spectral range from 0.9 to 45 μm with a spectral resolution of ~1 cm[-1] and a field of view of ~2 degrees that corresponds to a spatial resolution of about 10 km at pericentre. The main science objectives, which were to study temperature, aerosol structure and composition in the Venusian mesosphere (60–100 km) and to monitor trace gases in the lower atmosphere, are in jeopardy at the time of writing due to a jammed scan mirror. Efforts are underway to use special command sequences to free the mirror, or at least to force it into a position where it views the planet, but if this fails the other spectro-imaging experiments will be reprogrammed to partially recover the PFS scientific objectives.

SPICAV/SOIR (Spectroscopy for Investigation of Characteristics of the Atmosphere of Venus/Solar Occultation IR) combines three spectrometers for the UV and near-IR range to study the vertical structure and composition of the mesosphere (60–100 km) and lower thermosphere (100–200 km) in solar and stellar occultation, limb and nadir geometries. Occultation observations provide high sensitivity to the abundance of minor species like isotopes of CO_2, SO_2, COS, CO, HCl, and HF. Measurements of H_2O and HDO characterize the escape of D atoms from the upper atmosphere and give insight about the evolution of water on Venus. SPICAV/SOIR also searches for new trace gases, in particular hydrocarbons (CH_4, C_2H_2), nitrogen oxides (NO, N_2O), and chlorine bearing compounds (CH_3Cl, ClO_2).

The Venus Monitoring Camera (VMC) is a wide-angle camera for observations of the atmosphere and the surface in four narrow-band filters centred at 0.365, 0.513, 0.965, and 1.01 µm. The overall field of view is 17 degrees with 0.75 mrad/pixel, corresponding to a spatial resolution ranging from 0.2 km at pericentre to 50 km at apocentre. The main goal of the VMC is to investigate the cloud morphology and atmospheric dynamics by tracking the cloud features in the UV and near-IR ranges. Observations on the night side are used for thermal mapping of the surface and to search for spatial variations in water vapour in the lower atmosphere.

Two experiments focus on the analysis of the plasma environment of Venus and the interaction of the solar wind with the atmosphere. The Analyser of Space Plasmas and Energetic Atoms (ASPERA) comprises four sensors: two detectors of energetic neutral atoms, plus electron and ion spectrometers. It will measure the composition and fluxes of neutrals, ions and electrons to address how the interplanetary plasma and electromagnetic fields affect the Venus atmosphere and identify the main escape processes and help understand similarities and differences in the solar wind interaction with the other terrestrial planets, Earth and Mars. The magnetometer (MAG) has two fluxgate sensors to measure the magnitude and direction of the magnetic field in the magnetosheath, magnetic barrier, ionosphere, and magnetotail, with high sensitivity and temporal resolution and to characterize the boundaries between plasma regions. MAG can also search for lightning on Venus by measuring the strength of electromagnetic waves associated with atmospheric electrical discharges.

The Venus Express Radio Science Experiment (VeRa) uses signals emitted by the spacecraft radio system in the X- and S- bands (3.5 cm and 13 cm) to sound the structure of the neutral atmosphere and ionosphere with a vertical resolution of a few hundred meters. The experiment investigates the gravity field and surface properties of Venus, and the solar corona. An ultra-stable oscillator provides a high quality onboard reference frequency source for the spacecraft transponder. The same measurements will sound the abundance of H_2SO_4 vapor below the clouds (40–50 km), looking for variability that can be seen as a tracer for atmospheric motions.

In addition to the teams associated with these experiments, the project includes a number of Interdisciplinary Scientists and Supporting Investigators, who bring additional strengths in several categories, such as atmospheric radiative transfer calculations and energy balance models, inversion of spectroscopic and radiometric data to obtain temperature and species profiles and cloud parameters, and dynamical modeling of the Venusian atmosphere using general circulation models which also have application to the Earth and other planets. A long-term goal is to represent the climate process on Venus in a time-dependent model that will incorporate the results from the Venus Express investigations, leading to improved theories about the origin of the present state of Venus' climate, and informed speculations about its possible future evolution. A better understanding, not only of conditions on our planetary neighbor, but also of why they appear to diverge so much from conditions on the Earth, may then emerge.

VENUS EXPRESS SCIENCE OPERATIONS

After about 50 days of commissioning, the nominal science mission of Venus Express began on June 4, 2006 and is planned to extend until October 2, 2007, corresponding to a duration of slightly more than 2 Venus sidereal days of 243 Earth days each. The pericentre latitude of the highly eccentric orbit was initially at 78N and slowly drifts polewards; maneuvers are regularly performed to maintain the pericentre altitude in the 250–350 km range. Such a high inclination orbit gives good latitudinal coverage and, in particular, affords high-resolution views of the North polar region and extensive mapping and imaging of the Southern hemisphere.

The scheduled observations are characterized by high repeatability of measurements of dynamical phenomena, sufficient to make low-resolution 'movies' in which features can be tracked and their speeds and evolutionary characteristics identified. For operational purposes the Venus Express orbit is roughly divided in three parts: pericentre observations (23–2 hours orbital time, measured relative to periapsis), telecommunications (2–12 hours, during the descending branch of the orbit), and off-pericentre observations (12–23 hours, in the ascending branch). The orbital phase is maintained so that the ground station at Cebreros in Spain is always visible from the satellite between 2 and 12 hours orbital time for the downloading of data that is acquired during the previous orbit. During every orbit, the spacecraft and payload operate according to one of a series of pre-determined 'science cases', each of which specifies the data rate of each instrument as a function of time (c.f. Fig.2). Downlink and on-board data storage limitations generally mean that the remote sensing instruments must choose between periapsis, off-periapsis, and apoapsis viewing campaigns on any given orbit.

Each of these has its own particular advantage: for 97 minutes near periapsis, the altitude of the satellite is less than ~10,000 km, permitting high spatial resolution spectroscopic and imaging observations of the Northern high latitudes; during off-periapsis observations, the +Z axis of the spacecraft points to nadir or slightly off-nadir between approximately 15 and 23 hours to obtain a global view of the Southern hemisphere, enabling spectral imaging of the motions of mid-latitude cloud features for studies of atmospheric dynamics, while apoapsis mode emphasizes studies of the atmosphere in the South polar region. Special science cases are provided for less frequent use, including periods dedicated to VeRa bi-static sounding, which involves a slew maneuver by the spacecraft to maintain specular reflection geometry for a selected target on the Venusian surface; to stellar and solar occultations by SPICAV/SOIR; and to limb and radio occultation observations. Plasma and magnetic field measurements are obtained continuously on all orbits, to ensure maximum four-dimensional coverage of both the near-planet environment and the solar wind region. The modes of Venus Express operations as well as the nominal mission timeline are described in detail by Titov et al. (2006).

COMPARATIVE CLIMATOLOGY OF VENUS AND EARTH

The rationale for a new mission to Venus, and its detailed implementation, was based on the need to better understand the climate of Venus, the factors involved in maintaining its observed state, their commonality with known processes at work on the Earth, and why twin planets with a common origin now have such different atmospheric regimes. Progress will involve disentangling how Venus and Earth differ fundamentally, in bulk composition for example, or in the angular momentum state when they cooled initially, from differences due more to evolutionary factors, for instance,

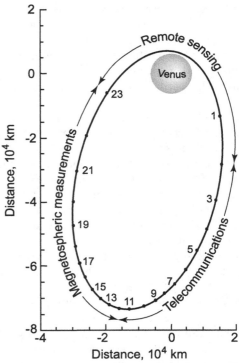

Figure 2. The Venus Express orbit, showing the time in hours relative to periapsis and the 8-hour segment reserved for transmitting the data back to Earth. The emphasis on different science goals is varied both during the orbit itself and from orbit to orbit. For instance, as shown here, remote sensing of the atmosphere might be emphasised while near the planet, with magnetospheric measurements given priority at greater distances. However, the time between 12 and 23 hours is not restricted to magnetospheric observations since distant observations of global dynamics are also important. Also most important plasma observations and occultations are performed close to the periapsis.

the possible early loss of an Earth-sized ocean from Venus. Had Venus and Earth been swapped at birth—that is, at the time when they had accumulated virtually all of their present mass but before their atmospheres were fully evolved—what would the inner solar system look like today? In this thought experiment, Venus is now at one astronomical unit from the Sun, and Earth a factor $1/\sqrt{2}$ closer. Venus still rotates slowly and has any bulk compositional differences it acquired in reality, for instance as a result of forming at the closer position to the centre of the protosolar cloud. The best guess based on the current paradigm is probably a hot, arid Earth and a temperate, oceanic Venus, but this is far from certain; it may be that Venus and Earth would have to swap rotation rates as well. The slow rotation of Venus may somehow account for the absence of an internal magnetic field on Venus, as we discuss below, and the lack of a magnetic shield from the solar wind could have been a key factor in the loss of water from early Venus.

Of course, we cannot observe Venus at any time other than the present and more complete answers to these grand questions of origin and evolution must be sought initially through a more complete understanding of the present conditions on Venus and Earth. Models might then be developed that show how Venus evolved with a massive, dry atmosphere compared to Earth, while still obeying the same physical laws acting within the known boundary conditions.

The similarities between Venus and Earth have often been noted, from their common origin in the inner solar system to their overall physical resemblance and, to first order at least, similar internal composition. The slow retrograde rotation of Venus is a marked difference that most likely originated in the collisional history of the planetesimals that aggregated to form the planets, while the lack of a planetary magnetic field at the inner of the twin planets remains difficult to explain, even in speculative terms. The latter difference is particularly surprising given that the mean density of Venus must mean that it has a large metallic core like the Earth, while evolutionary models and the apparently high level of modern volcanic activity both suggest that this core is still in a partially molten state. Why it is not associated with an internal dynamo is a mystery that neither Venus Express nor any of its precursors have been able to address; perhaps there is a connection with the slow rotation of the solid body of Venus, although why that should be is far from apparent, or perhaps Venus did have a field but it is currently in the null state between reversals similar to those seen in the terrestrial record. Perhaps the interior of Venus, lacking the cooling effect of efficient plate tectonics, is too hot to generate a field. The apparent suppression of plate tectonics may itself be a manifestation of the high temperature and extreme dryness, relative to Earth, of the crust. The return of core samples from the surface and interior of Venus may eventually shed light on this puzzle.

The most arresting contrast between Earth and Venus is the dramatic difference in surface temperature and pressure. Until the high surface temperature on Venus (around 730 K, which is higher than the melting point of the metals lead and tin) was first detected by ground-based radio astronomers and confirmed by Mariner 2 and Venera 9, it was widely expected that Venus would turn out to be a more tropical version of the Earth. However, once it became known that the Bond albedo of Venus is more than 2.5 times that of Earth (about 0.76 vs. about 0.3), so that Venus absorbs less radiative energy than Earth, despite its greater proximity to the Sun, it could well have been argued that Venus could be not warmer but cooler overall. The mean temperature on Venus is, in fact, lower than that on Earth if the comparison is made

at the atmospheric pressure of 1000 hPa that is characteristic of the surface of the latter.

The big difference, and the problem so far as the habitability of Venus' surface is concerned, is that the pressure, and hence the temperature, both continue to rise with depth below the 1 bar level. The profile roughly follows the hydrostatic and adiabatic formulae, as would be expected, leading to a temperature increase of about 10K for each km of depth below the 1 bar level. This amounts to some 450K altogether at Venus' surface pressure of 92 bars. If the Earth had such a high surface pressure, it too would be extremely hot, even without the increased proportion of greenhouse gases that is found on the Venus. About 96% of this is carbon dioxide, which, along with water vapor and other minor constituents, and some components of the ubiquitous cloud cover, drives the radiative energy balance at the surface in the direction of elevated temperatures (Titov et al., this book).

Typical measured temperature profiles for both Venus and Earth conform reasonably well to the predictions of simple radiative-convective model calculations (Figure 3). This confirms that the processes at work are basically the same in both cases and that, unlike many aspects of the climate on Venus, there are no mysteries, at least to first order. The factor that was so surprising when it was first discovered, the high surface temperature on Venus, is a consequence of the large mass of the atmosphere, rather than any mysterious thermal process. As discussed below, this may not be too surprising either, provided we can account for the history of water on Venus.

Enough sunlight diffuses through the cloud layers on Venus to provide about 17 watts per cm^2 of average surface insolation, about 12% of the total absorbed by Venus as a whole when the atmosphere is included. Most of the energy deposited at depth cannot escape as radiation but must instead be raised by convection along an approximately adiabatic temperature-pressure profile to a level near the cloud tops where it can radiate to space. Thus, the surface and lower atmospheric temperatures must rise to force enough infrared cooling at higher levels to balance the incoming solar energy. An airless body with the same albedo and heliocentric distance as Venus would reach equilibrium for a mean surface temperature of only about 230 K. This is close to the actual temperature at the Venusian cloud tops, as we should expect if they are the most important source of thermal infrared opacity in the tropopause region. Global measurements by the Pioneer Venus Orbiter of the net infrared emission and the total reflected solar energy (Schofield and Taylor, 1982) confirmed that the planet is in overall energy balance to within the accuracy of the measurement.

To first order, a plausible explanation for the apparent superabundance of CO_2 on Venus relative to Earth is not

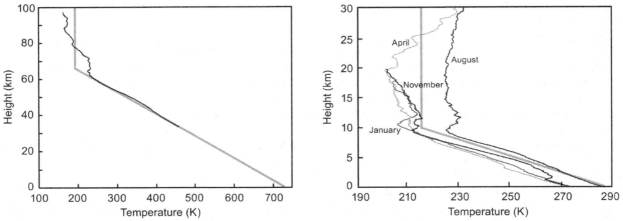

Figure 3. Simple radiative-convective equilibrium temperature profile models for Venus (left, Taylor, 2006) and Earth (right, Taylor, 2005) compared to representative measured profiles.

particularly difficult to find. It has been estimated that the carbonate rocks on the Earth hold the equivalent CO_2 amount (Kasting 1993), but since the conversion of atmospheric to crustal carbonate occurs much more efficiently in the presence of liquid water, in which the CO_2 first dissolves, the relatively water-depleted state of Venus may be responsible for so much of the gas remaining in the atmosphere. However, Venus has not always been so dry. The evidence from the D/H ratio, plus the cosmogonical argument that Venus should have accreted with similar amounts of H_2O to the Earth, could mean that Venus, too, was once covered by oceans to a considerable depth. How long this state survived is not known; nor is the abundance of carbonates in the component of Venus' crust that is, or has been, in contact with the atmosphere and hydrosphere (Donahue et al., 1997).

However, the crucial question of whether the current surface pressure on Venus is stable remains an interesting and important one. It is well known that the CO_2 abundance in Earth's atmosphere can vary, due to natural and anthropogenic factors, and that it is changing at the present time, with likely consequences for the global climate. If the climate on Venus is stable in the long term then it is likely that some mechanism provides a buffer that stabilizes the atmospheric carbon dioxide content. Since Urey (1952) proposed the exchange between atmospheric CO_2 and common minerals in the surface, it has been shown that the reaction ($CaCO_3$ + SiO_2 ⇔ $CaSiO_3$ + CO_2) reaches equilibrium at precisely the temperature and pressure found on the surface of Venus. Problems have been raised with this theory however (see, for instance, Hashimoto and Abe, 1997), including the question of how a sufficiently intimate contact between atmosphere and lithosphere is achieved. The answer is likely to be linked to the history of volcanism and the nature of the interior. The

existence of high mountains on Venus without the necessary (by analogy with Earth at least) plate tectonics that produced the large mountain ranges on Earth, and the relatively recent global resurfacing, speak of an intimate relationship between atmosphere and interior.

The thermosphere of Venus is cooler than Earth's, because of the greater abundance of carbon dioxide, which is very efficient at radiating heat to space. Above about 150 km, the temperature is approximately constant with height on the dayside at about 300K. The terrestrial thermosphere is the seat of rapid winds, up to 1000 m s⁻¹ or more, and this tends to redistribute energy originally absorbed from the Sun over the dark as well as the sunlit hemisphere. The result is a day-night difference of around 200 K about a mean temperature of 1000 K. On Venus however, the nighttime temperature in the thermosphere is very low, around 100 K. The transition from the day to night side values of temperature on Venus also show remarkably steep gradients (Keating et al., 1979) and modelers have great difficulty in reproducing both the minimum temperature and the short distance across the terminator with which it is attained.

The first-order differences between the atmospheric general circulation regimes on Venus and Earth (Figure 4) can be explained by the differences in the rotation rates of the solid bodies and in the optical depths of their atmospheres. The relative unimportance of Coriolis forces on Venus allow a single Hadley cell that extends much closer to the pole than on Earth, extending right to the edge of the polar collar without the intermediate Ferrel cell. Carbon monoxide measurements in the deep atmosphere by the NIMS experiment on the Galileo spacecraft (Collard et al., 1993), and now by Venus Express, are consistent with a deep Hadley circulation on Venus that extends from well above the clouds to the surface.

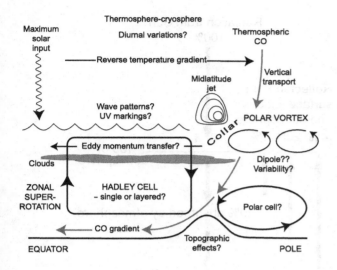

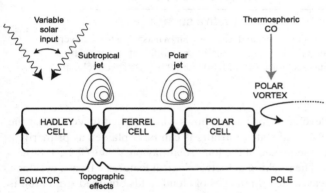

Figure 4. The main features of the atmospheric circulation on Venus and Earth. The existence of a Hadley circulation and polar vortices are common to both; Earth has a secondary cell at mid-latitudes that seems to be absent on Venus, probably because of the slower rotation. Venus exhibits rapid zonal super-rotation, which according to model experiments is a consequence of the extreme optical thickness of the atmosphere above a slowly-rotating surface.

On Earth, half of the radiant energy from the Sun is deposited at the surface (50%), with smaller proportions absorbed in the atmosphere (20%) or reflected back into space (30%). On Venus, however, the proportions are more like 3%, 21%, and 76% respectively, with the bulk of the energy absorbed by the planet deposited well above the surface in the principal cloud layers (Figure 5). GCM experiments (Lebonnois et al., 2005; Lee et al., 2005) show that global super-rotation develops in optically-thick atmospheres on slowly-rotating planets like Venus and Titan. However, the present state of model development (including the details of energy deposition profiles etc required in the model specification) is

deficient in that the predicted wind speeds are too slow, by a factor of 2 or more. The features seen in ultraviolet images rotate around the planet in a period of only 4 to 5 days, corresponding to wind velocities of more than 100 meters per second at the cloud tops, while the solid surface of Venus rotates at only about 2 meters per second, or once every 243 days. More information about cloud variability and wave modes in the atmosphere below the visible cloud tops, from repeated UV and IR mapping, should permit progress in understanding issues such as the role of the topography in maintaining or opposing the super-rotation and the role of waves or eddies in the transport of angular momentum.

Vortex behavior occurs in the polar region of any terrestrial planet, due to general subsidence of cold, dense air and the propagation of zonal angular momentum in the meridional flow. On Venus, the small obliquity and the large super-rotation lead to an extreme version of this effect, manifest by a sharp transition in the circulation regimes in both hemispheres at a latitude of about 65 degrees. There, the Hadley cell stops and we find the *circumpolar collar*, a belt of very cold air that surrounds the pole at a radial distance of about 2500 km and has a predominantly wavenumber-1 structure locked to the Sun (Plate 1). The vertical extent of the collar must be much less than its 5000 km diameter, and the indications from Pioneer Venus studies (Schofield and Taylor, 1983) and early Venus Express data are that it may be only about 10 km deep, with a complex vertical structure. The temperatures that characterize the collar are about 30°C colder than at the same altitude outside, so the feature generates pressure differences that would cause it to dissipate rapidly were it not continually forced by some unknown mechanism.

Inside the collar, the air at the center of the vortex must descend rapidly to conserve mass, and we expect to find a relatively cloud-free region at the pole, analogous to the eye of a terrestrial hurricane but much larger and more permanent. Interestingly, however, the 'eye' of the Venus polar vortex is not circular but elongated, and with brightness maxima (possibly corresponding to maximum in the downward flow) at either end of a quasi-linear feature connecting the two. This wave-2 characteristic gives the polar atmosphere a 'dumbbell' appearance in infrared images that use the thermal emission from the planet as a source, and has led to the name *polar dipole* for the feature. A dipole was first seen at the north pole by Pioneer Venus, and now a similar feature has been discovered and extensively studied at the south pole as well by Venus Express. The northern dipole was observed in successive images obtained in 1979–1980 to be rotating about the pole with a period whose dominant component, among several, was 2.7 Earth days (Schofield and Diner, 1983), i.e. with about twice the angular velocity

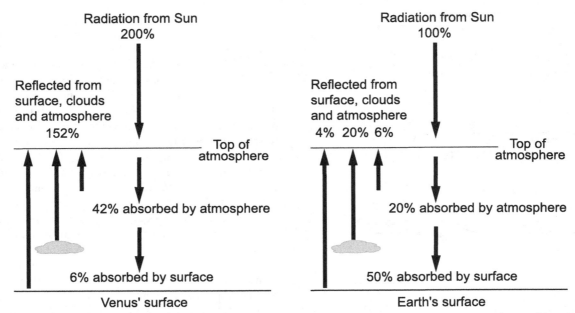

Figure 5. The different components of the radiative energy budgets of Venus and Earth are shown as planet-wide averages, taking the solar irradiance at the Earth as 100% and Venus as twice that. (Actually, the sunfall at Venus relative to that at Earth varies between 182% and 200% when the orbital eccentricities of 0.007 and 0.0167, respectively, are taken into account).

of the equatorial cloud markings. If angular momentum were being conserved by a parcel of air as it migrated from equator to pole the dipole might be expected to rotate five or six times faster. In fact, the ultraviolet markings are observed to keep a roughly constant angular velocity (solid body rotation) from the equator to at least 60° latitude, presumably accelerating poleward of this if the rotation of the dipole represents the actual speed of mass motions around the pole and not simply the phase speed of a wavelike disturbance superimposed on the polar vortex. At the time of writing, many new details of the dipole-collar structure are emerging from Venus Express VIRTIS maps, soundings, and movies that must, after detailed analysis, reveal much more of its true nature.

Interestingly, the thermal tide on Venus around the equatorial regions also has two maxima and two minima. (The thermal tide is simply the diurnal increase and decrease of temperature caused by the rising and setting of the Sun.) This does not seem to be directly connected with the polar dipole, since the two regions are separated by a narrow latitude band apparently free of planetary-scale waves, as well as by the predominantly wavenumber-one collar. The Earth's atmosphere has a small wavenumber-two component superposed on the familiar early-afternoon maximum to post-midnight minimum cycle, but this component dominates on Venus. For once, the dynamical theory of atmospheric tides, as

developed for Earth, shows when applied to Venus that the observed state of affairs is can be explained as primarily a consequence of the long solar day on Venus (Fels, 1986).

The tracking of meteorological features – fronts, cyclones, waves etc – in images of cloud fields obtained from orbit is of course a well-established technique for terrestrial research and forecasting. The Venusian equivalent was, for many decades, limited to the transient and quasi-permanent features seen in the ultraviolet images of the cloud-top region, where they revealed structures identified with Rossby and gravity wave activity (Belton et al., 1976). In the mid-1980s, this changed with the discovery of the near-infrared windows, which permitted imaging of the deep cloud structure. Before Venus Express, the best example of this was obtained by the Near Infrared Mapping Spectrometer (NIMS) on Galileo (Carlson et al., 1993) when it flew past Venus in February 1990 (Plate 2). The cloud patterns it imaged, with good spatial but very limited time resolution, were not obviously associated with the uv markings near the Venusian cloud tops, some 20 km higher in the atmosphere. Nor do they much resemble the familiar terrestrial global cloud patterns imaged daily by Earth weather satellites (Plate 2), although there must be similarities in the basic processes at work. Spectacular meteorological activity is clearly present in the deep atmosphere of Venus, and it is one of the main goals of the VIRTIS and VMC instruments on Venus Express

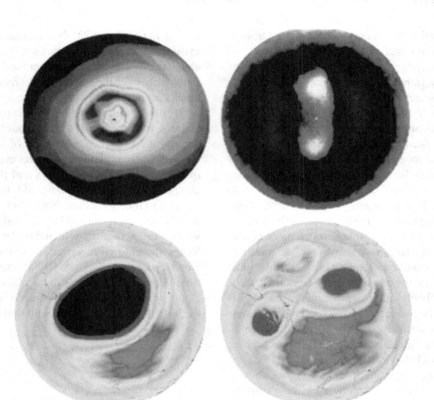

Plate 1. Polar vortices on Venus and Earth, shown respectively as radiance maps at a wavelength of 11.5 microns in the thermal infrared from Pioneer Venus Orbiter (Taylor et al., 1980) and as ozone maps from the Sciamatchy and GOME instruments on Envisat (ESA). The upper frames show a 72-day average of the Venus vortex (top left), showing the cold polar collar, and a similar average in a coordinate frame rotating every 2.7 days, showing the dipole structure. The lower frames show (left) the terrestrial ozone hole on 23 September 2003, when the polar vortex was fully developed, and (right) during its break-up in 2003, when a wavenumber-2 instability developed.

Plate 2. Meteorology on Venus and Earth as traced by cloud patterns, viewed in near-infrared and visible imaging, repectively, by the Galileo spacecraft during fly-by encounters with both planets on its journey to Jupiter.

and the UV and infrared cameras on Venus Climate Orbiter to track this in detail and hopefully elucidate its nature.

Certain minor atmospheric constituents on Earth and Venus are of primary importance to reveal key production and loss processes, act as tracers of the dynamics, and define the cloud chemistry. On Earth, these are water vapor, carbon monoxide and ozone; on Venus, water vapor, carbon monoxide and the SO_x family (i.e. sulfur dioxide and related compounds). The abundance of ozone on Venus is negligible but there is commonality between the two planets for the others.

The issue of the bulk abundances of water and carbon dioxide, where Venus appears to have lost most of the former but, as a result, retained in its atmosphere much more of the latter, has already been discussed. A measurable amount of water remains on Venus, as vapor above, within and below the clouds, plus an unknown quantity combined with sulphuric acid and probably other compounds in the cloud particles themselves. The indications from spectroscopic and entry probe data are, tentatively, that the H_2O abundance is fairly constant across the globe near the surface, but highly variable in the clouds and above (Koukouli et al., 2005). The water vapor measurements prior to Venus Express above, below and within the cloud layers show a baffling disparity that is presumably, by analogy with Earth, linked to cloud formation and dissipation processes and meteorological activity in Venus' atmosphere (Ignatiev et al., 1999; Koukouli et al., 2005). Systematic new measurements from a long-lived orbiter, sounding within and below the clouds for the first time, could radically improve our understanding of these.

Several of the other minor constituents in Venus's atmosphere also exhibit striking amounts of temporal and spatial variability, with glimpses of terrestrial analogies that can be followed up with new data. During the Galileo fly-by in 1991, NIMS near infrared measurements revealed an equator-to-pole gradient in the abundance of tropospheric carbon monoxide (Figure 6; Collard et al., 1993), which Taylor (1995) suggested was unlikely to be volcanic in origin but could be the result of a hemispherical Hadley circulation that extended from the lower thermosphere at around 120 km all the way down to the surface. While the Galileo data had large uncertainties and limited high-latitude coverage, early Venus Express data is confirming the equator-to-pole gradient seen by NIMS and the symmetry between hemispheres we would expect on a planet without seasons. The lower frame in Figure 6 shows measurements by the Improved Stratospheric and Mesospheric Sounder (ISAMS) instrument on the Upper Atmosphere Research Satellite of the seasonal CO profile on the Earth, which shows what seems to be a similar effect, in this case known to be due to the descent of

air rich in CO from CO_2 dissociation in the mesosphere. The main differences from Venus are the generally smaller CO abundances, and the fact that, enhanced values are found on Earth only over the winter pole, since the terrestrial vortex breaks up in the summer.

Of all the questions Venus Express was designed to address, near the top of the list would come the gathering of data on the existence and extent of active volcanism at the surface. There is plenty of indirect evidence, including radar images of massive lava flows and the profusion of sulfur compounds in the atmosphere. SO_2 has more than 100 times the abundance expected from chemical equilibrium with the surface, according to Fegley et al (1997). Sulfur dioxide was observed in ultraviolet measurements made by Pioneer Venus showing large variations in its mixing ratio near the cloud tops, which Esposito (1984) interpreted as evidence for time-dependent eruptions. It may be that parameters like the cloud opacity at solar and infrared wavelengths depend in the long term on volcanism, and that when the latter finally

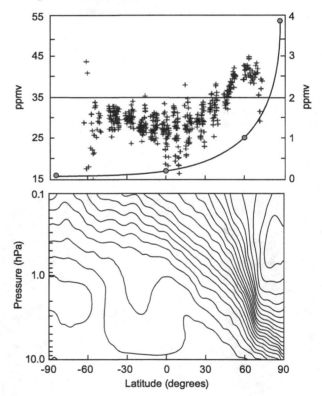

Figure 6. Carbon monoxide latitudinal profile on Venus (top. from Galileo NIMS, Collard et al., 1993) and Earth (December 1991, from UARS ISAMS, Allen et al., 1999). The curve in the upper figure is the latitudinal trend on Earth, taken from the lower figure, at the same approximate height (about 30 km) that applies to the Venus data, so the two can be compared (note the different abundance scales, for Venus on the left axis and for Earth on the right).

subsides the climate on Venus could change radically. On Earth, a single eruption like that of Pinatubo in 1992 had a measurable effect on the climate (Hansen et al., 1996). Venus Express can search for active volcanoes by searching for hot lava flows on the surface and estimate their contribution to the climate system by measuring the composition of sulfur-rich plumes of volcanic gases in the atmosphere. It may be possible to calculate approximately if the heat released from Venus' interior via volcanoes is comparable to that which the Earth discharges by convective and tectonic activity.

Another long-standing puzzle, this time without any obvious terrestrial analogue, is the origin of the contrasts in the cloud markings seen in ultraviolet images of Venus from space. Apparently, some ultraviolet-absorbing substance is non-uniformly dispersed through the clouds. Sulfur dioxide behaves in this way, and is definitely present in spectroscopic observations, but its spectrum does not match that of Venus precisely at all wavelengths. Some other material, probably another sulfur compound or even one of the allotropes of elemental sulfur, which also absorbs ultraviolet but not visible radiation, must be contributing also. The spectrometers on Venus Express are more advanced than any deployed previously and may shed light on this issue, and on aspects of the fairly mysterious question of Venus cloud chemistry in general.

On both planets, the surface is both a source and a sink for key atmospheric gases, and so is the boundary between the top of the atmosphere and space. An unknown amount of material, often assumed to be negligible but not necessarily so, arrives continuously as cometary and meteoritic debris. At the same time, molecules are lost to space by thermal escape and erosion by photon and particle erosion, mainly from the solar wind. A key question for the study of the induced magnetosphere around Venus is the role of the deflected solar wind in carrying off atoms and ions derived from atmospheric molecules, particularly the lighter elements, and especially hydrogen. Venus may once have had a massive ocean that was slowly removed by dissociation of water vapor in the upper atmosphere and subsequent loss of hydrogen. This scenario is supported, not just by the expectation that Venus was initially water-rich like the Earth, but also by the strong evidence of fractionation in the isotopes of hydrogen found on Venus, where the deuterium-to-hydrogen ratio is more than 100 times that found on Earth or in meteorites. The implications for the present high surface pressure and consequent extreme climate on Venus are obvious: the mass of the current atmosphere on any planet represents a balance between emissions from the crust by volcanism, the chemical recombination of atmospheric molecules with the surface, and escape to space. The details and relative proportions of these processes and budgets, current and historical,

remain quite unknown, making models of the origin and future evolution of Venus' atmosphere highly uncertain at present.

The Neutral Mass Spectrometer on Pioneer Venus Orbiter observed CO_2, CO, O, N_2, N, NO, He and H in the upper atmosphere of Venus. The UV spectrometer on the same mission established the presence of a corona of hot atoms, mainly H, O and C, around the planet. The homopause on Venus is at an altitude of about 135 km, leading to an increasing preponderance of the lighter species above this level. However, the lightest, atomic hydrogen and helium, are minor constituents in the atmosphere as a whole while atomic oxygen is produced in large quantities in the upper atmosphere by the photodissociation of carbon dioxide by solar UV, viz. $CO_2 + h\nu \rightarrow CO + O$. The net effect is that O atoms are the dominant species at altitudes above 170 km, especially during the daytime, followed by atomic hydrogen, helium, and molecular hydrogen (von Zahn et al., 1980).

Model calculations that take into account the distinctly different magnetospheric physics at Venus, compared to Earth (Figure 7) have compared the various loss processes for these species. A key question is whether the net loss rates, by all processes, for hydrogen and oxygen are in the ratio 2:1 as would be expected if the source molecule is water vapor and there are no large sinks of atmospheric oxygen on Venus's surface. Within large uncertainties, Lammer et al. (2005) find that this may indeed be the case for Venus, although not for Mars where the ratio is about 4 times larger and a surface sink for oxygen almost certainly must be invoked. The ASPERA and Magnetometer measurements by Venus Express should further elucidate this ratio and the loss rates for deuterium, and other species, and help improve evolutionary climate models.

SUMMARY AND LONGER-TERM GOALS

A new picture of the climate on Venus will emerge from the results expected from the entire suite of instruments on Venus Express. Of course, this one mission, even if completely successful, will not fully resolve the many issues touched on in the account given above. However, the payload is designed to advance the current state of planetary climatology, by the acquisition and dissemination of new knowledge about the Venusian climate and its place in our understanding of the climate regimes on all of the terrestrial planets (including Earth, Mars, and, for some purposes, Titan). With the successful arrival and deployment of the spacecraft and its payload (assuming eventual success at activating the Planetary Fourier Spectrometer), progress should certainly be achieved by Venus Express in the following key areas and objectives:

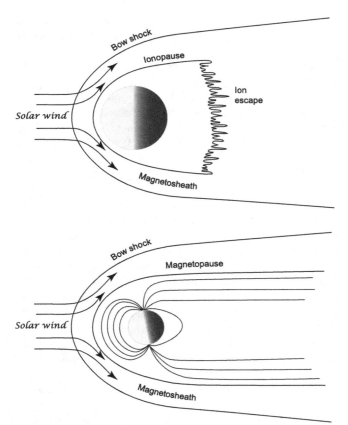

Figure 7. A simplified representation of the magnetospheres of Venus (top) and Earth (below). In both cases the solar wind, a supersonic plasma flow, produces a bow shock when decelerated by the planetary obstacle. At Earth, the particles are deflected by the magnetic field, while at Venus they impinge on the upper atmosphere at the ionopause, located where the solar wind dynamic pressure is balanced by the thermal pressure of the exosphere. Photo ionisation, charge exchange and electron impact ionisation lead to the removal of ionised atmospheric components by the plasma flow in a comet-like tail.

• detection of volcanic activity and better quantification of the volcanic gas inventory in the atmosphere

• improved knowledge of vertical cloud structure, microphysics and variability

• updated inventories of minor constituent abundances

• atmospheric temperature fields above, in and below the clouds

• new observational constraints from mapping on the general circulation and dynamical phenomena like the polar vortices and deep atmosphere 'weather'

• improved estimates of atmospheric loss rates for O, C, H, and D

• detection of any interannual and interhemispheric asymmetries and trends in all of the above.

Potential climate-related advances are not limited to these because of the exploratory and innovative aspects of the mission. Despite the fact that Venus Express will be the twenty-eighth spacecraft to arrive successfully at Venus since Mariner 2 in 1962, there is still considerable scope for serendipitous discoveries. These could derive, for example, from the fact that Venus Express will be the first mission to employ the near infrared transparency windows, discovered in the 1980s (see Taylor et al., 1997), from orbit, and so the first to carry out systematic remote sensing observations of the Venusian atmosphere below the clouds. Conversely, new findings will also pose new questions requiring further missions and new measurements to resolve.

Both expected and serendipitous findings may be used as a basis for:

• producing improved greenhouse models of the energy balance in the lower atmosphere

• validating and improving general circulation models of the atmosphere, with improved treatment of the zonal super-rotation, the meridional Hadley circulation, and the polar vortices

• generating new climate evolution models using simple physics constrained by measurements, and

• comparative studies in all three areas with the other terrestrial planets including Earth.

Venus Express will not address, let alone resolve, every one of the key questions about Venus that have accumulated as a result of exploration by the Venera, VEGA, Pioneer and Magellan missions. The knowledge gaps that will remain that can be predicted in advance are mostly in the area of atmospheric evolution (addressable by accurate measurements of noble gas isotopic ratios, for instance) and composition (a full understanding of surface-atmosphere interactions, cloud composition and chemistry, will require in-situ trace constituent abundance measurements, especially at the surface and in the clouds). Other areas that will be largely untouched by Venus Express are surface geology, geochemistry, and interior structure, and surface-atmosphere and surface-interior interactions. In this case, landed missions and sample return may be the optimum way forward.

REFERENCES

The Venus Express mission and the individual experiments and their objectives are described in a series of papers that can be found in two special issues of *Planetary and Space Science*, in press 2006, and these papers are not individually referenced here.

Similarly, detailed reviews of knowledge of Venus are given in the volumes *Venus* (1983) and *Venus 2* (1997), Univ. Ariz. Press, Tucson, AZ, and these papers are not individually referenced here either, except where they are directly quoted.

Allen, D. R., J. L. Stanford, M. A. Lopez-Valverde, N. Nakamura, D. J. Lary, A. R. Douglass, M. C. Cerniglia, J. J. Remedios and F. W. Taylor. Observations of Middle Atmosphere CO from the UARS ISAMS During the Early Northern Winter 1991/1992. *Journal of the Atmospheric Sciences*, 56, 563–583, 1999

Belton, M. J. S., G. R. Smith, G. Schubert, and A. D. Del Genio. Cloud patterns, waves and convection in the Venus atmosphere. *J. Atmos. Sci.* 33, 1394–1417, 1976.

Carlson, R. W., L. W. Kamp, K. H. Baines, J. B. Pollack, D. H. Grinspoon, Th. Encrenaz, P. Drossart, and F. W. Taylor, 1993. Variations in Venus cloud particle properties: a new view of the Venus' cloud morphology as observed by Galileo near-infrared mapping spectrometer. *Planet. Space Sci.* 41, pp. 477–485.

Collard, A. D., Taylor, F. W., Calcutt S. B., Carlson, R. W., Kamp L., Baines, K., Encrenaz, Th., Drossart, P., Lellouch, E., and Bézard, B. Latitudinal distribution of carbon monoxide in the deep atmosphere of Venus. *Planetary and Space Science*, 41, 7, 487–494, 1993.

Donahue, T. M., Grinspoon, D. H., Hartle, R. E., and Hodges R. R. Jr., 1997. Ion neutral escape of hydrogen and deuterium: evolution of water. In *Venus II*, Univ. Ariz. Press, Tucson, AZ, pp. 385–414.

Esposito, L. Sulphur Dioxide: Episodic Injection Shows Evidence for Active Venus Volcanism. *Science*, 223: 1072–1074, 1984.

Fegley, B. Jr, G. Klinghoefer, K. Lodders, and T. Widemann, 1997. Geochemistry of surface-atmosphere interactions on Venus. In *Venus II*, Univ. Ariz. Press, Tucson, AZ, pp. 591–636.

Fels, S. B. An Approximate Analytical Method for Calculating Tides in the Atmosphere of Venus. Journal of the Atmospheric Sciences: Vol. 43, No. 23, pp. 2757–2772, 1986.

Hansen, J., Mki. Sato, R. Ruedy, A. Lacis, K. Asamoah, S. Borenstein, E. Brown, B. Cairns, G. Caliri, M. Campbell, B. Curran, S. de Castro, L. Druyan, M. Fox, C. Johnson, J. Lerner, M. P. McCormick, R. L. Miller, P. Minnis, A. Morrison, L. Pandolfo, I. Ramberran, F. Zaucker, M. Robinson, P. Russell, K. Shah, P. Stone, I. Tegen, L. Thomason, J. Wilder, and H. Wilson 1996. A Pinatubo climate modeling investigation. In *The Mount Pinatubo Eruption: Effects on the Atmosphere and Climate* (G. Fiocco, D. Fua, and G. Visconti, Eds.). NATO ASI Series Vol. I 42, pp. 233–272. Springer-Verlag. Heidelberg, Germany.

Ignatiev N.I.; Moroz V.I.; Zasova L.V.; Khatuntsev I.V. Water vapour in the middle atmosphere of Venus: An improved treatment of the Venera 15 IR spectra. *Planetary and Space Science*, Volume 47, Number 8, 9 August 1999, pp. 1061–1075(15).

Kasting, J. F. Earth's Early Atmosphere, *Science,* 259, 5097, 920–926, 1993.

Keating, G. M., Taylor, F. W., Nicholson, J. Y., and Hinson, E. W. Short term cyclic variations of the Venus upper atmosphere. *Science,* 205, 62–65, 1979.

Koukouli, M. E., P. G. J. Irwin, and F.W. Taylor, 2005. Water vapour abundance in Venus' middle atmosphere from Pioneer Venus OIR and Venera-15 FTS measurements. *Icarus* 173, pp. 84–99.

Lammer, H., H. I. M. Lichtenegger, H. K. Biernat, N. V. Erkaev, I. Arshukova, C. Kolb, H. Gunell, A. Lukyanov, M. Holmstrom, S. Barabash, T. L. Zhang, and W. Baumjohann. Loss of hydrogen and oxygen from the upper atmosphere of Venus, submitted to Planetary and Space Science, 2005.

Lebonnois, S.; Hourdin, F.; Eymet, V.; Fournier, R.; Dufresne, J.-L. A new Venus General Circulation Model, in the context of the Venus-Express mission. American Astronomical Society, DPS meeting #37, #54.07, 08/2005.

Lee, C., Lewis, S.R., and Read, P.L. A numerical model of the atmosphere of Venus, Adv. Space Res., 36, 11, 214–2145, 2005.

Schofield, J.T. and Taylor, F.W.. Net global thermal emission from the Venus atmosphere. *Icarus,* 52, 245–262, 1982.

Schofield, J. T. & Diner D. J. Rotation of Venus's polar dipole. *Nature* 305, 116–119, 1983.

Schofield, J.T. and Taylor, F.W. Measurements of the mean solar-fixed temperature and cloud structure of the middle atmosphere of Venus. *Quarterly Journal of the Royal Meteorological Society,* 109, 57–80, 1983.

Svedhem, H., D.V. Titov, D. McCoy, J.-P. Lebreton, S. Barabash, J.-L. Bertaux, P. Drossart, V. Formisano, B. Häusler, O. Korablev, W.J. Markiewicz, D. Nevejans, M. Pätzold, G. Piccioni, T.L. Zhang, F.W. Taylor, E. Lellouch, D. Koschny, O. Witasse, M. Warhaut, A. Accomazzo, J. Rodriguez-Canabal, J. Fabrega, T. Schirmann, A. Clochet, M. Coradini. Venus Express – the first European mission to Venus. *Planetary and Space Science,* in press.

Taylor, F. W. Carbon Monoxide In The Deep Atmosphere Of Venus. *Adv. Space Res.,* 16, 6, 81–88, 1995.

Taylor, F. W., Beer, R., Chahine, M. T., Diner, D. J., Elson, L. S., Haskins, R. D. , McCleese, D. J., Martonchik, J. V., Reichley, P.E., Bradley, S.P., Delderfield, J., Schofield, J.T., Farmer, C.B., Froidevaux, L., Leung, J., Coffey, M.T., and Gille, J.C. Structure and meteorology of the middle atmosphere of Venus : infrared remote sounding from the Pioneer Orbiter. *J. Geophys. Res,* 85, 7963–8006, 1980.

Taylor, F. W., D. Crisp, and B. Bezard, 1997. Near-infrared sounding of the lower atmosphere of Venus, In *Venus II*, Univ. Ariz. Press, Tucson, AZ, pp. 325–351.

Taylor, F. W. *Elementary Climate Physics.* Oxford University Press, July 2005.

Taylor, F. W. Climate Variability On Venus And Titan, Space Science Series of ISSI, Vol.19, Solar Variability and Planetary Climates, Springer, in press 2006.

Taylor, F. W., 2006. Venus before Venus Express, *Planet. Space Sci.,* 54, 1249–1262, 2006.

Titov, D. V., H. Svedhem , D. Koschny, R. Hoofs, S. Barabash, J.-L. Bertaux, P. Drossart, V. Formisano, B. Häusler, O. Korablev, W.J. Markiewicz, D. Nevejans, M. Pätzold, G. Piccioni, T. L. Zhang, D. Merritt, O. Witasse, J. Zender, A. Accommazzo, M Sweeney, D. Trillard, M. Janvier, and A. Clochet. Venus Express science planning, *Planet. Space Sci.*, **54**, 1279–1297, 2006.

Titov, D. V., M. Bullock, D. Crisp, N. Renno, F. W. Taylor, and L. V. Zasova. Radiation in the atmosphere of Venus, this book.

Urey, H.C. *The Planets*, Yale University Press, New Haven, 1952.

Von Zahn, U., S. Kumar, H. Niemann, and R. Prinn (1983), Composition of the Venus atmosphere, in Venus, edited by D. M. Hunten, L. Colin, T. M. Donahue, and V. I. Moroz, pp. 299–430, Univ. Arizona Press, Tucson.

Experiencing Venus: Clues to the Origin, Evolution, and Chemistry of Terrestrial Planets via In-Situ Exploration of our Sister World

Kevin H. Baines[1], Sushil K. Atreya[2], Robert W. Carlson[1], David Crisp[1], David Grinspoon[3], Christopher T. Russell[4], Gerald Schubert[5], and Kevin Zahnle[6]

We review the current state of knowledge of (1) the origin and evolution of Venus and (2) the photochemical and thermochemical processes occurring in the middle and lower atmosphere there. For each, we review the promise of on-going and planned orbital observations by ESA's Venus Express and Japan's Venus Climate Orbiter missions. We review the need for future *in-situ* measurements for understanding Venus origin and evolution and present-day chemistry, and the implications for understanding the origin and history of the Earth and other bodies in the inner solar system, as well as for understanding terrestrial planets in other solar systems. We prioritize the goals remaining in the post Venus Express era, based on the Decadal Survey (National Research Council, 2003). Using past experience with Pioneer Venus, VEGAs, Veneras, and, most recently, Venus Express as guides, we suggest appropriate techniques and measurements to address these fundamental science issues.

1. INTRODUCTION

The exploration of Venus is experiencing a renaissance as space agencies in Europe and Japan develop, launch, and execute missions to Earth's sister planet. On April 11, 2006,

[1] Jet Propulsion Laboratory, California Institute of Technology, Pasadena, California, USA.

[2] Department of Atmospheric, Oceanic, and Space Sciences, The University of Michigan, Ann Arbor, Michigan, USA.

[3] Department of Space Sciences, Denver Museum of Nature and Science, Denver, Colorado, USA.

[4] Institute of Geophysics and Planetary Physics, University of California, Los Angeles, California, USA.

[5] Dept. of Earth and Space Sciences, Institute of Geophysics and Planetary Physics, University of California, Los Angeles, California, USA.

[6] NASA/Ames Research Center, Moffett Field, California, USA.

Exploring Venus as a Terrestrial Planet
Geophysical Monograph Series 176
Copyright 2007 by the American Geophysical Union.
10.1029/176GM11

the Venus Express orbiter was placed into orbit about Venus by the European Space Agency. As a rapid-development follow-on to Mars Express, this relatively small (~ $300 M), fast (~ 3 years from selection to launch) mission uses the Mars-Express-based spacecraft bus and an array of instruments inherited from the Rosetta and Mars Express missions to provide the most systematic, thorough and complete investigation of Venus's complex atmosphere yet achieved.

During its 16-month prime science mission, which began in June 2006, the Venus Express orbiter circles the planet daily in a polar elliptical orbit, regularly imaging and sounding the planet from the upper atmosphere, through its various middle-level cloud layers, down to the depths of the lower atmosphere and surface (c.f., Baines *et al.*, 2006; Taylor, 2006; and Titov *et al.*, 2006 for mission summaries). A variety of UV, visible, near-IR, mid-IR, and radio techniques map clouds and chemically-reactive species throughout the atmosphere, enabling a new understanding of Venus's active photo- and thermochemical processes. In addition the nadir-viewing mapping capabilities of the Visible and Infrared Thermal Imaging Spectrometer (VIRTIS) provide the potential for mapping near-IR surface properties and a search for outgassing and thermal emissions indicative of active volcanism. The Venus Radio experiment (VeRa) periodically acquires detailed radio occultation profiles down to the ~ 36 km level. It also probes the surface in its bistatic radar mode, using Earth-based radio telescopes to receive the echoes of VeRa signals directed at the surface. In the upper reaches of the atmosphere, a sophisticated fields and particles experiment—known as Analyser of Space Plasmas and Energetic Atoms (ASPERA)—measures the solar wind and the flux of ionized species leaving Venus, thereby elucidating the present-day rate of loss of hydrogen and other species from the planet. Currently, after six months in orbit, savings in the fuel needed to maintain a stable orbit suggests that the mission is on course for a planned 16-month mission extension, with potential for a similar second extension, thus providing up to ten Venus solar days of on-orbit observations by June, 2010.

As currently planned, in that same year, 2010, the Japan Aerospace Exploration Agency (JAXA) will launch the Venus Climate Orbiter (VCO). Its suite of four cameras will repeatedly image the planet's cloud systems at two levels and surface from an equatorial orbit. Together, as currently planned, Venus Express and VCO will globally map both the observed winds and theoretically-derived equilibrium wind field of Venus throughout more than a dozen Venus days, thereby providing information extremely useful for determining the nature of dynamical processes powering the planet's baffling world-wide super-rotating zonal winds.

Such missions are also providing fundamental new data on the nature of the planet's surface. In particular, near-infrared maps in the 0.9–1.2-μm spectral range by VIRTIS on Venus Express are beginning to reveal in detail the spatial variability of thermal surface emission, due to variations in elevation and surface composition. As first discovered by Allen (1984) in groundbased observations and further revealed by the Galileo and Cassini flybys (Carlson *et al.*, 1991; Baines *et al.*, 2000), several useful spectral "windows" between CO_2 bands exist in the 0.9–1.2-μm range which allow thermal emission from Venus's hot (~ 740K) surface to be readily viewed under nighttime conditions wherein the strong glint from sunlit clouds is suppressed. The adiabatic lapse rate of 8K per km altitude ensures dramatic variations of surface thermal flux with elevation, as previously demonstrated by the flyby mapping of the Near Infrared Mapping Spectrometer (NIMS) on-board the Galileo spacecraft enroute to Jupiter (Carlson *et al.*, 1991, 1993a) and by ground-based observations (Lecacheux *et al.*, 1993; Meadows and Crisp, 1996). Venus Express and VCO will exploit these windows to globally map surface properties at a spatial-resolution of ~ 50 km, as limited by the diffusion of upwelling radiation caused by atmospheric

scattering. Incorporation of known elevations from Magellan and Pioneer Venus Orbiter radar measurements in the analysis is likely to enable the separation of elevation effects from effects due to variations in the emissivity of surface materials, thus providing meaningful constraints on the distribution of surface materials over the globe (Hashimoto and Sugita, 2003). In addition, both missions plan to conduct searches for active volcanism by watching for unusual spatial and/or temporal anomalies in surface brightness (Hashimoto and Imamura, 2001) or low-altitude abundances of volcanic gases (such as SO_2). The discovery and characterization of active volcanism—the first for any terrestrial planet beyond Earth—would provide a fundamental key to understanding the planet's geology, interior, atmospheric evolution, and climate, as well as provide a deeper perspective on the geologic nature of terrestrial planets throughout the universe.

Additional impetus for the new renaissance in Venus exploration has recently come from the U. S. planetary science community. In 2003, under the auspices of the Space Studies Board of the National Research Council, the community's Solar System Exploration Survey (SSES) produced its report "New Frontiers in the Solar System: An Integrated Exploration Strategy" (National Research Council, 2003). A major finding of this "Decadal Survey", as this report is informally known, is the need for direct, *in-situ* sampling of Venus, of both its atmosphere and surface. Despite the intense orbital scrutiny provided by Venus Express and VCO, there are major questions dealing with the evolution and chemistry of the planet, as well as its dynamics, that can only be appropriately addressed via *in-situ* measurements. These include measurements of the abundance of inert noble gases and their isotopes—which are readily accessible and bear the scars of Venus' origin and evolution—as well as a wide range of chemically-active gases important for understanding Venus' present-day meteorology and chemical cycles. None of the noble gases and just a handful of reactive gases can be measured remotely from orbiting spacecraft. Thus, as recognized by the Decadal Survey, *in-situ* atmospheric sampling is the essential next step. Many measurements can only be made by *in situ* sampling, and *in situ* sampling provides "ground truth" for the gases sounded remotely by Venus Express, VCO, and ground-based spectroscopy.

In this paper, based largely on the findings of the Decadal Survey (NRC, 2003), we review the need for *in-situ* sampling of the Venus atmosphere, concentrating on the scientific objectives that can be met by relatively small missions, such as may be affordable under NASA's Discovery Program. We then expand the discussion to cover all of the high-priority Venus science objectives identified by the Decadal Survey, and summarize viable mission architectures that can accomplish each of these objectives.

2. THE NEED FOR *IN-SITU* EXPLORATION OF VENUS

More than any other pairing of terrestrial planets, Venus and Earth are virtual twins. To within 20%, they have the same radius, volume, mass, and gravity. They also have similar bulk compositions. At the tops of their major clouds, they also have similar temperatures and pressures. Furthermore, based on a simple, first-order analysis of the solar flux and the deposition of solar energy as determined from their respective solar distances and global geometric albedos, they also have the same effective temperature.

Yet, Venus does not follow our expectations. Its mean surface temperature is nearly 2.6 times that of Earth (740 K vs 288 K terrestrial mean). Despite a rate of rotation more than 100 times slower than Earth's, its solar-powered 1-bar-level winds flow globally at an average speed greater than the sustained speed of the Earth's mightiest jet streams. Both the surface and atmosphere of Venus are exceedingly dry (~30 ppm water vs ~ 30,000 ppm in Earth's atmosphere). Its ubiquitous clouds are composed not of water, but of sulfuric acid, indicative of a complex cycle of sulfur-based corrosive chemistry throughout the planet's extensive 94-bar thick atmosphere and surface.

Thus, despite their bulk similarities, Venus is not at all like Earth. It is an alien place, seemingly inhospitable to the types of long-duration, *in-situ* exploration thus far accomplished on Earth, the Moon and Mars. The strange, alien nature of Venus is linked to it history. Due to both cataclysmic and gradual events in the past, Venus proceeded along a distinctly different evolutionary path than Earth. Answering fundamental questions of Venus's history—as can only be done with *in-situ* sampling of ancient materials imbedded within the planet's atmosphere and surface—will undoubtedly reveal key lessons for the past, present, and future of Earth's environment and climate, as well as all terrestrial planets.

As noted by the Decadal Survey (NRC, 2003), direct sampling of atmospheric constituents is key to understanding how the planet formed and evolved as well as understanding the complex chemistry coupling the surface and atmosphere. In particular, the heavy noble gas elements xenon, krypton, and argon and their isotopes provide an accessible historical record of ancient events, pertinent to planetary formation and early evolutionary processes. Radiogenic isotopes of helium, argon and xenon also provide dating constraints on geologic processes that over the eons may have delivered materials from the deep interior to the surface.

The remarkable property that renders the noble gases and their isotopes so valuable in determining ancient events is their stability against chemical alterations. Unfortunately, this stability against chemical reactions manifests itself as well in an absolute stability against photonic interactions. Thus no spectral features exist. Consequently, the abundances of these elements cannot be readily assessed by remote sensing techniques, as from the Venus Express orbiter. Only *in-situ* sampling can do it.

Beyond sampling of noble gases for understanding the historical record, the sampling of the plethora of other constituents present in Venus's atmosphere is extremely valuable for understanding how Venus works today. Precise, simultaneous sampling of a variety of sulfur-bearing constituents and related molecules over a wide range of altitudes, latitudes, longitudes, and time can yield important insights into cloud formation/dissipation, the sulfur cycle, meteorology, and climate. Such insights can be significantly enhanced by simultaneous measurements of dynamically related processes such as pressure and temperature, vertical and horizontal winds, and cloud properties (*e.g.,* particle sizes and distributions). Such precise sampling at specific altitudes can only be achieved *in-situ*. Ultimately—since atmospheric sulfur is the product of either volcanism and/or tectonic activity– these atmospheric measurements bear directly on understanding the current state and recent history of volcanic and geologic processes.

3. THE EVOLUTION OF VENUS: EVIDENCE FROM BULK PROPERTIES AND RADAR IMAGERY

The similarity of Venus and Earth in size, mass and distance from the Sun suggests that Venus and Earth had similar early histories of accretion and internal differentiation. Nevertheless, Venus and Earth proceeded along different evolutionary paths, as is evident from a comparison of the planets' satellites, magnetic fields, tectonics, rotational states, and atmospheres.

Venus does not have a moon, but the reason for this is not understood. Of course, it could simply be by chance, but Venus was likely subject to major impacts in its early evolution. Such an impact with Earth apparently produced its moon. The lack of a Venusian moon has been attributed to inward spiraling and collision with the planet of a satellite that might once have existed. It has also been explained by satellite escape. However, neither explanation is plausible. The inward spiraling scenario is unlikely since post giant impact rotation and orbital motion tend to be in the same sense, contrary to the direction of Venus's rotational spin today. The escape scenario is unlikely for reasonable Q values and moon masses (Stevenson, 2006).

Venus does not have an intrinsic magnetic field (Russell, 1980; Phillips and Russell, 1987; Donahue and Russell 1997, Nimmo, 2002), a fact that has at least two reasonable explanations both connected with the planet's thermal evolution. On the assumption that Venus was initially hot and differentiated, it has long possessed a liquid metallic core. Hence it is possible that the planet had an early dynamo-generated magnetic field. Whether such a field existed and persisted for some length of time depends on the core cooling history. The rate of core cooling, controlled by the ability of the mantle to extract heat from the core, must be sufficient to support convection and dynamo action in the core. This suggests one explanation for the absence of a present day magnetic field. Venus underwent a transition during its evolution from efficient core cooling by a plate tectonic-like mode of mantle convection to an inefficient mode of stagnant or sluggish lid mantle convection that was unable to cool the core rapidly enough to support dynamo activity. Nimmo and Stevenson (2000) have proposed a similar explanation for the absence of a present day Martian magnetic field. A potential test of the existence of an ancient dynamo is the detection of remanent magnetization from this epoch. However the high surface temperature on Venus makes this detection unlikely but not necessarily impossible.

A second explanation for the lack of a magnetic field is also related to inefficient core cooling by the mantle and the lower pressure in Venus' core compared with the pressure in Earth's core due to Venus's smaller mass. It is possible that Venus's core has not yet cooled sufficiently to initiate inner core growth, but has cooled enough to prevent the operation of a purely thermally driven dynamo (Stevenson et al., 1983). It is believed that dynamo action in Earth's core is substantially facilitated by compositional convection driven by inner core solidification. The absence of plate tectonics on Venus is consistent with a style of mantle convection that is inefficient in cooling its core. That Venus might have a liquid core at present is indicated by Doppler radio tracking data of the Magellan and Pioneer Venus spacecraft which have provided an estimate of the tidal Love number k_2 (Sjogren et al., 1997). The Love number k_2 is determined from the time variations in the gravitational coefficients C_{22} and S_{22} at the solar period. Konopliv and Yoder (1996) have found $k_2 = 0.295 \pm 0.066$, a value consistent with a liquid core (Yoder, 1997).

Magellan radar global imagery strongly suggests that Venus lacks plate tectonics, a conclusion supported by the almost total absence of a global system of ridges and trenches similar to those found in the oceans on Earth (Kaula and Phillips, 1981). At the present time, Venus is a one-plate planet. Mantle convection in Venus must therefore be of the sluggish or stagnant lid mode, a relatively inefficient way to cool the core (Schubert et al. 1997). The absence of plate tectonics on Venus could have something to do with the planet's lack of water. If Venus had water early in its history

it could have had a style of mantle convection similar to plate tectonics. With a plate tectonic-like mantle convection regime there might also have been a magnetic field. It is therefore possible that atmospheric evolution, by way of an intense greenhouse effect and loss of water, had a controlling influence on the evolution of the entire solid planet, shutting down plate tectonics, core cooling, and dynamo action. The atmosphere, a relatively minor part of Venus by mass, could have controlled the thermal and rotational history of the entire planet.

The absence of plate tectonics on Venus and its implications for the relatively inefficient removal of heat from the planet's interior could account for a hypothesized major event in Venus's evolution. It has been proposed that the relatively young age of Venus's surface (about 500–800 Myr according to studies of the impact crater population, McKinnon et al. 1997, Herrick et al. 1997) was set in a global volcanic resurfacing event and that relatively little volcanism has occurred since (Schaber et al. 1992, Basilevsky et al., 1997). The resurfacing event could be the means by which Venus expels its heat. One way this could happen is by the global foundering of a thick, relatively cold and heavy lithosphere and its replacement by the relatively hot underlying mantle. Such events might have occurred episodically throughout Venus's history. Between such events the lithosphere would thicken but Venus would have no efficient way, like plate tectonics on Earth, to expel its heat. Instead the heat building up in the interior during the quiescent period would be lost in the mantle overturn when the lithosphere thickened enough to become gravitationally unstable. The initiation of such an event might be evident today on Venus's surface in the form of large coronae. McKenzie et al. (1992) and Sandwell and Schubert (1992a, b) have argued that the perimeters of several large coronae on Venus, specifically Artemis, Latona, and Eithinoa, resemble terrestrial subduction zones in both planform and topography. Artemis chasma has a radius of curvature similar to that of the South Sandwich subduction zone on the Earth. Sandwell and Schubert (1992a) proposed that the large coronae are incipient circular subduction zones. The foundering lithosphere is replaced by ascending hot mantle in a manner similar to back-arc spreading on the Earth. A consequence of a global resurfacing event would be the emplacement of mantle volatiles into the atmosphere. Of course, the resurfacing event and its consequences do not necessarily have to occur abruptly on the geologic time scale, but could take place over several hundred million years or even longer.

Although Venus is a one-plate planet at present, Magellan radar imagery has shown that there are tectonic features on its surface that resemble major tectonic features on Earth. Beta Regio, a volcanic highland, has many of the features of a continental rift on Earth. It has a domal structure with a diameter of about 2000 km and a swell amplitude of about 2 km. It has a well defined central rift valley with a depth of 1–2 km and there is some evidence for a three-armed planform (aulacogen). Alta, Eistla, and Bell Regiones have similar rift zone characteristics (Grimm and Phillips, 1992; Senske *et al.*, 1992). Aphrodite Terra with a length of some 1500 km is reminiscent of major continental collision zones on Earth, such as the mountain belt that extends from the Alps to the Himalayas. Ishtar Terra is a region of elevated topography with a horizontal scale of 2000–3000 km. A major feature is Lakshmi Planum, which is an elevated plateau similar to Tibet with a mean elevation of about 4 km. This plateau is surrounded by linear mountain belts—Akna, Danu, Freyja, and Maxwell Montes—which reach elevations of 10 km, similar in scale and height to the Himalayas (Kaula *et al.*, 1997).

Venus rotates in a retrograde sense with a spin period of 243 days. How this spin state came about is unknown, but it might be maintained by a balance between solid body tides raised on Venus by the Sun's gravity and a frictional torque on the solid planet exerted by the super-rotating atmosphere (Gold and Soter 1969, 1979, Kundt 1977, Ingersoll and Dobrovolskis 1978, Dobrovolskis and Ingersoll 1980). Such a balance would require the source of atmospheric angular momentum to be the solar torque on the atmospheric thermal tide. Alternatively, the super-rotating atmosphere could derive its angular momentum from the solid planet, resulting in a secular variation in the planet's spin. The angular momentum of Venus' atmosphere is 1.6×10^{-3} of the solid body angular momentum, making the possibility of angular momentum exchanges, both long term and short term, highly significant for both the atmosphere and solid planet; these momentum exchanges could be manifest in measurable length of day variations (Schubert, 1983).

As noted earlier, the atmosphere of Venus bears no resemblance to Earth's atmosphere. The atmosphere of Venus is almost 100 times as massive as Earth's, with extreme surface pressure and temperature conditions: about 94 bars and 740K (Seiff 1983). Venus's atmosphere is predominantly CO_2, roughly equivalent to the amount of CO_2 tied up in carbonate rocks on Earth, and is extremely dry, with about 30 ppm of H_2O, corresponding to ~ 10^{-3} of the water content of Earth's atmosphere. Consequently, Venus's ubiquitous cloud system is comprised not of water, but of sulfuric acid, indicative of a complex cycle of sulfur-based corrosive chemistry throughout the atmosphere. How did the atmosphere evolve to its present state if it was initially similar to Earth's atmosphere with a substantial complement of water? A strong possibility is that an intense greenhouse effect led to the vaporization of all water on early Venus, followed by photodissociation

of the water vapor at high altitudes and the escape of the released hydrogen into space. Without liquid water on its surface, CO_2 could then not be sequestered in the surface rocks. This change in Venus's atmosphere, if it occurred, would have triggered changes in the planet reaching all the way to its core. The loss of water would stop plate tectonics. This would have changed the style of mantle convection, slowed the rate of core cooling, and turned off the magnetic field.

If Venus once had an intrinsic magnetic field, it is possible, though perhaps not very likely, that some remanent magnetization of the crust might have survived to the present. However, while the surface temperature of Venus is below the blocking temperature of common minerals, the prolonged exposure of rocks to high temperature over hundreds of millions of years or longer would tend to destroy their magnetic domains.

How can these evolutionary story lines be checked? One way, as discussed in detail below, is the measurement of minor constituents and their isotopic compositions.

4. NOBLE GASES: THE KEY TO THE PAST

Since noble gases are thermochemically and photochemically inert, they provide a variety of keys to unlocking the history of planets, including Earth and Venus. Non-radiogenic noble gas isotopes preserve records of the materials that originally formed the planets and the evolution of their atmospheres. In contrast, radiogenic components are the daughters of parent molecules. Because noble gases are enormously depleted in planets compared to other elements, radiogenic noble gas isotopes produced by decay of even rare radioactive parents can be relatively abundant and readily measured. Consequently, the abundances of radiogenic isotopes provide a variety of clocks for understanding the formation and evolutionary history of Venus and Earth, as well as other planetary bodies such as the Moon and Mars. How these planets initially formed, the cataclysmic events in their early histories and the nature of major geologic events throughout their evolution are examples of basic insights potentially provided through accurate sampling of noble gases and their isotopes (cf., Plate 1). Accurate measurements for Venus, in particular, would help elucidate the history of the entire inner Solar System, as we now discuss.

In the Beginning: The Origin of Venus as Revealed by Neon, Xenon, Krypton and Argon

Neon Isotopic Ratios: Earth and Venus as Fraternal Twins? On Earth, neon in the atmosphere has an isotopic composition that is mass-fractionated with respect to neon

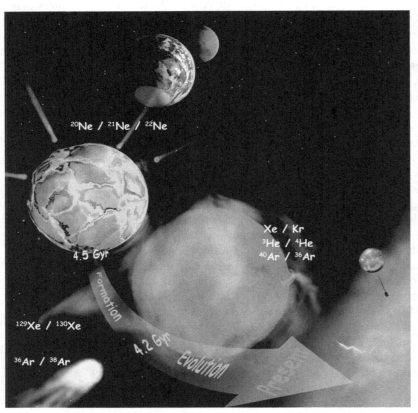

Plate 2. Near -Infrared Mapping Spectrometer (NIMS) map of hemispherical asymmetry in Venus cloud particle distribution: Light-grey-colored aerosols in the north are 10 times larger by volume than dark-colored particles in the south (Carlson et al., 1993b).

Plate 1. Clues to the formation and early history of the terrestrial planets are locked in the abundances of noble gases and their isotopes on Venus. The abundances of xenon, krypton and other tracers of Venus's past await in-situ atmospheric sampling by a future mission, such as by a probe or balloon-borne science station, depicted on the right.

in the mantle, with the atmosphere incorporating a larger proportion of the heavier neon isotopes. This suggests that neon has escaped from Earth's atmosphere. Further evidence of neon escape comes from the ~ 100 times less neon/argon ratio seen on Earth today compared to parent nebular values. Sasaki and Nakazawa (1988) have modeled both the bulk and fractionated neon, and have shown that, in principle, Earth could lose the neon from a solar nebula composition atmosphere while generating the observed amount of fractionation.

Since neon has three stable isotopes, in principle we can tell whether neon on Earth is related to that on Venus by mass fractionation of a common ancient neon located in a common part of the parent nebula, or whether Venus and Earth accreted their neon from different sources. Specifically, if the $(^{22}Ne/^{20}Ne, ^{21}Ne/^{20}Ne)$ ratios for Venus and Earth fall on the mass fractionation line predicted by escape processes, it would imply the two planets began as neon twins, sharing the same source of noble gases (and perhaps other volatiles), and then evolved via escape to the present-day ratios. On the other hand, if the observed ratios don't both fall on the fractionation/escape line, then the two planets likely accreted their neons from disparate sources, thus indicating that a variety of formation processes and realms in the parent nebulae helped to create the inner planets. Such information has broad implications as well for the formation of extrasolar planets within other parent nebulae.

Xenon and Krypton: Coded Messages from Ancient Cataclysms. Xenon (Xe) is of special interest because, on both Earth and Mars, its abundance and isotopic composition do not closely resemble any known solar system source materials, which together point to major events in the early histories of planets. There are three major differences. First, xenon's nine non-radiogenic isotopes are strongly mass fractionated (with a gradient of ~4% per amu) compared to the solar wind or any of its plausible solar system sources (Pepin, 1991). Second, the bulk abundance of non-radiogenic xenon is depleted with respect to krypton (Kr). The depletion is a factor of ~20 when comparing Earth to typical chondritic meteorites. The strong mass fractionation of the non-radiogenic isotopes implies that what we see today on Earth and Mars are ragged remnants of atmospheric escape that occurred during the formation processes of these planets.

The third major difference involves radiogenic isotopes of xenon, which are also markedly depleted compared to the measured abundances of their radioactive parents in primitive Solar System materials. The bulk and radiogenic component depletions together add up to the "missing Xe problem." The important point is that, because xenon is the heaviest gas in the atmosphere, xenon loss implies that

escape processes affected all the gases of the atmosphere. All three xenon abundance characteristics indicate a variety of cataclysmic events on both Earth and Mars late in the planetary formation process.

Important sources of radiogenic xenon are the decay of short-lived isotopes of iodine (^{129}I, half-life 15.7 Myr) and plutonium (^{244}Pu, half-life 82 Myr). Iodine produces just a single isotope, ^{129}Xe; thus, iodine-decay-produced xenon is relatively easy to measure. Some 7% of Earth's ^{129}Xe is from ^{129}I, while fully 50% of the ^{129}Xe in Mars's atmosphere is from ^{129}I. Yet these abundances are much less than expected for planetary atmospheres if there had been no escape. Earth, for example, retains only ~0.8% of its cosmic complement of ^{129}I-generated ^{129}Xe produced within its accreted material, while Mars retains just 0.1% (Porcelli and Pepin, 2000; Ozima and Podosek, 2002). In other words, more than 99% of the ^{129}Xe produced from the radioactive decay of iodine has been lost from Earth and Mars. Given ^{129}I's 15.7 million year half-life, much of this escape must have occurred from planet–sized bodies during accretion. But, as well, its half-life is long enough that much of it must have also been lost after Earth was a planet, and some of it after the Moon-forming impact. The degree to which Venus resembles the other terrestrial planets provides a measure of the similarity of planetary evolution during the time of giant impacts in the planet formation process.

Somewhat similar arguments apply to radiogenic xenon born of spontaneous fission of ^{244}Pu. Plutonium is especially important because its 82 million year half-life probes the early evolution of the planet after accretion. Fission xenon has been detected in Earth's mantle, and there appears to be some in Earth's atmosphere. How much is uncertain, but the upper limit is about 20% of what Earth's primordial plutonium should have generated. Whatever process removed fissogenic xenon from the atmosphere (or prevented its venting into the atmosphere) was one that took place some 200 million years after Earth formed.

Four general scenarios can be invoked to explain the loss and fractionation of xenon. In one scenario, the terrestrial planets experienced different degrees of atmospheric blow-off. The cause was hydrogen escape driven by solar EUV radiation more than one hundred times stronger than today. As the hydrogen escaped, it dragged other gases with it, especially the lighter isotopes of the noble gases. What remained behind became isotopically heavy (Zahnle and Kasting, 1986; Hunten *et al.*; 1987; Sasaki and Nakazawa, 1988; Zahnle *et al.*, 1990a; Pepin, 1991). A variant attributes escape to the Moon-forming impact (Pepin, 1997). In either of Pepin's (1991, 1997) models, escape was followed by degassing of the lighter noble gases from the interior, thus replenishing Earth's atmosphere with the amounts of lighter noble gases seen today. In either model there is relatively little escape from Venus, so that Venus holds a larger and less altered portion of its original complement.

A related model ascribes most of the xenon loss to atmospheric erosion caused by impacts during accretion. Impact erosion is expected to treat all gases equally, so that the only fractionation is between gases and condensed materials. Thus one expects impact erosion to expel noble gases much more readily than water. This may have been the means by which planets lost their radiogenic xenon. But impact erosion cannot account for the mass fractionation of xenon's isotopes. Thus impact erosion is not the whole story.

In a third scenario, the xenon isotopic pattern seen on Earth today is the signature of a common external source such as large planetesimals or comets. This hypothesis then requires that much of Earth's primordial xenon—including radiogenic products from plutonium and iodine decay—be sequestered deep in the planet's interior. Large planetesimals can fractionate isotopes by gravitational segregation (Ozima and Podosek, 2002; Zahnle *et al.*, 1990b). Extremely porous cold bodies immersed in nebular gases, such as large comets, are required. Comets are a possible source primarily because their compositional make-up vis-a-vis noble elements is not well-constrained. Formed in the cold reaches of the outer solar system, comets could perhaps incorporate xenon isotopic signatures substantially different from those in the inner nebula. Such bodies could then deliver these signatures to the Earth (and likely other nearby bodies) to produce the enigmatic isotopic xenon distribution seen today.

A fourth scenario invokes properties unique to xenon among the noble gases. An example is that Xe alone among the noble gases is easier to ionize than hydrogen. Consequently, when embedded in a partially ionized hydrogen wind that is flowing upward to space, xenon tends to be present as an ion while the other noble gases are neutral. The Coulomb force between xenon ions and protons is big and thus xenon ions are relatively easily dragged away by the escaping wind, while the lighter noble gases, being neutral, go nowhere. This describes a variant on atmospheric blowoff that leaves the bulk of the atmosphere intact while allowing xenon to escape.

Sampling xenon and its isotopes on Venus can help resolve which mechanism is responsible for the xenon signature observed on Earth, elucidating a crucial part of Earth's early history. A common fractionation pattern would indicate a common source of xenon isotopes. This would strengthen the hypothesis that comets or large planetesimals were a major source of volatiles throughout the inner solar system. On the other hand, the solar EUV blowoff theory would be strengthened if the fractionation of xenon isotopes in Venus were found to be different than found on Earth, since the power of

hydrogen blowoff might be expected to differ between Venus and Earth. The Earth-blowoff hypothesis would be strengthened further if radiogenic ^{129}Xe were found in large amounts on Venus, rather than the small amounts (~7%) found on Earth, thus indicating that a blowoff took out virtually all of the iodine-generated xenon from the early terrestrial atmosphere. Abundant radiogenic xenon on Venus would also indicate that impact erosion was relatively unimportant. Unfortunately, no isotopic abundance information for Venus exists today. Only a crude bulk Xe abundance upper limit of <4 ppb has been assigned (Donahue and Russell, 1997).

Isotopic measurements of Venus's xenon also provide a fundamental test of the "U-Xe" hypothesis for Earth, which asserts that terrestrial xenon is the sum of mass fractionated primordial U-Xe and a small dose of heavy isotopes obtained from the spontaneous fission of ^{244}Pu. The hypothetical primordial U-Xe is depleted in heavy isotopes with respect to the solar wind and other known sources. Indeed, with a low ^{136}Xe/^{130}Xe ratio of 1.66, some 8% less than the solar wind, U-Xe represents an extreme end-member of postulated fractionation patterns, and has yet to be detected in the solar system.

If U-Xe is in fact a good description of Earth's primordial xenon, then it must have come from a source that was isotopically distinct from the solar nebula as a whole. It is probable that Venus also accreted its xenon from this source. In fact, the atmosphere of Venus is perhaps the one place in the solar system where one could expect to find U-Xe, as predicted by the U-Xe hypothesis (Pepin, 1991). If it turns out that U-Xe is indeed present on Venus, as signified, in particular by a ^{136}Xe/^{130}Xe ratio about 8% lower then in the solar wind, the implications for the accretion of solid bodies throughout the inner solar system would be profound.

To resolve these various scenarios, xenon isotopic ratio measurements of at least 5% accuracy are desired. This accuracy well resolves the ~20% fractionation characteristic of terrestrial and martian xenon. To measure the difference in ^{136}Xe/^{130}Xe between xenon derived from hypothetical U-Xe and solar wind xenon (8%), or to measure abundance of radiogenic ^{129}Xe (7% on Earth), requires a somewhat higher precision, at least ~3%. Fissogenic xenon is perhaps more problematic: its reported detection in Earth's atmosphere at the 4% level is model-dependent. But in principle as much as 10% of the ^{136}Xe on Venus could be fissogenic, and a measurement accuracy of ~3% in the ^{136}Xe/^{130}Xe ratio would likely provide valuable insight.

Additional information on the source and nature of planetary material comes from the bulk ratios of various heavy noble gases. The uniform enhancement of xenon, krypton, and argon seen in Jupiter by the Galileo probe implies that "solar composition" planetesimals ("cold comets") were abundant in the solar system. Kr/Xe and Ar/Kr on Venus thus provide tests on whether these solar-like planestimals also reached Venus and contributed to the volatile inventories of the terrestrial planets. However, existing measurements of krypton on Venus differ by more than an order of magnitude (Von Zahn *et al.*, 1983). If the lower krypton estimate is correct, then Kr/Xe and Ar/Kr on Venus more closely resemble the solar wind and Jupiter's atmosphere, thus strengthening the solar-like planetesimal hypothesis. If the higher estimate is correct, then the Venus Ar/Kr ratio instead resembles a number of other objects, including meteorites, Earth, Mars, and lab measurements of gases trapped in cold ice (Notesco *et al.*, 2003), and the Kr/Xe ratio can then be used to discriminate between the different sources to clarify the nature and distribution of materials that formed the inner planets. Thus, determination of the bulk krypton abundance at the 5% level would provide fundamental insight into the origin of Venus, specifically providing valuable insights into the commonality of its volatile materials with other solar system objects.

Argon and Neon: Evidence for a Single Large Impact. A major discovery by the Pioneer Venus probes was that noble gases on Venus do not obey the "planetary" pattern generalized from Earth, Mars, and many meteorites. The striking result is that Venus has vastly too much neon and primordial (non-radiogenic) argon (^{36}Ar and ^{38}Ar), indicating an unusual source of argon not seen in the other inner planets.

There are three leading theories for the high abundance of neon and argon, two of which involve an anomalous large impact event not experienced by the other inner planets. The first hypothesis is that the solar wind implanted noble gases into meter-size particles in the neighborhood of Venus that later assembled into a large, km-scale body that eventually struck Venus (McElroy and Prather, 1981; Wetherill, 1981). Thus the source of enhanced neon and argon is the Sun, delivered via a very large, "solar-contaminated" impactor. Sasaki (1991) showed that this model can work in the more realistic context of an optically thick, vertically extended dust disk generated by collisions between planetesimals.

A second theory holds that Venus experienced an impact of a very large, very cold comet from the outer solar system. If this comet had the same argon-rich composition of the hypothetical pollutants of Jupiter's atmosphere, it would need to have been >~200 km diameter to supply the enhanced argon content of Venus. Relatively recent work by Gomes *et al.* (2005) estimates that the late bombardment delivered ~10^{23}g of cometary material to Earth and Venus; i.e., the mass of a dozen 200 km diameter comets. Thus, if such cold comets existed—and enhanced noble gas abundances on Jupiter suggest that they probably did—the Gomes *et al*

(2005) result implies that there is no difficulty supplying enough argon from the outer solar system.

A third theory is that argon was captured gravitationally from the solar nebula on both Venus and Earth, but survived better on Venus because this planet's thicker atmosphere provided more protection against catastrophic loss by giant impacts (Genda and Abe, 2005).

In-situ measurement of ^{36}Ar/^{38}Ar in Venus' atmosphere helps to distinguish between these theories by, in particular, testing whether the solar wind ratio (currently under analysis by the Genesis mission, e.g. Grimberg *et al.*, 2007) is observed on Venus. The Pioneer Venus probes measured the ^{36}Ar/^{38}Ar ratio on Venus to only 10%, too crude to determine if the planet's signature is that of the solar wind, based on early Genesis results. New measurements at the 1–3% level could prove valuable in separating out these scenarios once the Genesis mission results are finalized.

Venus Geologic Activity Through Time: Interior Degassing Revealed by Radiogenic Argon and Helium

On Earth the most powerful probes of mantle degassing are radiogenic argon and radiogenic and non-radiogenic helium. Venus has about 25% as much atmospheric radiogenic ^{40}Ar (from potassium, ^{40}K decay, half-life 1.3 Gyr) as Earth. This strongly implies that Venus either (1) is less degassed than Earth, or (2) has significantly less potassium (Turcotte and Schubert, 1988). A plausible scenario would be a large decrease in the degassing rate of Venus one billion years or so after formation. The low ^{40}Ar abundance coupled with the very high ^{36}Ar content results in a ^{40}Ar/^{36}Ar ratio near unity. In contrast, the large ^{40}Ar/^{36}Ar ratios measured for Earth, Mars, and Titan (~ 150–2000, Owen, 1992; Atreya *et al.*, 2006a, 2006b) has been taken to indicate active geologies.

Helium (He) offers another probe of planetary degassing through time. Helium has two stable isotopes: the abundant ^{4}He and the rare ^{3}He. More than 90% of the ^{4}He on Earth is radiogenic, chiefly made by the decay of uranium and thorium (with relevant half-lives of 0.7, 4.5, and 12 billion years), whilst most of the ^{3}He on Earth is primordial (i.e., non-radiogenic). On Earth the continents, which are granitic and greatly enriched in thorium and uranium, are a major source of ^{4}He, while the mantle is a source of both isotopes in the ratio of 90,000:1. Degassing of ^{3}He indicates that Earth's interior still retains a noble gas imprint that dates to its formation. Similarly, measurements of ^{3}He and ^{4}He on Venus could give insight into the degree of mantle degassing and associated active geology there.

Helium escapes rapidly from Earth's atmosphere but apparently does not currently escape from Venus (Donahue and Russell, 1997). If there really has been no atmospheric escape, about twice as much radiogenic ^{4}He would be expected as radiogenic ^{40}Ar (assuming an Earth-like K/U ratio). What is actually observed seems to be less ^{4}He. If ^{4}He escaped, then the corresponding ^{3}He escape rate should be greater, thus reducing the ^{3}He/^{4}He ratio. On the other hand, a high ^{3}He/^{4}He ratio in the current atmosphere could imply either (1) a relatively small helium escape rate (as expected, e.g., Donahue and Russell, 1997), (2) a relatively large and relatively recent degassing event from the mantle (perhaps on the time scale of water degassing events discussed below), or (3) a relatively small amount of near-surface ^{4}He-producing granitic material (Turcotte and Schubert, 1988). Better information on both the escape rates and isotopic ratios of helium (the latter to a measurement accuracy of approximately 20%) would provide new insight into the geology, outgassing history, and evolution of Venus. The Venus Express ASPERA experiment will directly measure the helium escape rates. What remains, then, is atmospheric measurements of the helium isotopic abundances, which can only be provided by *in-situ* observations.

5. BEYOND NOBLE GASES: ATMOSPHERIC LOSS AND CLIMATE THROUGH TIME

Due largely to the intense greenhouse effect, the atmosphere of Venus today is vastly different from that at Venus' formation. The large ratio of deuterated water, HDO, relative to H_2O—some 150 times that found on Earth—attests to the loss of most of Venus' water during the planet's evolution. The hypothetical sources and mechanisms responsible for the higher D/H ratio are currently controversial and range from the loss of a primordial ocean to steady state mechanisms, wherein water supplied by cometary infall and volcanic outgassing is lost by atmospheric hydrogen escape and oxidation of Fe-bearing crustal minerals (Grinspoon, 1993). A key question is the amount of water in the past, estimated to be equivalent to a global ocean between 5 and 500 m in depth. The large uncertainty arises from a lack of precise measurements of the D/H ratio, as well as uncertainties in the current loss rate of H and O. Another source of uncertainty is the fractionation factor between H and D loss, and how this factor has varied with time, as differing loss processes have dominated. The D/H ratio ranges from 0.013 to 0.025 based on PVO mass spectrometer and IR spectroscopy data, considering the full range of measurement uncertainties. If escape fluxes are currently in the upper end of the range suggested by current (pre-Venus Express) observations, $\approx 3 \times 10^7$ cm^{-2}s^{-1}, then the greatly enhanced (over terrestrial) D/H ratio must reflect loss over the last 0.5 Gyr, and would mask a primordial signature (Grinspoon, 1993; Donahue 1999).

Additional information on atmospheric loss comes from the isotopic ratio of nitrogen, ^{15}N/^{14}N. Currently this ratio ($3.8 \pm 0.8 \times 10^{-3}$) is known to ±20%, at which precision the ratio is comparable to the terrestrial value (3.7×10^{-3}) and broadly similar to nitrogen in meteorites. However, the value is significantly different than in the atmospheres of Mars ($5.9 \pm 0.5 \times 10^{-3}$) or Jupiter ($2.3 \pm 0.3 \times 10^{-3}$). The usual scenario is that Earth and Mars accreted their nitrogen from a common meteoritic or chondritic source, and that heavy nitrogen on low-gravity Mars means that the light nitrogen preferentially escaped, while light nitrogen on Jupiter means that Jupiter's nitrogen was supplied by a cometary or nebular source. Because Venus has roughly the same amount of nitrogen as Earth, the expectation is that Venus accreted its nitrogen from the same common source that Earth and Mars did. Comparison of the nitrogen isotopes between Venus and Earth should help determine whether nitrogen escape was significant on at least one of the two planets. In summary, *in-situ* measurements of the isotopic ratios of ^{15}N/^{14}N and HDO/H_2O to ~3% would provide data fundamental to understanding (1) present and past atmospheric loss rates, (2) the nature of the ancient climate of Venus, and (3) the role of comets in providing volatiles to the inner planets.

Among the light gases, the most important to understand from the perspective of planetary composition and accretion are the stable oxygen isotopes. Because there are three such isotopes, it is possible to distinguish fundamentally different compositions from mere isotopic fractionation. For example, the Earth and Moon share the same distinctive oxygen composition, which is is clearly distinct from those of nearly all meteorites, including the meteorites that come to us from Mars and Vesta (e.g., Stevenson, 2005). Specifically, the ^{17}O/^{16}O ratio of Vestan meteorites lies about 0.02% below the Earth-Moon mass-dependent fractionation line, and Mars lies 0.04% above that line. That the Earth and Moon share the same oxygen isotopes is currently viewed as a profound mystery, because according to the best available simulations of the Moon-forming impact, the Moon forms almost wholly from the planet that struck the Earth (Canup, 2004), and hence ought to be as different from Earth as Mars is.

Similar isotopic measurements of oxygen for Venus would be valuable in understanding the relationships of the materials comprising Earth, Moon, and Venus. If Venus is distinct from the Earth and Moon, we would learn that Venus accreted from a different pool of planetesimals than the Earth did. If it is the same as Earth, it would not only show Venus accreted from the same reservoir of materials as Earth, but it would also suggest that the equivalence of the Earth and Moon is not so mysterious.

To be useful, ^{17}O/^{16}O ratios need to be measured to significantly better than 0.01%. This level of precision may be available with tunable laser spectroscopy techniques (e.g., Christensen *et al.*, 2007a, 2007b) applied to isotopic carbon-dioxide. Adequate vertical sampling would be desired to effectively address issues arising from photochemical fractionations produced by atmospheric photochemistry.

6. VENUS TODAY: UNDERSTANDING ACTIVE GEOLOGIC, CHEMICAL, AND DYNAMICAL PROCESSES

Tracers of Active Volcanism

In-situ measurements of SO_2 and OCS provide constraints on present-day volcanic activity and geologic processes. Temporal variations in atmospheric content are especially intriguing. Indeed, a global decrease in SO_2 on a decadal scale was first observed in the atmosphere above the clouds by the Pioneer Venus Orbiter (Esposito, 1984). This decrease may have been due to the dissipation of a volcanic eruption of SO_2 prior to the arrival of the spacecraft (*ibid*). Alternatively, the SO_2 decrease may have been due to dynamical processes, including decadal-scale variations in the circulation within the Venus mesosphere where both SO_2 and CO exhibit steep vertical gradients (Clancy and Muhleman, 1991; Bézard *et al.*, 1993). More refined measurements of SO_2, CO, and OCS obtained *in-situ* lower down in the atmosphere, and with more extensive spatial and temporal coverage, would provide significantly more meaningful constraints on volcanism than previously available.

Additional information on interior degassing during recent geologic times—primarily by volcanism—are potentially provided by the isotopic ratios of other light gases, in particular $^{33}S/^{32}S$, and $^{34}S/^{32}S$, which are fixed in volcanic and interior processes. However, sulfur isotopes are susceptible to modifications via surface chemistry involving iron sulfides and via the atmospheric sulfur cycle, including fractionation processes involving photochemistry. Indeed, Laboratory experiments at 193 nm obtain anomalous fractionations—deviations from simple mass-dependent fractionation—as large as 7% (Farquhar *et al.* 2004). But in practice natural samples that have not been biologically fractionated show conventional fractionations smaller than 1% and anomalous fractionations of a few tenths of a percent at most. The biggest anomalous fractionation reported in a Martian meteorite is 0.1% (Farquhar *et al.*, 2007).

Useful measurements of the $^{34}S/^{32}S$ and $^{33}S/^{32}S$ isotopes ratios occur about the 1% level, which would yield quantitative constraints on anomalous isotopic effects in photochemical hot spots. More realistically, 0.1% precision is needed to discriminate reaction pathways. A long-duration mobile mission (e.g. a balloon) utilizing the high-precision afforded by tunable laser spectroscopy (e.g., Christensen *et al.*, 2007a, 2007b) would be especially useful since good characterization of sulfur isotopes requires multiple samples obtained in many environments.

Another potential tracer of volcanic activity is water and its isotope HDO. Measurement of the abundances of magmatic H_2O and HDO released in a volcanic eruption would help to distinguish recent volcanic sources from exogenic or primordial sources, thus yielding valuable insights into the evolution of the H_2O-poor atmosphere and the efficacy of present theories of global tectonics, insights into the history and current rate of volcanic activity, and constraints on the oxidation rate of the crust.

Chemistry, Composition, and Transport

Venus atmospheric chemistry involves complex and varied chemical cycles—H_2SO_4 cloud formation from SO_2 and H_2O, CO generated by photochemistry, OCS and HCl produced by thermochemistry, and SO_2, H_2O, and HCl in volcanic gases (Prinn and Fegley, 1987). By influencing radiative balance, these gases and clouds play a large role in establishing and maintaining the current climate. As well, the temporal and spatial variability of SO_2 and OCS give clues to the current rate of volcanic outgassing and the state of geologic activity on the planet. In addition, a fuller understanding of the radiative roles of these molecules and their connection with current and past resurfacing and outgassing would provide important clues to possible past climates on Venus.

Previous *in-situ* probe missions—lasting less than an hour in the atmosphere—provided sparse sampling of these important chemical constituents. The longer-lasting VEGA balloons lacked instruments capable of measuring composition. As a result, the roles of chemically-active species in maintaining radiative balance and climate, and their connection to volcanic activity, are poorly understood. A long-duration *in-situ* mission, perhaps balloon-borne, can make precise compositional measurements repeatedly over a range of altitudes and latitudes, to help resolve these important issues.

The Sulfur Cycle. Sulfur dioxide (SO_2) plays key roles in the planet's radiative balance, atmospheric chemistry and surface-atmosphere interactions (cf., Prinn and Fegley,1987; Moroz *et al.*, 1990) . It is the third most abundant gas below the clouds, the feedstock for the global H_2SO_4 clouds, the second most important greenhouse gas, and, as noted earlier, historically the most suspected tracer of volcanic activity. In addition, by constraining models of climate evolution, SO_2 measurements provide key data for understanding the origin of the Venus greenhouse effect, with implications for understanding the future climate evolution of Earth.

The stability of the atmosphere is intimately intertwined with SO_2 and its role in the sulfur cycle. The existence of H_2SO_4 clouds and ultraviolet absorbing haze (suspected to be comprised of sulfur allotropes) are largely due to SO_2 and the sulfur cycle. As part of this process, the low abundances of H_2O and O_2 are partly due to their removal by SO_2 in the formation of H_2SO_4. The non-uniform height profile of CO is also due to sulfur chemistry. Simply put, oxidation of SO_2 is one of the most important phenomena in the Venus atmosphere.

There are three main Venus sulfur cycles. First, the geologic cycle occurs at and below the surface, and is the ultimate source of atmospheric sulfur. Under the pressure-cooker environment of the hot, CO_2-laden near-surface atmosphere, OCS and H_2S gases are released from pyrite (FeS_2), particularly during active volcanism. Next, 10–40 km above the surface, the slow atmospheric cycle occurs, transforming these species into elemental and allotropic sulfur and SO_3. The latter then produces a small amount of H_2SO_4, by combining with H_2O. The slow cycle is thought to be responsible for producing UV-absorbing elemental and allotropic S_n, the putative major UV absorber observed on Venus. In this process, OCS is converted into CO and S_n.

Finally, the fast atmospheric cycle occurs primarily above the main cloud deck, and to some extent, in the lower atmosphere, from 40 to 80 km altitude. This chemical cycle—also involving CO—utilizes photochemical as well as thermochemical reactions, and leads to the primary production of SO_2 and bulk H_2SO_4 clouds on Venus.

Comprehensive, *in-situ* measurements of SO_2, H_2O, and CO in the middle and upper atmosphere over an extended period and range of altitudes and latitudes, together with measurements of the mass density and abundances of H_2SO_4 particles, could lead to a new understanding of cloud formation and the variability of greenhouse gases. Together, these measurements and results would provide a fundamental key dataset for understanding thermal structure, radiative balance, chemistry, and climate on Venus.

CO: A Tracer of Atmospheric Transport. CO is produced by photochemistry in the upper atmosphere as well as by thermochemistry in the middle atmosphere. Previously observed spatial variations in the CO abundance appear indicative of global-scale dynamics (Collard *et al.*, 1993). A 35% CO enhancement north of 47°N observed by Galileo/NIMS has been attributed to concentration by the polar descending branch of an equator-to-pole cell (Taylor, 1995). *In-situ* measurements over a wide range of latitudes could map CO directly, providing another means to assess latitudinal and vertical circulation.

Water. Water is the third most important greenhouse gas in Venus' atmosphere, is intimately involved in forming the global H_2SO_4 clouds, and is a potential tracer of volcanism. Furthermore, the observable H_2O inventory is ~10^5 times lower than on Earth, an important clue for understanding the different evolutionary paths of the two planets.

The VORTEX experiment on Pioneer Venus detected spatial variability of the cloud-top H_2O vapor abundance (Schofield *et al.*, 1982). On the night side, the water abundance was below the detection limit (6 ppm) and the equatorial mid-afternoon was the wettest (up to 100 ± 40 ppm vs. <6–30 ppm elsewhere). This enhancement may have been generated by vertical uplift of deeper, moister air via convection and Hadley circulation. To validate the role of water in cloud formation, search for additional evidence of volcanic activity, and refine existing theories of atmospheric evolution, *in-situ* missions should sample H_2O over a wide range of latitudes and longitudes, to an absolute abundance accuracy of 15%.

Nitrogen. N_2 is the second most abundant species in the Venus atmosphere. Since the lifetime of N_2 on Venus is >100 Myr, a uniform mixing ratio is expected. Yet, past gas chromatograph measurements by the Pioneer Venus and Venera spacecraft are surprising, suggesting a height-dependent mixing ratio. Specifically, the N_2 abundance varies widely with altitude, from a low of 2.5% abundance near 22 km to a high of 4.6% at 52 km (e.g., Table V of von Zahn *et al.*, 1983). This apparently strong altitude-dependent behavior of nitrogen implies that either the Pioneer Venus and/or Venera measurements are in error, or we do not completely understand the physico-chemical processes in the interior and atmosphere of Venus. *In-situ* measurements are needed to measure the N_2 abundance to 5% over altitude, sufficient to determine the validity of the Pioneer Venus and Venera measurements and whether N_2 is truly a well-mixed component of the Venus atmosphere.

Clouds, Meteorology, and Lightning. In-situ measurements of the spatial and temporal variability of middle-level clouds and their constituent gaseous cloud-forming material can provide quantitative constraints on cloud evolution, including the growth and dissipation rates of cloud mass densities and constituent particles. The need for such measurements was highlighted by the glimpse of Venus provided by the Galileo spacecraft enroute to Jupiter.

The Near-Infrared Mapping Spectrometer (NIMS) experiment onboard Galileo found large variations in the mean particle size of cloud particles, with marked hemispherical asymmetry (Carlson *et al.*, 1993b; cf., Plate. 2). Particles were ten times larger by volume in the northern hemisphere. Explanations for such marked hemispherical differences in cloud particle sizes are uncertain, but likely involve spatial variations in dynamical properties such as temperature, eddy diffusion (turbulence), and strengths of up/downdrafts that bring cloud-forming gases (principally SO_2 and H_2O) into the region. The NIMS results indicate that if cloud particle size is due to mixing of vertically stratified source regions (*e.g.,* photochemical and condensation source mechanisms), then mixing must be coherent over very large spatial scales, in turn implying relatively small variations in small-scale dynamical regimes. Yet, the distinct regional character of particle sizes may indicate sharp regional variations in the strength of dynamical mechanisms (*e.g.,* turbulence, up/downwelling).

Measurements by an *in-situ* airborne mission can scrutinize cloud evolution in unprecedented detail. In particular, *in-situ* measurements of (1) cloud particle sizes, acquired simultaneously with measurements of (2) H_2O, SO_2, and other species involved in the formation of H_2SO_4 clouds, can be correlated as well with (3) the measured vertical velocities and (4) temperature variations, to distinguish among mechanisms hypothesized for the distinct regional particle-size differences seen by NIMS.

Lightning—both an indicator of dynamics and an enabler of non-local thermodynamic equilibrium—was detected by the Venera 11–14 landers (Ksanfomaliti, 1979, 1983) and by Pioneer Venus (Taylor *et al.*, 1979; Russell, 1991) as electromagnetic impulses. Venera data indicate that the rate of lightning discharges is possibly greater than on Earth (Russell, 1991). The locations of the flashes of the two optical investigations to have successfully observed lightning (Krasnopolsky, 1983; Hansell *et al.*, 1995) are shown in Figure 1. Both plasma wave and optical observations show frequent occurrence in the dusk terminator region and on the

equator similar to the local time and latitudinal asymmetries in the distribution on Earth. A long-term *in-situ* monitor of lightning within the convective cloud layer could provide key observations of lightning frequency and strength over all local times and over a wide range of latitudes and elucidate its role in the chemistry of the Venus atmosphere. Lightning is a potentially important source of NOx compounds (NO, NO_2, NO_3) in the middle atmosphere. The existence of NOx species is indirect evidence for lightning, as it is the only source of NO in the lower atmosphere. If NOx species exist at a level of 10 ppb or greater, they have a potentially significant impact on (a) recycling of CO_2 through NOx-HOx catalytic reactions, and (b) catalytic oxidation of SO_2 to H_2SO_4. Krasnopolsky (2006) has recently reported an NO abundance of 5.5 ± 1.5 ppb from high-resolution spectra at 5.3 μm obtained with the TEXES spectrograph at the NASA/IRTF facility on Mauna Kea, Hawaii. The results correspond to a global flash rate of 87 s^{-1} and 6 km^{-2} y^{-1}, if the flash energy on Venus is similar to that on Earth ($\sim 10^9$ J).

In-situ measurements of NO-bearing species can be used to better understand middle-cloud-level chemistry and further elucidate the importance of lightning in the Venus atmosphere. The results could be compared with measurements of lightning emission rates and strengths to further clarify the role of lightning in shaping the chemistry of the Venus atmosphere.

7. IN-SITU EXPLORATION OF THE VENUS ATMOSPHERE: POTENTIAL MISSION ARCHITECTURES

The Decadal Survey (NRC, 2003) and the accompanying white paper on Venus (Crisp *et al.,* 2002) have specified the community-wide consensus of science priorities for Venus. Missions involving direct, *in-situ* measurements of atmospheric properties figured heavily in the recommended priorities, most notably resulting in the recommendation of a Venus atmospheric and surface sampling mission as a high-priority New Frontiers mission. Here, we briefly re-visit the Venus science objectives recommended by the Decadal Survey, with an eye to identifying the minimum mission architectures that can effectively address each objective.

Tables 1–3 list each of the science themes, questions, and priority science investigations for Venus, as determined by the Inner Planets Panel of the Decadal Survey (cf., Table 2.1 of the Decadal Survey report (NRC, 2003). Each table covers a particular Theme promoted by Decadal Survey—Venus Past, Present, and Future. We have then listed, in bold, the simplest, least-expensive, *in-situ* mission architectures which we deem can meaningfully achieve a significant portion of these recommended priority science investigations. Both

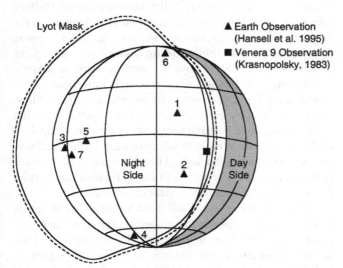

Figure 1. Locations of visible lightning flashes seen from a terrestrial telescope by Hansell et al. (1995) and from orbit by Krasnopolsky (1983).

Table 1. Effective Mission Architectures for Discovering Venus's Past

Theme	Questions	Priority Science Investigations	In-Situ Methods (Simplest Mission Architecture)	Orbiter Methods
Past: What led to the unique character of our home planet?	a. What are the bulk compositions of the inner planets and how do they vary with distance from the Sun?	1. Determine elemental and mineralogic surface compositions 2. Measure noble gas compositions of atmospheres 3. Measure oxygen isotopic ratios of the unaltered surface and atmosphere. 4. Determine interior (mantle) compositions.	**Lander** **Atmospheric Probe Lander** **Surface Stations**	 Volcanic emissions via spectra?
	b. What is the internal structure and how did the core, crust, and mantle of each planet evolve?	1. Determine horizontal and vertical variations in internal structures. 2. Determine the compositional variations and evolution of crusts and mantles 3. Determine major heat-loss mechanisms and resulting changes in tectonic and volcanic styles. 4. Determine characteristics of Fe-rich metallic cores (size, liquid and solid components).	**Geophysical Network** **Geophysical Network** **Geophysical Network** **Geophysical Network**	High-resolution Radar? High-resolution Radar? High-resolution Radar?
	c. What were the history and role of early impacts?	1. Determine large-impactor flux in the early solar system and calibrate the lunar impact record. 2. Determine the global geology of the inner planets. 3. Investigate how major impacts early in a planet's history can alter its evolution and orbital dynamics.	Sample Return Low-Alt Aerostats Geophysical Network	Radar Radar
	d. What is the history of water and other volatiles and how did the atmospheres of inner planets evolve?	1. Make high-precision measurements of noble gases and light stable isotopes. 2. Determine the composition of magmatic volatiles.	**Atmospheric Probe** Aerostats, Geophysical Network	 NIR spectroscopy (?)

potential cost-effective orbiter and *in-situ* missions are listed. Table 4 summarizes the *in-situ* mission architectures themselves. From top to bottom in Table 4, these missions become progressively more complex and costly as their capabilities in exploring the depths of Venus increase.

Fundamental science goals are attainable by each mission class, from ~$450M Discovery-class atmospheric sampling missions to ~$2000M Flagship-class, long-duration missions to the planet's lower atmosphere and surface. We note in particular that many high-priority science measurements and objectives appear attainable with low-cost, relatively low-risk Discovery-class mission architectures, including (1) simple probes similar to the Pioneer Venus vehicles, and (2) balloon missions building on the VEGA experience. Such missions can readily capitalize on recent improvements in sampling instrumentation—including light-weight gas chromatographs/mass spectrometers (GCMS) and tunable laser spectrometers—and precise interferometric radio and Doppler tracking of *in-situ* vehicles, to provide effective exploration of Venus at relatively low cost and low risk.

If they are to last for more than a few hours, missions to the relatively hostile lower atmosphere and surface require developments in new technologies dealing with high-temper-

ature operations. The mean surface temperature of ~470 C requires either (1) the development of a full suite of electronic components and instruments able to work under these high-temperature conditions, or (2) the development of effective refrigeration techniques to preserve benign temperatures for the most sensitive systems, at least. An alternative scheme would be to develop air-borne technologies enabling frequent "bobbing" between the hostile lower atmosphere/surface and the relatively benign environs near 55 km altitude. This would allow rapid descents to investigate the lower atmosphere followed by quick ascents, which together limit the craft's exposure to the high-temperature environment while enabling the achievement of effective near-surface and surface science. Such technological developments seem possible in the next decade or so, but need strong financial support by the world's space agencies to achieve success.

Among the major questions about Venus listed in Tables 1–3 are ones relating to the planet's interior. Clearly, we need to do more than just explore the atmosphere if we are to understand Venus and how it evolved to its present state. A possible source of information about the interior structure of Venus would be the variations that might exist in Venus' length of day or pole position. We have already

Table 2. Effective Mission Architectures for Discovering Venus's Present

Theme	Questions	Priority Science Investigations	In-Situ Methods (Simplist Mission Architecture)	Orbiter Methods
Present: What common dynamic processes shape Earth-like planets?	a. What processes—stabilize climate?	1. Determine the general circulation and dynamics of the inner planets' atmospheres.	**High-Altitude Aerostat**	Cloud tracking, Thermal Mapping
		2. Determine the composition of the atmospheres, especially trace gases and isotopes.	**High-Altitude Aerostat**	Near-IR spectroscopy
		3. Determine how sunlight, thermal radiation, and clouds drive greenhouse effects.	**High-Altitude Aerostat**	Visible, NIR, Mid-IR
		4. Determine processes and rates of surface/atmosphere interaction.	**Aerostats, Lander**	NIR, Volcanic emissions?
	b. How do active internal processes shape the atmosphere and surface environments?	1. Characterize current volcanic and/or tectonic activity and outgassing.	Aerostats? Geophysical Net.	NIR spectral imaging
		2. Determine absolute ages of surfaces.	**Sample Return**	
		3. Characterize magnetic fields and relationships to surface, atmosphere, and the interplanetary medium	Geophysical Net.	Magnetospheric fields and particles expmts
	c. How do active external processes shape the atmosphere and surface environments?	1. Quantify processes in the uppermost atmospheres of the terrestrial planets.		UV spectra, occultations, INMS

Table 3. Effective Mission Architectures for Discovering Venus's Future

Theme	Questions	Priority Science Investigations	In-Situ Methods (Simplist Mission Architecture)	Orbiter Methods
Future: What fate awaits Earth's environment and those of the other terrestrial planets?	a. What do the diverse climates of the inner planets reveal about the environment?	1. Characterize the greenhouse effect through meteorological observations.	**High-Altitude Aerostat**	Cloud/gas structure/distribution via NIR spectroscopy
	b. How do varied geologic histories enable predictions of volcanic and tectonic activity?	1. Assess the distribution and age of volcanism on the terrestrial planets.	Low-Altitude Aerostats? Geophys Net, Sample Return	Radar
		2. Search for evidence of volcanic gases in inner planet atmospheres.	**Probe(s)**, Aerostat(s)	UV and NIR spectroscopy
	c. What are the consequences of impacting particles and large objects?	1. Determine the recent cratering history and current flux of impactors in the inner solar system	Low-Altitude Aerostats Geophysical Network	Radar
	d. What are the resources of the inner solar system?	1. Assess volatile resources.	Probe(s), Aerostat(s)	NIR spectroscopy (?)
		2. Assess mineral resources.	**Lander(s), Geophysical Net. Sample Return**	

noted that length of day variations on Venus are likely due to atmosphere-solid body interactions and possibly interior processes. Whereas length of day variations on Earth are of order milliseconds, length of day variations on Venus could be of order hours (Schubert, 1983). Length of day variations on Venus are potentially observable from Earth through long term radio interferometric measurements.

The size and physical state of the core, the locations and nature of mantle phase transitions, and the thickness of the crust could be determined by a seismic/geodetic network that could directly probe the interior. Knowledge of the physical state of the core, in particular whether Venus has a solid inner core, would enable understanding of why Venus does not have an intrinsic magnetic field and how dynamos

Table 4. Potential *In-Situ* Mission Architectures

Mission Type	Class	Duration of Science Operations	Spatial Coverage	Measurement Objectives	Instrument Technique
Atmospheric Probe (From 100 km to surface)	Discovery	~ 1 hour	Single point	Gas abundances Atmospheric Structure	GCMS, Tunable Laser Spectroscopy P/T measurements Nephelometer Accelerometer
High-Altitude Aerostat (~ 56 km altitude,~20C)	Discovery	Weeks	Hemisphere	As above, plus: Meterology, dynamics circulation, volcanos	As above, plus: Radio Tracking, Doppler Tracking, Pressure-level Tracking
Low-Altitude Aerostat (~ 5 km altitude, ~ 440 C)	New Frontiers?	Weeks	Regional	As above, plus: Surface geology, Surface composition	As above, plus: Vis-NIR imager/spectroscopy, NIR mapping spectrometer
Lander	New Frontiers	~ 3 hours	Single point	As for Probe, plus: Elemental geochemistry Surface composition	As for Probe, plus:XRD, GRS, LASAR Ablation Spectroscopy, NIR mapping spectrometer, Imaging microscope
Surface Station	Flagship	Weeks	Single point	As for Lander, plus: Seismometry Core sampling Surface Meteorology	As for Lander, plus: Seismometer, Surface Wind Monitor
Geophysical Network (Multiple Surface Stations)	Flagship	Weeks	Near global	As for Surface Station, plus: High-Res Seismometry	Same as for Surface Station
Sample Return (Lander + Aerostat + Return Capsule)	Flagship +	~ 3 hours	Single poin	As for Lander and Low-Altitude Aerostat, plus: Detailed surface mineralogy and geological and planetary evolution	Same as for Lander and Low-Altitude Aerostat

work. A geophysical network could provide measurements of the surface heat flow, an important indicator of the thermal state of the interior and the planet's complement of radiogenic elements. More information about the variations in composition and mineralogy of the surface rocks by remote sensing or *in situ* sampling or sample return to Earth or an orbiting spacecraft laboratory would clarify how differentiation has produced the Venusian crust and if there are fundamental compositional differences between the crusts of the volcanic and plateau highlands similar to the differences between oceanic and continental crusts on Earth. Finally, the radiogenic element content of surface rocks provides important constraints on Venus' thermal state and evolution.

Acknowledgments. We wish to thank David J. Stevenson for valuable comments and suggestions. Much of the work described in this paper was carried out at the Jet Propulsion Laboratory, Pasadena, CA, under contract with the National Aeronautics and Space Administration. Funding for KHB, SKA, RWC, DC, DG and CTR was largely provided by NASA for their participation in support of ESA's Venus Express Mission.

REFERENCES

Allen, D.A., and Crawford, J.W., (1984). Cloud structure on the dark side of Venus. *Nature* **307**, 222–224.

Atreya, S. K., Mahaffy, P. R., and Wong, A-S (2006a). Methane and related trace species on Mars: Origin, loss, implications for life, and habitability. *Planet. Space Sci.,* in press.

Atreya, S. K., Adams, E. Y., Niemann, H. B., Demick-Montelara, J. E., Owen, T. C., Fulchignoni, M., Ferri, F., and Wilson, E. (2006b). Titan's methane cycle. *Planet. Space Sci.* 54, 1177–1187.

Baines, K. H., Bellucci, G., Bibring, J.-P., Brown, R. H., Buratti, B. J., Bussoletti, E., Capaccioni, F., Cerroni, P., Clark, R. N., Cruikshank, D. P., Drossart, P., et al. (2000). Detection of submicron radiation from the surface of Venus by Cassini/VIMS. *Icarus* **148**, 307–311.

Baines, K. H. Atreya, S., Carlson, R. W., Crisp, D., Drossart, P., Formisano, V., Limaye, S. S., Markiewicz, W. J., and Piccioni, G. (2006). To the depths of Venus: Exploring the deep atmosphere and surface of our sister world with Venus Express. *Planetary and Space Sci.* **54**, 1263–1278.

Basilevsky, A. T., Schaber, G. G., Strom, R. G. (1997). The resurfacing history of Venus, in *Venus II.* Edited by S. Bougher, W., Hunten, D. M., and Philips,R.J.. Tucson, AZ : University of Arizona Press, pp.1047—1084.

Bézard, B., de Bergh, C., Fegley, B., Maillard, J. P., Crisp, D., Owen, T., Pollack, J. B., and Grinspoon, D. (1993). The abundance of sulfur dioxide below the clouds of Venus. *Geophys. Res. Lett.*, **20**, 1587–1590.

Canup R.M. (2004). Simulations of a late lunar-forming impact. *Icarus* **168**, 433–456.

Carlson, R. W., Baines, K. H., Encrenaz, Th., Taylor, F. W., Drossart, P., L., Kamp, L. W., Pollack, J. B., Lellouch, E., Collard, A. D., Calcutt, S. B., *et al.* (1991). Galileo infrared imaging spectroscopy measurements at Venus. *Science* **253**, 1541–1548.

Carlson, R.W., Baines, K.H., Girard, M.A., Kamp, L.W., Drossart, P.,Encrenaz, Th., and Taylor, F.W. (1993a). Galileo/NIMS near-infrared thermal imagery of the surface of Venus. *Proc. of XXIV LPSC.* p.253.

Carlson, R. W., Kamp, L. W., Baines, K. H., Pollack, J. B., Grinspoon, D. H., Encrenaz, Th., Drossart, P., and Taylor, F. W. (1993b). Variations in Venus cloud particle properties: A new view of Venus's cloud morphology as observed by the Galileo Near-Infrared Mapping Spectrometer. *Planetary and Space Science* 41, 477–485.

Christensen, L. E., Brunner, B., Truong, K. N., Mielke, R. E., Webster, C. R., and Coleman, M. (2007a). Measurement of sulfur isotope compositions by tunable laser spectroscopy of SO_2. *Analytical Chem.*, in press.

Christensen. L. E., Webster, C. R., and Yang, R. Q. (2007b). Aircraft and balloon in situ measurements of methane and hydrochloric acid using interband cascade lasers, *Applied Optics* 46, No. 7, 1132–1138.

Clancy, R.T. and Muhleman , D. O. (1991). Long term (1979–1990) changes in the thermal, dynamical, and compositional structure of the Venus mesosphere as inferred from microwave spectral line observations of $C_{12}O$, $C_{13}O$, and CO_{18}. *Icarus* **89**, 129–146.

Collard, A. D., Taylor, F. W., Calcutt, S. B., Carlson, R. W., Kamp, L. W., Baines, K. H., Encrenaz, Th., Drossart, P., Lellouch, E., and Bézard, B. (1993). Latitudinal distribution of carbon monoxide in the deep atmosphere of Venus. *Planetary and Space Science* 41, 487–494.

Crisp, D., Allen, M. A., Anicich, V. G., Arvidson, R. E., Atreya, S. K., Baines, K. H., Banerdt, W. B., Bjoraker, G. L. Bougher, S. W., Campbell, B. A., Carlson, R. W., *et al.* (2002). Divergent evolution among Earth-like planets: The case for Venus exploration. In *The Future of Solar System Exploration, 2003–2013* (M. Sykes, Ed.). ASP Conference Series, pp. 5–34.

Dobrovolskis, A. R. and Ingersoll, A. P. (1980). Atmospheric tides and the rotation of Venus. I.Tidal theory and the balance of torques. *Icarus* **41**, 1–17.

Donahue, T. M., and Russell, C. T. (1997). The Venus atmosphere and ionosphere and their interaction with the solar wind: An overview. In *Venus II* (S. W. Bougher, D. M. Hunten, and R. Phillips, Eds.), Univ. of Arizona Press, Tucson. pp. 3–31.

Donahue, T.M. (1999). New analysis of hydrogen and deuterium escape from Venus. *Icarus* **141**, 226–235.

Esposito, L. W. (1984). Sulfur dioxide: Episodic injection shows evidence for active Venus volcanism. *Science* **223**, 1072–1074.

Farquhar, J., Kim, S.-T., Masterson, A. (2007). Sulfur isotope analysis of the Nakhla meteorite: Implications for the origin of sulfate and the processing of sulfur in the meteorite parent. LPSC XXXVIII, Contribution No. 1338.

Farquhar, J., Johnston, D.T., Calvin, C., Condie, K. (2004). Implications of Sulfur Isotopes for the Evolution of Atmospheric Oxygen. LPSC XXXV, Contribution No. 1920.

Fegley, B. Jr. (2004). Venus, in *Treatise on Geochemistry*, Volume 1 (A. M. Davis, ed.), Elsevier Press.

Gelman, B. G., Zolotukhin, V. G., Lamanov, N. I., Levchuk, B. V., Muhkin, L. M., Nenarokov, D. F., Okhothikov, B. P., Rotin, V. A., and Lipatov, A. N. (1979). Gaschromatographical analysis of chemical composition of the atmosphere of Venus done by the Venera 12 probes. *Plasma Astron. Zh.* **5**:217; *NASA Tech. Mem.* TM-75476.

Genda, H., and Abe, Y. (2005). Enhanced atmospheric loss on protoplanets at the giant impact phase in the presence of oceans. *Nature* **433**, 842–844.

Gold, T. and Soter , S. (1969). Atmospheric tides and the resonant rotation of Venus. *Icarus* **11**, 356–366.

Gold , T. and Soter, S. (1979). Theory of Earth-synchronus rotation of Venus. *Nature* **277**, 280–281.

Gomes, R., Levison, H., Tsiganis, K., Morbidelli, A. (2005). Origin of the cataclysmic Late Heavy Bombardment period of the terrestrial planets. *Nature* **435**, 466–469.

Grimberg, A., Baur, Burnett, D. S., H., Bochsler, P., and Wieler, R. (2007). The depth distribution of neon and argon in the bulk metallic glass flown on Genesis. *Lunar and Planetary Sci. Conf. XXXVIII*, #1270.

Grimm, R. E. and Phillips, R. J. (1992). Anatomy of a Venusian hot-spot—Geology, gravity, and mantle dynamics of Eistla Regio. *J. Geophys. Res. –Planets* **97**, E10, 16035–16054.

Grinspoon, D. H. (1993). Implications of the high D/H ratio for the sources of water in Venus' atmosphere. *Nature* **363**, 428–431.

Hansell, S. A., Wells, W. K., and Hunten, D. M. (1995).Optical detection of lightning on Venus. *Icarus* **117**, 345–351.

Hashimoto, G. L., and T. Imamura, T. (2001). Elucidating the rate of volcanism on Venus: Detection of lava eruptions using near-infrared observations. *Icarus* **154**, 239–243.

Hashimoto, G. L. and Sugita, S. (2003). On observing the compositional variability of the surface of Venus using nightside near-infrared thermal radiation. *JGR-Planets* **108** (E9), 5109, doi:10.1029/2003JE002082.

Herrick, R. R., Sharpton, V. L., Malin, M. C., Lyons, S. N., and Feely, K. (1997). Morphology and morphometry of impact craters, in *Venus II* . Edited by Stephen W. Bougher, D.M. Hunten, and R.J. Philips. Tucson, AZ : University of Arizona Press. pp.1015–1046.

Hunten, D. M., Pepin, R. O., and Walker, J. C. G. (1987). Mass fractionation in hydrodynamic escape. *Icarus* **69**, 532–549.

Ingersoll, A. P. and Dobrovolskis, A. R. (1978). Venus rotation and atmospheric tides. *Nature* **275**, 37–38.

Kaula, W. M. and Phillips, R. J (1981). Quantitative Tests for plate-tectonics on Venus. *Geophys. Res. Letters* **8**, 1187–1190.

Kaula, William M., Owen, T., Runcorn, S. K., Tozer, D. C. (1994). The Tectonics of Venus and Discussion. Philosophical Transactions: Physical Sciences and Engineering. 349, 345–355.

Kaula, William M., Lenardic, A., Bindschadler, D. L., and Arkani-Hamed, J. (1997) Ishtar Terra, in *Venus II*, S. W. Bougher, D. M. Hunten, and R. J. Phillips, editors, The University of Arizona Press, pp 879– 900

Konopliv, A. S. and Yoder, C. F. (1996). Venusian k(2) tidal Love number from Magellan and PVO tracking data. *Geophys. Res. Letters* **23**, 1857–1860

Krasnopolsky, V. A. (1983). Lightning on Venus according to information obtained by the satellites Venera 9 and 10. *Kosmich. Issled.* **18**, 429–434.

Krasnopolsky, V. A. (2006). A sensitive search fro nitric oxide in the lower atmospheres of Venus and Mars: Detection on Venus and upper limits for Mars. *Icarus* **182**, 80–91.

Ksanfomaliti, L. V. (1979). Lightning in the cloud layers of Venus. *Kosmich. Issled.* **17**, 747–762.

Ksanfomaliti, L. V. (1983). Electrical activity in the atmosphere of Venus. I. Measurements on descending probes. *Kosmich. Issled.* **21**, 279–296.

Kundt, W. (1977). Spin and atmospheric tides of Venus. *Astron. Astrophysics.* **60**, 85–91.

Lecacheux, J., Drossart, P., Laques, P., Deladierre, F. and Colas, F. (1993). Detection of the surface of Venus at 1.0 um from gro-pund-based observations. *Planet.Space Sci.* **41**, 543–549.

Meadows, V. S. and Crisp, D. (1996). Ground-based near-infrared observations of the Venus nightside: The thermal structure and water abundance near the surface. *J. Geophys. Res.* **101**, 4595–4622.

McElroy, M. B. and Prather, M. J. (1981). Noble gases in the terrestrial planets. *Nature* **293**, 535–539.

McKenzie, D. Ford, P. G. Johnson, C., Parsons, B., Sandwell, D., Saunders, S., and Solomon, S. C. (1992). Features on Venus generated by plate boundary processes. *J. Geophys. Res.-Planets* **97**, E8, 13533–13544.

McKinnon, W. B., Zahnle, K. J., Ivanov, B. A., Melosh, H. J. (1997). Cratering on Venus: Models and Observations, , in *Venus II*. Edited by S. W. Bougher, D.M. Hunten, and R.J. Philips. Tucson, AZ : University of Arizona Press, pp. 969– 1014.

Moroz, V. I. *et al.*, (1990). Water vapor and sulfur dioxide abundances at the Venus cloud tops from the VENERA 15 infrared spectrometry data. *Adv. Space Res* **10**: 77.

National Research Council (NRC) of the National Academies, Solar System Survey Space Studies Board (2003). New Frontiers in the Solar System: An Integrated Exploration Strategy. The National Academies Press. Washington, DC.

Nimmo, F. (2002). Why does Venus lack a magnetic field?, *Geology* **30**, 987–990.

Nimmo, F. and Stevenson, D. J. (2000) Influence of early plate tectonics on the thermal evolution and magnetic field of Mars, *J. Geophys. Res.* **105**, 11,969–11,979.

Notesco, G., Bar-Nun, A., and Owen, T. (2003). Gas trapping in water ice at very low deposition rates and implications for comets. *Icarus* **162**, 183–189.

Owen, T. C. (1992). The composition and early history of the atmosphere of Mars. In *Mars* (ed., H. Kieffer, *et al.*). Univ. of Arizona Press, pp.818.

Ozima , M. and Podosek, F. A. (2002). *Noble Gas Geochemistry*, Cambridge University Press, New York.

Pepin, R. O. (1991). On the origin and early evolution of terrestrial planet atmospheres and meteoritic volatiles. *Icarus* **92**, 2–79.

Pepin, R. O. (1997). Evolution of the Earth's noble gases: Consequences of assuming hydrodynamic loss driven by giant impact. *Icarus* **126**, 148–156.

Phillips, J. L., and Russell, C. T. (1987). Upper limits on the intrinsic magnetic field of Venus. *J. Geophys. Res. –Space Phys* **92**, 2253–2263.

Prinn, R.G., and Fegley, B. (1987). The atmospheres of Venus ,Earth and Mars—a critical comparison. *Ann. Rev. Earth and Planet. Sci.,* **15**, 171–212, 1987.

Porcelli , D., and Pepin, R. O. (2000).Rare gas constraints on early Earth history. In *Origin of Earth and Moon*, eds. R. M. Canup and K. Righter. University of Arizona Press, Tucson. pp. 435–458,

Russell, C. T. (1991). Venus lightning. *Space Sci Rev* **55**, 317–356.

Russell, C. T. (1980). Planetary magnetism. *Reviews of Geophysics.* **18**, 77–106.

Sandwell, D. T., and Schubert, G.(1992a). Flexural ridges, trenches, and outer rises around coronae on Venus. *J. Geophys. Res.-Planets.* **97**, E10,16069-16083.

Sandwell, D. T. and Schubert, G. (1992b). Evidence for retrograde lithospheric subduction on Venus. *Science* **257**, 766–770.

Sasaki, S. (1991). Off-disk penetration of ancient solar wind. *Icarus* **91**, 29–38

Sasaki, S. and Nakazawa, K. (1988). Origin of isotopic fractionation of terrestrial Xe: Hydrodynamic fractionation during escape of the primordial H_2-He atmosphere. *Earth Planet. Sci. Lett.* **89**, 323–334.

Schaber, G. G., Strom, R. G., Moore, H. J.,Soderblom, L. A., Kirk, R. L., Chadwick, D. J. Dawson, D. D., Gaddis, L. R., Boyce, J. M., Russell, J., and Schaber, G. (1992). Geology and distribution of impact craters on Venus—What are they telling us? *J. of Geophy. Res.-Planets* **97**, E8, 13257–13301

Schofield,J. T., Taylor, F. W., and McCleese, D. J. (1982). The global distribution of water vapor in the middle atmosphere of Venus. *Icarus* **52**, 263–278.

Schubert, G. (1983). General Circulation and the Dynamical State of the Venus Atmosphere, in *Venus*, Edited by D. M. Hunten, L. Colin, T. M. Donahue, and V. I. Moroz, Tucson, AZ: The University of Arizona Press, pp. 681–765.

Schubert, G., Solomatov, V. S., Tackley, P. J., and Turcotte, D. L. (1997). Mantle convection and the thermal evolution of Venus, in

Venus II. Edited by S. W. Bougher, D.M. Hunten, and R.J. Philips. Tucson, AZ : University of Arizona Press, pp.1245– 1287.

Seiff, A. (1983). Thermal structure of the atmosphere of Venus, in *Venus*, eds. D. M. Hunten, L. Colin, T. M. Donahue, and V. I. Moroz, (Tucson: Univ. Arizona Press) pp. 154–158.

Senske, D. A., Schaber, G. G., and Stofan, E. R. (1992). Regional topographic rises on Venus—Geology of Western Eistla Regio and comparison to Beta-Regio and Atla-Regio. *J. Geophys. Res.-Planets* **97**, E8, 13395–13420.

Sjogren, W. L., W. B. Banerdt, P. W. Chodas, A. S. Konopliv, G. Balmino, J. P. Barriot, J. Arkani-Hamed, T. R. Colvin, and M. E. Davies (1997) The Venus gravity field and other geodetic parameters, in *Venus II*, S. W. Bougher, D. M. Hunten, and R. J. Phillips, editors, The University of Arizona Press, Tucson, pp 1125–1161.

Stevenson, D. J. (2005). The oxygen isotope similarity between the Earth and Moon—Source region or formation process? *Lunar and Planetary Sci. Conf. XXXVI*, #2382.

Stevenson, D. J. (2006) Evolution of Venus: Initial conditions, internal dynamics and rotational state. Paper presented at the AGU Chapman Conference on *Exploring Venus as a Terrestrial Planet*, Key Largo, Florida, 13–16 February, 2006.

Stevenson, D. J., Spohn, T., Schubert, G. (1983). Magnetism and thermal evolution of the terrestrial planets. *Icarus* **54**, 466–489.

Taylor, F. W. (1995). Carbon monoxide in the deep atmosphere of Venus. *Adv. Space Res.* **16**, (6 81–88.

Taylor, W. W. L., Scarf, F. L., Russell, C. T., and Brace, L. H. (1979). Evidence for lightning on Venus. *Nature* **282**, 614–616.

Taylor, F. W. (2006). Venus before Venus Express. *Planetary and Space Sci.* **54**, 1249—1262.

Titov, D. V., Svedhem, H., Koschny, D., Hoofs, R., Barabash, S., Bertaux, J. –L., Drossart, P., Formisano, V., Häusler, B., Korablev, O., Markiewicz, W. J., Nevejans, D., Pätzold, M., Piccioni, G., Zhang, T. L., Merritt, D., Witassee, O., Zender, J., Accomazzo, A., Sweeney, M., Trillard, D., Janvier, M., and Clochet, A. (2006) Venus Express science planning. *Planetary and Space Sci.* **54**, 1279—1297.

Turcotte, D. (1993) An episodic hypothesis for Venusian tectonics. *J. Geophys. Res.-Planets* **98**, E9, 17061–17068.

Turcotte, D. L. and Schubert , G. (1988). Tectonic implications of radiogenic noble gases in planetary atmospheres *Icarus* **74**, 36–46.

Von Zahn, U., Komer, S., Wiemann, H., and Prinn, R. (1983). Composition of the Venus atmosphere. In *Venus* (D. M. Hunten, L. Collin, T. M. Donahue, and V. I. Moroz, Eds.), Univ. of Arizona Press, Tucson. pp. 297–430.

Wetherill, G. W. (1981). Solar wind origin of ^{36}Ar on Venus. ***Icarus*** **46**, 70–80.

Yoder, C.F. (1997).Venusian Spin Dynamics, in *Venus II* . Edited by Stephen W. Bougher, D.M. Hunten, and R.J. Philips. Tucson, AZ : University of Arizona Press, pp. 1087–1124.

Yoder, C. F. and Russell, C. T. (1997). The Venus atmosphere and ionosphere and their interaction with the solar wind: An overview, in *Venus II*. Edited by S. W. Bougher, D.M. Hunten, and R.J. Philips. Tucson, AZ, University of Arizona Press., p.1087.

Zahnle K. and Kasting, J. F. (1986). Mass fractionation during transonic escape and implications for loss of water from Mars and Venus. *Icarus* **68**, 462–480.

Zahnle, K., Kasting, J., and Pollack, J. (1990a). Mass fractionation of noble gases in diffusion-limited hydrodynamic hydrogen escape. *Icarus* **84**, 502–527.

Zahnle, K., Pollack, J. B., and Kasting, J. F. (1990b). Xenon fractionation in porous planetesimals. *Geochimica et Cosmochimica Acta* **54**, 2577–2586.

Kevin H. Baines, Jet Propulsion Laboratory, California Institute of Technology, M/S 183-601, 4800 Oak Grove Drive, Pasadena, California 91109, Ph: (818) 354-0481, FAX: (818) 393-5148, Email: kbaines@aloha.jpl.nasa.gov.

Astrobiology and Venus Exploration

David H. Grinspoon

Department of Space Sciences, Denver Museum of Nature & Science, Denver, Colorado, USA

Mark A. Bullock

Department of Space Studies, Southwest Research Institute. Boulder, Colorado, USA

For hundreds of years prior to the space age, Venus was considered among the most likely homes for extraterrestrial life. Since planetary exploration began, Venus has not been considered a promising target for Astrobiological exploration. However, Venus should be central to such an exploration program for several reasons. At present Venus is the only other Earth-sized terrestrial planet that we know of, and certainly the only one we will have the opportunity to explore in the foreseeable future. Understanding the divergence of Earth and Venus is central to understanding the limits of habitability in the inner regions of habitable zones around solar-type stars. Thus Venus presents us with a unique opportunity for putting the bulk properties, evolution and ongoing geochemical processes of Earth in a wider context.

Many geological and meteorological processes otherwise active only on Earth at present are currently active on Venus. Active volcanism most likely affects the climate and chemical equilibrium state of the atmosphere and surface, and maintains the global cloud cover. Further, if we think beyond the specifics of a particular chemical system required to build complexity and heredity, we can ask what general properties a planet must possess in order to be considered a possible candidate for life. The answers might include an atmosphere with signs of flagrant chemical disequilibrium and active, internally driven cycling of volatile elements between the surface, atmosphere and interior. At present, the two planets we know of which possess these characteristics are Earth and Venus.

Venus almost surely once had warm, habitable oceans. The evaporation of these oceans, and subsequent escape of hydrogen, most likely resulted in an oxygenated atmosphere. The duration of this phase is poorly understood, but during this time the terrestrial planets were not isolated. Rather, due to frequent impact transport, they represented a continuous environment for early microbial life. Life, once established in the early oceans of Venus, may have migrated to the clouds which, on present day Venus, may represent a habitable niche. Though highly acidic, this aqueous environment enjoys moderate temperatures, surroundings far from chemical equilibrium, and potentially useful radiation fluxes. Observations of unusual

Exploring Venus as a Terrestrial Planet
Geophysical Monograph Series 176
Copyright 2007 by the American Geophysical Union.
10.1029/176GM12

191

chemistry in the clouds, and particle populations that are not well characterized, suggest that this environment must be explored much more fully before biology can be ruled out. A sulfur-based metabolism for cloud-based life on Venus has recently been proposed (Schulze-Makuch et al., 2004). While speculative, these arguments, along with the discovery of terrestrial extremophile organisms that point toward the plausibility of survival in the Venusian clouds, establish the credibility of astrobiological exploration of Venus. Arguments for the possible existence of life on Mars or Europa are, by convention and repetition, seen as more mainstream than arguments for life elsewhere, but their logical status is similar to plausibility arguments for life on Venus.

With the launch of COROT in 2006 and Kepler in 2008 the demographics of Earth-sized planets in our galaxy should finally become known. Future plans for a Terrestrial Planet Finder or Darwin-type space-based spectrograph should provide the capability of studying the atmospheric composition and other properties of terrestrial planets. One of the prime rationales for building such instruments is the possibility of identifying habitable planets or providing more generalized observational constraints on the habitable zones of stellar systems. Given the prevalence of CO_2 dominated atmospheres in our own solar system, it is quite likely that a large fraction of these will be Venus-like in composition and evolutionary history. We will be observing these planets at random times in their evolution. In analogy with our own solar system, it is just as likely that we will find representatives of early Venus and early Earth type planets from the first 2 billion years of their evolution as it is that we will find "mature Venus" and "mature Earth" type planets that are roughly 4.5 billion years old. Therefore, in order to be poised to use the results of these future observations of extrasolar planets to make valid, generalized inferences about the size, shape and evolution of stellar habitable zones it is vital that we obtain a much deeper understanding of the evolutionary histories and divergence of Earth and Venus.

The Mars Exploration Rover findings of evidence for aqueous conditions on early Mars have intensified interest in the possible origin and evolution of life on early Mars. Yet the evidence suggests that these deposits were formed in a highly acidic and sulfur-rich environment. During this phase, Mars may well have had sulfuric acid clouds sustained by vigorous, sulfur-rich volcanism. This suggests that a greater understanding of the chemistry of the Venusian atmosphere and clouds, and surface/atmosphere interactions, may help to characterize the environment of Mars when life may have formed there. In turn, if signs of early life are found on Mars during the upcoming decades of intensive astrobiological exploration planned for that planet, it will strengthen arguments for the plausibility of life in an early and gradually acidifying Venusian environment. Of our two neighboring planets, Venus and Mars, it is not yet known which held on to its surface oceans, and early habitable conditions, for longer.

1. INTRODUCTION: A BRIEF HISTORY OF VENUS EXOBIOLOGY

The atmosphere of Venus was discovered by Mikhail V. Lomonosov at the Saint Petersburg Observatory during a transit of the sun in 1761 (Lomonosov, 1955). From that time up until the eve of the space age, Venus was widely viewed as a planet likely to have surface conditions closely resembling those of Earth, and consequently was seen as one of the most promising locations for extraterrestrial life. For centuries astronomers have been aware of the proximity of Venus to Earth in the solar system, the obvious similarity in bulk properties between Venus and Earth, and the cloud-filled sky. The clouds were widely interpreted before the 1970's to be condensed water, and were often held to suggest the presence of a large surface reservoir of water. The current consensus of astrobiology, that life is most likely to be found on planets with liquid water reservoirs (Pace, 2001; Irwin and Schulze-Makuch, 2001), has long been held by speculative scientists, who often believed Venus to be a likely location for life. For example, the distinguished British astrophysicist Sir Arthur Eddington, in a book published in 1928, summarized these arguments and declared that "Venus, so far as we know, would be well adapted for life similar to ours... If transplanted to Venus we might perhaps continue to live without much derangement of habit" and that Mars, by contrast, "has every appearance of being a planet long past its prime" (Eddington, 1935).

This view was first seriously contradicted by data in 1956 when Earth-based observations at 3.15 cm yielded a radio brightness temperature of 620 ± 100 K (Mayer et al., 1957). This suggested to some researchers that the surface temperature of Venus was much greater than that of Earth, but alternative interpretations involved a high-altitude, non-thermal source of radiation, preserving the possibility of moderate surface temperatures. Thus, at the inception of planetary exploration the question of whether or not Venus maintained habitable conditions at the surface was still in doubt. In a 1961 paper, Sagan (1961) summarized conflicting current views on possible environmental conditions on Venus, stating: "The state of our knowledge of Venus is amply illustrated by the fact that the Carboniferous swamp, the wind-swept desert, the planetary oil field, and the global Seltzer ocean each have their serious proponents, and those planning eventual manned expeditions to Venus must be exceedingly perplexed over whether to send along a paleobotanist, a mineralogist, a petroleum geologist, or a deep-sea diver."

This question was definitively resolved with the microwave radiometer on Mariner 2, one of the first experiments at another planet from any spacecraft, which showed a distinctive signature of limb darkening at a wavelength of 1.9 cm (Sagan 1969), demonstrating that the source was a surface much hotter than the stability range of liquid water. Sagan's hypothetical manned mission would most fruitfully include the mineralogist although, as will be discussed herein, there may yet be work for the paleobotanist.

The Mariner 2 results led to a precipitous fall from grace for Venus as a home for life in both the scientific and popular literature. A New York Times editorial in February, 1963, entitled "Venus Says No", described the Mariner 2 results as "disheartening, disillusioning", and declared that "The message from Venus may mark the beginning of the end of mankind's grand romantic dreams".

By 1971, after the Venera 7 and 8 landings, it was clear that Venus possessed a surface temperature of ≈ 735K and a surface pressure of ≈ 90 bars (Marov et al. 1973). Subsequently, the subject of life on Venus received little mention in the literature. However, Sagan and Morowitz (1967) proposed the possibility of multicellular organisms generating hydrogen and utilizing float bladders in the Venusian clouds.

More recently, the possibility of surface life in early Venusian history has received support from the prevailing view (still unsupported by definitive empirical evidence) that Venus possessed surface oceans early in its history, by accumulating evidence for a fast origin of life on the young Earth, by dynamical and microbiological support for the notion that meteorites with viable microorganisms were likely exchanged by the young terrestrial planets, and by the continuing discovery of terrestrial extremophile organisms inhabiting a wide range of physical conditions and employing a wide range of metabolic strategies.

Extensions of these same arguments, plus observations of the surprising adaptability of terrestrial life to even seemingly unlikely niches where liquid water is available in some form, has lead some authors to argue recently for the possibility that life on Venus, rather than going extinct when surface water was lost, may have migrated to the global cloud deck, and may possibly persist there today (Grinspoon, 1997; Schulze-Makuch et al, 2004).

In the last decade, exobiology, now astrobiology, has flourished as a new meta-discipline (Morrison, 2001). A major focus of this field is the effort to develop a generalized understanding of terrestrial planet evolution in order to understand the locations, demographics, morphologies, and longevity of stellar habitable zones and the types of terrestrial planets that may inhabit them. Exploration of Venus is a central and indispensable part of this effort.

2. HABITABLE CONDITIONS ON EARLY VENUS

While early planetary exploration definitively revealed that surface conditions on Venus are now incompatible with organic life, subsequent developments in both planetary science and astrobiology have supported the belief that earlier in its history Venus may have supported life.

Here we will not attempt a complete review of ideas about the formation and early evolution of Venus, but merely point out some aspects of our contemporary understanding of these topics which are most relevant to astrobiology.

The original water abundance of Venus is highly unconstrained. The high D/H ratio observed, 2.5×10^{-2} or ~150 times terrestrial (Donahue et al. 1997) has often been cited as evidence of a large primordial water endowment (Donahue et al. 1982). However, theoretical work has shown that given the likelihood of geologically recent water sources and the large uncertainty in the modern and past hydrogen and deuterium escape fluxes, the large observed D/H may result from the history of escape and re-supply in the most recent $\sim 10^9$ years of planetary evolution, and does not reliably or uniquely preserve a signal of primordial water escape (Grinspoon 1997, 1998, 1993; Donahue, 1997).

Thus, at present the best arguments for a sizable early Venusian water endowment are those based on consideration of likely modes of planet formation and early volatile delivery, along with the constraint that Earth received at least one FTO (full terrestrial ocean) of water during formation. Indeed, most models of water delivery to early Earth involve impact processes that would have also supplied Venus with abundant water. Whether the Earth's water was primarily delivered within rocky material from the inner solar system (Morbidelli et al., 2000) or icy material from the outer solar system (Grinspoon, 1987; Ip and Fernandez, 1998), gravitational scattering of objects from these source regions would have ensured that Venus received a roughly similar amount. It is worth noting that the stochasticity of the volatile accretion process for terrestrial planets is still largely unknown. It has even been suggested that Earth may have received most of its water in one large planetesimal impact (Morbidelli et al., 2000). Stochastic processes could have created large inequities in original volatile inventory among neighboring planets. However, given the great similarity in bulk properties between Venus and Earth, their close proximity in the Solar System, and present understanding of accretional processes, the best assumption at present is that Venus and Earth started with similar water endowments.

Given the faint young sun (Kasting and Grinspoon, 1991) and the almost certain presence of liquid water oceans on the young Venus, the interest for astrobiology is obvious: According to our current understanding of the required conditions and timescale for the origin of life, Venus is a place where this should have occurred. Even if a separate origin of life did not occur on Venus (or on Earth) there is good reason to believe that interplanetary transfer by impact ejecta would have distributed life among the young terrestrial planets (Melosh, 1988).

In the scientific literature, much more attention has been given to the possibility of life on early Mars during a period in which that planet may have been substantially warmer and wetter than at present. This bias is in part due to the fact that the most ancient Martian surface geology is well-preserved and observable with spacecraft, whereas nearly all the observable surface of Venus is probably younger than 1 GY (McKinnon et al., 1997). Abundant evidence exists for many kinds of water-formed features and deposits on the Martian surface. However, paradoxically, models of the early

Martian climate under the influence of the faint young sun do not produce surface temperatures warmer than the freezing point of water. This is at present an unsolved problem. Surface water on Mars may have been a transient or localized phenomenon, perhaps allowed by heating due to stochastic large impacts (Segura et al. 2002). Alternatively, the steady-state climate may have been warmer through mechanisms that are presently not understood. It is safe to say that the longevity of habitable surface conditions on Mars is currently not known and that therefore it is not known whether habitable conditions persisted for longer on Mars or Venus.

If one accepts that Venus started with a large, albeit currently unconstrained, water inventory, the first order question of Venus' evolution to its current state then becomes the question of water loss. Models of early atmospheric evolution have supported the idea that Venus may have lost most of its original hydrogen inventory through a phase of hydrodynamic escape fueled by the enhanced EUV flux from the early sun during a phase of runaway or "moist" greenhouse (Kasting et al, 1984; 1988) followed in turn by loss of hydrogen through various thermal and nonthermal escape mechanisms, including Jeans escape, charge exchange, collisional ejection (Kumar et al., 1983), and an electric field driven flow of ions in the night-side hydrogen bulge (Donahue et al., 1997).

The timescale for this loss of water has been suggested as several hundred million years (Kasting et al., 1984, 1988), but remains highly uncertain. Kasting et al. (1988) in many ways optimized his calculations to get rid of oceans quickly. These calculations provide an upper limit on surface temperatures during the moist greenhouse phase. This yields an upper limit on water fluxes to the upper atmosphere, and an upper limit to the resulting hydrogen escape fluxes. Therefore, it yields a lower limit on the lifetime of the ocean. Indeed, although the work of Kasting et al. (1984, 1988) currently represents the state of the art in modeling escape from early Venus, all of the uncertainties mentioned in this work add up to an overall uncertainty in the lifetime of the Venusian ocean that is larger than the lifetime of the solar system. Thus these calculations serve to show that it is possible, with known processes, to remove an FTO worth of water from Venus, but they do not provide a well-constrained estimate of the timescale for water loss.

One of the most difficult aspects of modeling the climate history and timescale for water loss on Venus is understanding the effects of clouds on the early environment. Cloud feedback is not included in any of the currently published models of the moist greenhouse. Yet, qualitatively, cloud feedback is expected to stabilize surface temperatures with rising solar flux, as cloud coverage and thickness increase with increasing temperature, and therefore albedo increases as the planet warms. Therefore cloud feedback would be expected to extend the lifetime of the moist greenhouse.

Plate 1 shows the results of some very simple calculations of the effects of clouds on temperature structure on a planet with a CO_2-H_2O atmosphere. In order to test the notion that clouds could have played an important, even dominant role in the radiation balance of a warm, wet Venus we have developed simple 1-D, semi-gray, 2-component, 2-stream model of a wet Venus atmosphere. Lapse rates in the convective lower atmosphere are controlled by a wet adiabat, and water clouds are allowed to form where the atmosphere becomes saturated with water. We use simple analytic expressions to calculate the albedo and scattering properties of these clouds. We considered cases with 0, 50 and 100% cloud cover. To illustrate the effects of clouds, we have considered a 1 bar N_2 atmosphere with 350 ppm CO_2. Atmospheric H_2O is allowed to vary over many orders of magnitude, from 6 mbar to 60 bars. The full radiative-convective equilibrium temperature profiles, along with the locations and extent of clouds are shown in Fig. 1 for atmospheres with 6 mbar, 60 mbar, and 2 bar of H_2O. For 6 mbar, we obtain something like the surface temperature of present-day Earth, with clouds existing between 2 and 10 km. With 10 times this amount of water, the surface temperature is 100 K higher, and clouds inhabit the atmosphere at about 20 km. For the 2-bar H_2O case, the tropopause rises to about 50 km and surface temperatures are about the same as present-day Venus. For all these cases, we assume 100% cloud cover.

The results of these somewhat crude calculations suggest that clouds will indeed act to significantly cool the surface during the moist greenhouse phase. Since hydrogen escape rates depend sensitively on upper atmosphere water abundance, which is in turn strongly dependent on surface temperature, these results lead us to believe that cloud-albedo feedback will significantly prolong the lifetime of early Venusian oceans. In this light, it is interesting to note that there are no strong geological constraints on the longevity of Venus' oceans, beyond the observation that there are no obvious signs of surface water seen in the ~1 GY record recorded by the Magellan radar maps. Thus it is possible that the lifetime of Venus' oceans was measured in billions of years, rather than hundreds of millions. These considerations motivate a much more rigorous calculation of climate during the moist-greenhouse phase.

Based on these results, Grinspoon and Bullock (2003) suggested that with the increased stability of surface oceans provided against the warming sun by cloud-albedo feedback, it is quite possible that oceans may have persisted for substantially longer than the oft-quoted 600 MY, perhaps as long as 2 GY. Thus the surface of Venus may have been a habitable environment for a substantial fraction of solar

system history. The rapid rate of dissociation of water and subsequent escape of hydrogen may have resulted in an oxygen-rich atmosphere during this time. This is an intriguing possibility for astrobiology given the likelihood that the late evolution of multicellular life on Earth may have been due to the timescale for atmospheric oxygenation.

3. VENUS AND COMPARATIVE PLANETARY ASTROBIOLOGY

In the last decade upwards of 200 planets have been discovered orbiting other stars (Reid, 2007). Observational selection has resulted, as of this writing, in an absence of terrestrial planets with masses close to that of Earth among those observed. However, formation and dynamical models predict a large number of terrestrial planets (Raymond *et al.* 2006). As far as we know, with the possible exception of the existence of life on Earth, there is nothing very unusual about the solar system, so discovery and remote sensing of many extrasolar terrestrial planets is widely anticipated. These predictions will soon be testable, as data from the COROT and Kepler missions begin to reveal the demographics of terrestrial planets in our galaxy and more advanced observational programs make possible the study of the spectra and light curves of such bodies.

Given this expected harvest of terrestrial planet data over the coming decades, the importance of studying these worlds for putting Earth, and life, in context, and the certainty that knowledge of these planets will be restricted to remote sensing for the foreseeable future, it is vital that we have "ground truth" in the form of terrestrial planet studies that combine remote sensing with *in situ* exploration. Extrasolar terrestrial planets to be observed can be expected to sample a complete suite of evolutionary states representing early, mature and late phases of planetary history. This increases the importance of comparative studies of the current states and evolutionary histories of Venus, Earth and Mars. For understanding the possible evolutionary histories of Earth-sized planets, the Venus-Earth comparison represents a unique opportunity.

If experience of extrasolar planet discovery thus far is any guide, then the variety of terrestrial planets is likely to be large and surprising. Making sense of this diversity with such a small baseline of well-known local examples of terrestrial planet evolution seems like a daunting task. However, to the extent that we can reliably piece together the evolutionary histories of the local terrestrial planets, this task is eased by the fact that we then, in a sense, have more than three local examples. Exploration of the current terrestrial planets provides us with an increasingly detailed snapshot of planetary evolution at one moment of geological time around a 4.55 billion year old G-type star. In the first billion years of solar system evolution, Venus, Mars and Earth were all very different from their current states, in ways that would be easily observable through remote sensing. To the extent that we can understand, with some confidence, the likely past and future states of local terrestrial planets, we can expand our knowledge base to more than the three examples provided by the current states of these planets.

In turn, when we have built up a sizeable database of observable characteristics of extrasolar planets around stars of different ages, it is likely that patterns will emerge which will help us to constrain our models of planetary evolution. Eventually we will have a database that will allow us to discern common classes of evolutionary pathways for terrestrial planets. As we discover and observe extrasolar terrestrial planets, we will see the full range of evolutionary stages and end states. We will undoubtedly see planetary systems of a wide variety of ages, from newly formed systems to "middle aged" systems like our own around stable main-sequence stars, to older planetary systems around late stage stars. These observations will help us to refine our understanding of the early histories and distant fates of Venus, Earth and Mars.

As our understanding of terrestrial planet evolution has increased, the importance of water abundance as a substance controlling many evolutionary factors has become increasingly clear. This is true of biological evolution, as the presence of liquid water is widely regarded as the key to the possibility of finding "life as we know it" on other worlds (Pace, 2001; Benner et al., 2004). It is also true of geological and climatic evolution. Water is among the most important climatically active atmospheric gasses on the terrestrial planets. It is also a controlling variable for tectonic style and geologic processes (Bercovici, 2003), as well as a mediator of surface-atmosphere chemical reactions (Walker et al., 1981). Thus, understanding the sources and sinks for surface water and characterizing the longevity of oceans and the magnitude of loss mechanisms on terrestrial planets of differing size, composition and proximity to stars of various stellar types, and the range of physical parameters which facilitates plate tectonics is key to defining stellar habitable zones.

As we learn enough through exploration and further modeling, to better characterize the evolutionary history of Venus, we will build a context for interpreting observations of extrasolar terrestrial planets. In particular, given the likelihood that Venus and Earth started out with similar surface conditions, and that Venus underwent loss of potentially one or several FTOs worth of water, the semi-controlled "experiment" of the apparently divergent histories of Venus and Earth is of particular interest for characterizing the histories and fates of Earth-sized worlds, and understanding

their dependence on initial conditions including stellar type, stellarcentric distance and initial volatile abundance.

It is striking that of the three local terrestrial planets, two have lost their oceans either to a subsurface cryosphere or to space, and one has had liquid oceans for most of its history. It is likely that planetary desiccation in one form or another is common among extrasolar terrestrial planets near the edges of their habitable zones.

On Venus, the very low abundance of water in the atmosphere and crust, combined with ongoing volcanism, have led to a sulfur-rich environment (Prinn 1985). This is most obvious in the globally-encircling sulfuric acid cloud layers, but there are strong experimental and observational reasons to believe that sulfur gases in the atmosphere interact vigorously with the surface (Fegley and Prinn 1989; Bullock and Grinspoon 2001; Prinn 2001).

Perhaps one of the most important discoveries about the history of water on Mars has been the very recent detection of massive layered sulfates by the Opportunity Mars Exploration Rover in Meridiani Planum (Squyres et al 2004a). These record not only the past existence of flowing surface water, but also the chemical nature of the water that was there last (Squyres et al. 2004b, 2005; Grotzinger et al. 2005). Meridiani sedimentary rocks are about 40% sulfate, with embedded nodules of hematite and significant quantities (a few percent) of the iron sulfate mineral jarosite (Clark et al. 2005). This implies that the sediments were laid down in highly acidic water, with a pH of between 2 and 4, when conditions were also oxidizing (Elwood Madden et al. 2003). The layered deposits in Meridiani, as seen from the Mars Observer Camera aboard the MGS spacecraft, are about 1 km thick and cover a region about the size of the Colorado Plateau (Hynek et al. 2002). It is very likely that they are the chemical remnants of an acidic sea. The history of Meridiani is probably not unique on Mars, as layered sulfate deposits are also seen in much larger quantities within Valles Marineris by the Omega spectrometer on the Mars Express Spacecraft (Gendrin et al. 2005). The lack of carbonate deposits of any kind strongly argues for a sulfur-rich, acidic environment as the last of Mars' surface water disappeared (Bullock and Moore 2005).

For large quantities of water to maintain a low pH on early Mars, they must have been supplied with sulfur, most likely from volcanism. Phillips et al. (2001) showed that the entire Tharsis bulge is a large igneous construct, emplaced mostly during the late Noachian era. The formation of Tharsis alone could have supplied enough sulfur to Mars' atmosphere to acidify the oceans and produce the chemical record that we see today. The early Martian atmosphere would very likely have hosted sulfuric acid/water clouds or hazes, much like Venus does today.

Both Venus and Mars have sulfur-rich environments, and may have experienced a watery past followed by an acidic phase as they desiccated. Current observations of Mars are thus leading to the increasing relevance of studying Venus as an early Mars analog. If there was an origin of life on Mars, it would most likely have taken place in acidic, sulfur-rich conditions.

Astrobiology, with its predecessor exobiology, has been accused of being a "science without a subject". This is not strictly true, as the subject includes the history of Earth, its comparison with other planets, and the origin, evolution and future of life. Nevertheless, the subject of life and its potential distribution in the universe is not informed by a large number of known examples. In fact, arguably, this is one of the few sciences based on a single example of a complex phenomenon (along with cosmology and perhaps evolutionary biology). Thus, it is important to question all assumptions about biology's universal characteristics which are based on the potentially unique qualities of Earth's biosphere, lest we slip into a kind of scientific solipsism, convinced through our ignorance and lack of perspective, rather than through actual evidence, that life elsewhere must share our own exact characteristics. Our planetary exploration, with an increasing focus on astrobiology, is designed to "follow the water". This is a reasonable strategy but it is based, at best, on an educated guess about life's universals. Such an approach is necessary, for now, for us to proceed with astrobiological exploration in the absence of other biological examples. Yet, in our current ignorance, all conclusions as to what is "universal" about life must be regarded as provisional. Given this philosophical problem, any alternative means of evaluating the habitability of planets should be given serious consideration. The most common means of evaluating planetary habitability begins and ends with the assumption that other planets, in order to be suited for life, must possess a similar environment to that of Earth, at least similar enough for liquid water to be present (see for example Schulze-Makuch and Irwin, 2003).

One alternative that has been suggested is the "Living Worlds Hypothesis" of Grinspoon (2003). According to this idea, life is not merely a local phenomenon that requires certain chemical or thermodynamic conditions in specific micro-environments to exist, but is best thought of as a planetary property. The only inhabited world we know of is so pervasively inhabited that it is difficult to find a terrestrial environment that has not been occupied by life, and probably impossible to find an environment that has not been modified by life. It is also true that the natural history of Earth is so deeply bound up with the history of life that it is almost meaningless to discuss the history of the "nonliving part" of Earth as a separate topic from the history of terrestrial life (Margulis and Lovelock, 1974). It cannot be argued, based on a statistical sample of

one, that because these are characteristics of life on Earth, that they should be universal characteristics of inhabited planets. That is, unless the longevity and stability of life on Earth are causally related to life here being a global property of the planet. If this were the case, then for example, the existence of Martian life persisting in isolated remnants of a former global biosphere which still retain liquid water and energy sources is much less likely (Grinspoon, 2003).

This viewpoint focuses attention on global properties, such as rate of exchange of matter and energy between interior and surface, surface energy gradients, available sources of free energy, magnitude of energy gradients, or atmospheric disequilibrium properties, that may be required for life to persist for geological timescales. In this light it is interesting to note that Venus and Earth share some global properties that may be relevant for this type of consideration. The global distribution of surface ages may serve as a crude proxy for total level of ongoing endogenous planetary activity. Although we have only a rough knowledge of the surface age distribution of Venus, it is clear that almost all surface units are younger than 1 GY, as is also true of Earth's surface. This is in marked contrast to Mars, the Moon, and Mercury, where almost all surface units are older than 3 GY. If we think beyond the specifics of a particular chemical system required to build complexity and heredity, we can ask what general properties an inhabited planet must possess. Judging from our sample of one inhabited planet, the answers might include vigorous geological activity continuing over billions of years, an atmosphere with signs of flagrant chemical disequilibrium and active, internally driven cycling of volatile elements between the surface, atmosphere and interior. At present, the two planets we know of which possess these characteristics are Earth and Venus.

In addition to the study of life's evolution and distribution, astrobiology also considers questions about the future of life. Does Venus have an astrobiological future? Given the slowly brightening sun and the continued exospheric loss of hydrogen it would seem that left to its own devices, the biological future of Venus is not a promising one. But will Venus always be left to its own devices? Several researchers have considered the possibilities of planetary engineering or terraforming Venus, or a Venus-like planet, altering its surface environment to be more conducive to organic life (Fogg, 1987; Dyson, 1989; Adelman, 1992; Pollack and Sagan, 1994). While an exploration of this speculative topic is beyond the scope of this paper, the interested reader is encouraged to pursue the available literature.

4. THE POSSIBILITY OF EXTANT LIFE

The common exercises of astrobiology each contain some, perhaps unavoidable, degree of geocentrism. Most reviews of the possibility of life on Venus begin with an examination of terrestrial extremophiles which exhibit resistance to environmental extremes associated with the Venusian environment (Cockell, 1999). These include acidophiles, hyperthermophiles, organisms with unusual osmotic capabilities, pressure tolerance, radiation resistance, ability to survive unusual atmospheric compositions, and so forth. This approach is pragmatic in that it allows us to use data from known organisms to assess the habitability of extraterrestrial environments, but it is worth reminding ourselves that any organism found on another planet will most likely not be a terrestrial extremophile and the idea that alien organisms can only inhabit environments proscribed by the limits of terrestrial organisms is an untested and extremely conservative assumption.

Given the high danger of solipsistic, geocentric reasoning, it is a useful exercise to consider life in non-obvious locations, including those where it is generally considered unlikely. In this spirit, some researchers have persisted in arguing that extant life on Venus cannot be ruled out given our present ignorance about life's range and limits, and the current constraints on our knowledge of the Venusian environment (Grinspoon, 1997; Schulze-Makuch and Irwin, 2002; Schulze-Makuch et al., 2004).

These arguments have focused on the possibility of life in the global cloud decks of Venus. Properties of the Venus clouds which are potentially hospitable to life include the following: (1) the clouds form an aqueous environment. While water is scarce, water vapor concentrations reach several hundred ppm in the lower clouds, (2) the cloud region has temperatures of 300–350K and pressures around 1 bar. (3) the clouds are much larger, more continuous and stable than clouds on Earth (Grinspoon et al. 1993). At some latitudes the cloud particles do not fall at all, as vertical transport velocities are higher than fall velocities (Imamura and Hashimoto, 1998). Thus the particle lifetimes will be approximately equal to the Hadley circulation timescale of 70–90 days. This is several orders of magnitude longer than the typical division time of bacteria. ; (4) the atmosphere is in chemical disequilibrium, with H_2 and O_2, and H_2S and SO_2 coexisting; (5) the "mode 3" particles in the lower cloud, which dominate the mass of the cloud deck, may be non-spherical, and may contain an unknown, non-absorbing core material which comprises up to 50% by volume of the particles (Cimino, 1982; Grinspoon et al. 1993) and are comparable in size to microbes on Earth; (6) the superrotaton of the atmosphere enhances the potential for photosynthetic reactions by producing a day-night cycle of 4–6 days, compared with 117 Earth days on the surface. (7) the unknown UV absorber has some properties in common with a photosynthetic pigment (Grinspoon, 1997).

There are several obvious objections to the idea of life in the Venusian clouds, although each depends on unproven assumptions made from extrapolating the properties and limits of terrestrial life. One objection has been summarized as "Why aren't the clouds green on Earth?". In other words, if cloud-based life were possible, surely on Earth, where life is nearly ubiquitous and has demonstrated its exhaustive evolutionary opportunism, the clouds would show obvious signs of inhabitation. However, Grinspoon (1997) points out that on Earth clouds are a comparatively transient and fragmented niche, with typical particle lifetimes of days, as compared to months on Venus. Further, it has not been established that clouds on Earth are not an inhabited niche, and some researchers have found evidence for terrestrial cloud-based bacterial colonies. In experiments conducted at the Sonnblick Observatory at an altitude of 3106 m, Sattler et al. (2001) analyzed condensing clouds and found growth and reproduction of microbes in super-cooled cloud droplets. They concluded that the limiting factor for the persistence of microbial life in cloud droplets is residence time in the atmosphere. Thus, for water-based life it may be that, in terms of stability and continuity, the discontinuous and cold clouds of Earth represent a much more extreme environment for life than those of Venus.

Another common objection to the idea of life in the Venus clouds is that, although they are an aqueous medium, the cloud droplets are strongly acidic. All known terrestrial acidophile organisms rely on active trans-membrane proton pumps to maintain their cytoplasm at a pH between 4 and neutral. How strong an impediment to life is the high acidity of the sulfuric acid clouds? Supporters of possible Venusian cloud life point to the ever expanding range of known terrestrial acidophile organisms.

Interestingly, the first known representatives of extreme acidophiles (growing near pH 0) that were found were eukaryotes rather than prokaryotes (Hoover, 2006). *Cyanidium caldarium* is a red algae (a rhodophyte) that contains both chlorophyll "a" and C-phycocyanin in its chloroplasts. It colonizes most of the hot, acid soils and waters on Earth. Three other extremely acidophilic eukaryotes are fungi: *Aconitium cylatium*, *Cephalosporium sp.*, and *Trichosporon cerebriae* (Schleper et al., 1995). The green alga *Dunaliella acidophila* can survive pH 0, but the maximum for growth is at pH 1 (Pick, 1999).

The archaeon *ferroplasma acidarmanus* thrives at pH 0 (Edwards et al., 2000), This eubacterium has no cell wall and it grows at pH 0 in acid mine drainage in Iron Mountain in California. The cell membrane is the only barrier between the cytoplasm and concentrated sulfuric acid infused with high concentrations of copper, arsenic, cadmium and zinc. This is the current low pH record for bacteria.

Although even some chemistry textbooks state that negative pH is not possible, this is in error. The error comes from the common use of the hydronium (H_3O^+) ion to measure pH, and the consequent statement in many popular science textbooks and some textbooks that pH can only exist within a range between 0 and 14. However, pH, commonly defined as pH = -log[H^+] (negative the logarithm of the hydrogen ion molarity) is better defined as pH = -log a_{H^+} (negative the log of the hydrogen ion activity). In very strong acid solutions, where there are few water molecules per acid formula unit, the influence of the hydrogen ions in the solution is enhanced. Thus the effective concentration, or the activity, of the hydrogen ions is much higher than the actual concentration. This commonly leads to negative values of pH. For example, commercially available concentrated HCl solution (37% by mass) has pH $\approx$ -1.1. (Lim, 2006). Figure 1, from Nordstom et al. (2000) shows the dependence of pH on sulfuric acid concentration, when the hydrogen activity coefficient is properly included. Hot springs near Ebeko volcano, with naturally occurring HCl and H_2SO_4, have pH of $\approx$ -1.7, and acid drainage at the Richmond Mine at Iron Mountain, CA have been measured at -3.6. The latter are the most acidic waters found on Earth so far.

As seen in Plate 2, the relatively water-rich aerosols in the upper cloud have a small range of positive pH, from 0.3 to 0.5. In the lower cloud, with its larger and more water-poor particles, pH can be as low as -1.3. The aerosol H_2SO_4 concentrations were calculated using the cloud model of Bullock and Grinspoon (2001), constrained by Pioneer Venus data, with corrections for high activities from Nordstrum et al. (2000).

There are now many terrestrial organisms known that can survive in negative pH conditions. The most extreme hyperacidophiles known to date belong to the archaea. For example, *Picrophilus oshimae* and *P. torridus* are able to grow at pH −0.2. They grow within the moderately thermal regime and were discovered in Japan near hydrothermal springs with solfataric gases (Schleper et al., 1996).

There are settings on Earth with acidity that exceeds that in the Venus clouds. For example, at Iron Mountain, California, a superfund cleanup site where pyrite mining has produced waters which are extremely concentrated in sulfuric acid, a pH of -3.6 has been measured (Nordstrum et al., 2000). As the water flowing from this site mixes with fresh water it produces a gradient of increasing pH in the downstream direction. This creates an interesting experiment in extremophile biology. How far upstream, in the direction of decreasing pH, does life exist? Unfortunately, the answer is not yet known. Organisms existing in water at pH ~0 have been cultured. However, it is not easy to search for life in the more acidic waters in the negative pH zone of

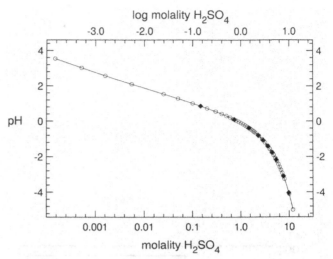

Figure 1. After Nordstom et al (2000). The dependence of pH on sulfuric acid concentration, when the hydrogen activity coefficient is properly included, calculated using the Pitzer ion interaction theory.

this stream, as ordinary culture mediums would simply dissolve in this water (Nordstrum, 2005). New, acid-resistant, culture mediums will have to be created in order to test for life in the most acidic waters. Thus, the low pH limit of terrestrial life is currently not known.

Another environmental constraint often mentioned as potentially prohibitive for life in the Venus clouds is the high UV flux, which would prove fatal for contemporary terrestrial surface organisms, as UV radiation is damaging to biological macromolecules. However, adaptation to high UV fluxes under selection pressure has been achieved by some terrestrial organisms. One strategy is the development of UV absorbing pigments such as carotenoids and scytonemin (Wynn-Williams et al. 2002). Other organisms have used layers of soil or water for protection (Pierson et al. 1993; Wynn-Williams and Edwards, 2000), or shielding by organic compounds from dead cells (Marchant et al. 1991). Still others, such as *Deinococcus radiodurans*, take a more active approach toward restoring UV damaged DNA and UV-sensitive proteins, repairing and resynthesizing macromolecules as needed (Battista, 1997; Ehling-Schulze and Scherer, 1999). The fungus *Fusarium alkanophylum* exhibits optimal growth under high does of UV radiation, provided that the growth medium contains sulfur-rich proteins (Marcano *et al.* 2002).

Given the evolutionary history of UV flux at the Earth's surface during the history of life, and the ability of many organisms to adopt to high UV fluxes, it is unlikely that the UV flux presents an impediment to life in the Venus clouds. Cockell (1999, 2000) points out that the UV flux in the

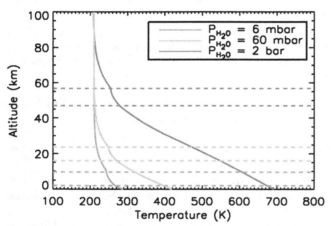

Plate 1. Radiative-convective equilibrium temperature profiles, as described in the text, along with the locations and extent of clouds for atmospheres with 6 mbar, 60 mbar, and 2 bar of H_2O.

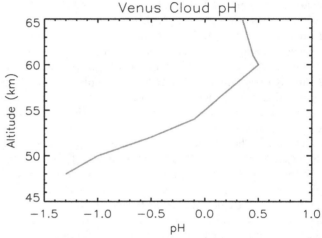

Plate 2. The pH of Venus' clouds as a function of altitude. The relatively water-rich aerosols in the upper cloud have a small range of positive pH, from 0.3 to 0.5. In the lower cloud, with its larger and more water-poor particles, pH can be as low as -1.3. Aerosol H_2SO_4 concentrations were calculated using the cloud model of Bullock and Grinspoon (2001), constrained by PV data. Correction for high activities is from Nordstrum et al. (2000).

upper clouds of Venus is comparable to the surface flux on Earth during the archean, before the build-up of atmospheric ozone but after the origin of life. Thus the UV flux cannot be considered to be a limit to the habitability of the cloud environment. In fact, it has been proposed that appropriately adapted organisms might use this flux as an energy source to power metabolism (Grinspoon, 1997; Schulze-Makuch et al. 2004).

Schulze-Makuch et al. (2004) proposed that cycloocta-sulfur (S_8) could be employed as a UV shield, and might also be employed as a UV-powered photosynthetic electron donor. They point out that S_8 is the most stable form of elemental sulfur within the clouds, and does not react with sulfuric acid. It absorbs strongly in the UV, with an absorption maximum at 285 nm, near the UV wavelengths most damaging to cells ($\approx$240–280 nm). The absorption is so strong that a few monolayers of S_8 would provide a significant UV shield for putative organisms. S_8 has been proposed as a UV shield on the early Earth (Pavlov et al. 2002). S_8 also reemits at visible wavelengths, providing another possible mechanism for photosynthetic exploitation of UV radiation.

Elemental sulfur might also be microbially produced by a photosynthetic reaction in which hydrogen sulfide is oxidized to elemental sulfur and carbon dioxide is reduced to organic carbon (equation 1)

$$2H_2S + CO_2 + light \rightarrow \text{``}CH_2O\text{''} + H_2O + 2\,S \qquad (1)$$

Many terrestrial organisms using this reaction thrive in marine sediments and hot springs (e.g. Vethanayagam, 1991, Bryantseva et al., 2000). The oxidized sulfur could then be polymerized to S_8. Several photosynthetic microbes, which use H_2S as their electron source, deposit sulfur outside the cell including Green Sulfur bacteria, some Purple Sulfur bacteria and some cyanobacterial species (e.g. Pierson et al., 1993; Tortora et al., 2001). Putative Venusian microbes might deposit elemental sulfur on the outside of their cells to convert potentially harmful UV radiation to electromagnetic frequencies that are usable for photosynthesis, or harvest the energy of UV photons through conversion to appropriate electron donors. Alternatively, they may just utilize sulfur allotropes as "sunscreen" and utilize visible light for photosynthesis. Schulze-Makuch et al. (2004) point out that while on Earth selection has favored photosynthetic mechanisms that operate at neutral pHs and are non-functional at high acidity, on Venus forms of life may have evolved different molecular machinery to deal with the progressively acidifying conditions.

Another concern about the habitability of the Venus cloud environment is that there is not sufficient access to the "biogenic elements". The biogenic elements are generally given as C,H,O,N,S,P as well as several trace metals, such as Co, Mn, and Mo. It seems dangerously geocentric to expect extraterrestrial life to require the exact same list of elements as life that has evolved in the near-surface environment of Earth, where these elements are available. Nonetheless, it can be pointed out that among the elements known to exist in the cloud region of Venus are C, H, O, N, S, P, Cl, F and most likely Fe. Among those suspected to exist are I, Br, Al, Se, Te, Hg, Pb, Al, Sb and As (Marov and Grinspoon, 1998). Thus the availability of a rich assortment of elemental building blocks, including most of the "biogenic elements" is not an impediment to possible life.

The previous arguments have been offered in defense of the proposition that life, once established, might be maintained in the clouds. It is harder to imagine an origin of life occurring in the cloud environment (though this may also be due to geocentric bias). Yet, as previously discussed, the surface of Venus, early in planetary history, may have had the necessary liquid water, organic rich, stable environment to support an origin of life, or survival of life that may have been transplanted from Earth or Mars through impact panspermia. As surface conditions became hostile, life could have adapted to an atmospheric niche under directional selection.

If we accept the premise that the present day clouds of Venus might be habitable, and that cloud life may have been implanted by lithopanspermia or migration from an ancient habitable surface, another question immediately presents itself. Have the clouds been continuously habitable over billions of years? It is not known whether the clouds have been a permanent feature of Venus. If Venus has been cloud free for any period of time then, in the absence of a habitable surface environment, life depending on the cloud niche could not have persisted. Today, the clouds form one continuous habitat and if they temporarily disappeared from the planet, there would be no refuge. Bullock and Grinspoon (2001) found that the lifetime of the global cloud deck against destruction by reaction of sulfur with surface minerals is 30 MY in the absence of replenishment of atmospheric SO_2 by volcanic outgassing. Thus, it is quite possible that Venus has gone through cloud-free epochs. In this case, a cloud-based biota would not have survived over planetary history. On the other hand, it is possible that the clouds have not disappeared but have at times become thinner, discontinuous and fragmented, before returning to their thick, globally continuous state. Such a history could theoretically have played a productive role in the evolution of Venusian biota by creating geologically significant time periods during which temporarily isolated environments would have facilitated increased speciation. Future missions to Venus and theoretical modeling of atmospheric and cloud evolution will illuminate the global history of outgassing and the question of whether the clouds have been a continuous, stable environment over planetary history.

5. PLANETARY PROTECTION

At the dawn of the Space Age, as contamination of other planets with terrestrial biota (forward contamination) and contamination of Earth by alien microorganisms (back contamination) by vehicles of space exploration became a theoretical possibility, the scientific view of Venus was in transition. Attempts were made to interpret the anomalous microwave radiation thought (correctly) by some to indicate a surface temperature outside the range of liquid water. This is illustrated by the 1960 paper in *Science* "Exobiology: Approaches to Life Beyond the Earth", in which biologist Joshua Lederberg introduced the term "Exobiology" to the wider scientific community: "The habitability of Venus is connected with its temperature, a highly controversial subject. Perhaps the most useful first contribution to the exobiology of Venus would be a definitive measurement of its temperature profile. Even should the surface be unbearably hot, this need not preclude a more temperate layer at another level." In this seminal paper, Lederberg argued that "...we must *rigorously* exclude terrestrial contaminants from our spacecraft." and that the risk of back contamination, while apparently slight, must also be taken seriously, given the large magnitude of the potential consequences of error.

In the wake of the Mariner 2 results indicating surface conditions incompatible with organic life, planetary protection arguments and requirements for Venus missions were relaxed. In 1970, a study by the Space Science Board (the predecessor of the current Space Studies Board) of the National Research Council included quantitative estimates of the probability of contamination of Venus by an interplanetary spacecraft, and concluded that "A slight possibility exists that terrestrial organisms could grow on airborne particles near to the cloud tops of Venus... ...Life on Venus is no more than a remote contingency, but the possibility of contamination by terrestrial organisms must be considered." This study recommended that only minimal precautions be applied to Venus bound spacecraft due to planetary protection considerations (Space Science Board, 1970).

A later study in 1972 (Space Science Board, 1972) similarly concluded that the probability of surface contamination of Venus by terrestrial spacecraft was effectively zero, and that the probability of growth of terrestrial microorganisms in the more temperate regions of the atmosphere was $\leq 10^{-6}$.

In 2005, in light of advances in astrobiology, including the discovery of extremophile organisms, new ideas about the possible viability of cloud-based life on Venus, and the prospect of a new generation of Venus spacecraft, NASA's Office of Planetary Protection, under the direction of Planetary Protection Officer John Rummel, decided to revisit the question of planetary protection protocols for Venus. The Space Studies Board's Committee on Origin and Evolution of Life (COEL) was asked to provide advice on planetary protection concerns related to missions to and from Venus, and a Task Group on Planetary Protection Requirements for Venus Missions was formed and heard expert testimony at several meetings. In particular, the COEL was asked to:

1. Assess the surface and atmospheric environments of Venus with respect to their ability to support Earth-origin microbial contamination, and recommend measures, if any, that should be taken to prevent the forward contamination of Venus by future spacecraft missions;
2. Provide recommendations related to planetary-protection issues associated with the return to Earth of samples from Venus; and
3. Identify scientific investigations that may be required to reduce uncertainty in the above assessments.

After hearing testimony and reviewing the relevant issues, the Task Group concluded, in summary, that:

- No significant risk of forward contamination exists in landing on the surface of Venus.
- No significant risk exists concerning backward contamination from Venus surface sample returns.
- No significant forward contamination risk exists regarding the exposure of spacecraft to the clouds in the atmosphere of Venus.
- No significant risk exists concerning atmospheric sample returns from the clouds in the atmosphere of Venus.

The task group thus recommended that the previous COSPAR Category II planetary protection classification of Venus be retained (see Attachment 2 in Space Science Board, 2006, available online at http://www.nap.edu/catalog/11584.html for full descriptions of these categories). Category II includes all types of missions to those target bodies where there is "significant interest relative to the process of chemical evolution and the origin of life, but where there is only a remote chance that contamination carried by a spacecraft could jeopardize future exploration." For category II bodies, the legal requirements are only for simple documentation. This required documentation includes a short planetary protection plan, primarily to outline intended or potential impact targets; brief pre-launch and post-launch analyses detailing impact strategies; and a post-encounter and end-of-mission report providing the location of inadvertent impact, if such an event occurs.

For planetary protection concerns, the relevant question is ultimately not the probability of any habitable niche existing on present day Venus, but the likelihood that such a niche, if it does exist, possessing physical conditions which overlap the conditions under which terrestrial organisms can survive, grow and reproduce. The judgment of the ad hoc Task Group was that the chance of such overlap is too slight to significantly impact planning for future Venus missions. This is, in a sense, good news for the future astrobiological exploration of Venus as it means that such missions are not likely to be burdened with the cost and complexity of meeting complex planetary protection requirements.

6. FUTURE INVESTIGATIONS AND EXPLORATION

There are some investigations that can be undertaken on Earth to deepen our knowledge of possible life on Venus or Venus-like terrestrial planets. These would include further study of extremophile organisms, particularly those in highly acidic environments and environments requiring extreme osmoregulation for survival. Knowledge of the specific biochemical mechanisms that make these extremophilities possible, and the energetic costs of these adaptations would be particularly valuable in extrapolating these abilities to putative organisms in alien environments. Follow-up experiments must be conducted to further explore the provocative claims of Sattler et al. (2001) of finding reproducing bacteria in cloud environments on Earth, as well as thorough biological investigation of permanently cloudy environments on Earth, such as tropical cloud forests.

Another potentially revealing avenue of investigation might be attempts to grow terrestrial extremophiles in simulated Venusian clouds. The acidity of these droplets could be varied, as a simulation of the likely evolution of the Venusian cloud deck, in response to billions of years of volcanic outgassing, from a higher pH when the planet was dominated by water clouds during early escape of a steam-dominated atmosphere, to the extremely acidic current conditions. It would be very instructive to learn when, in the history of Venus history, the clouds became uninhabitable to terrestrial microorganisms, if indeed they are. These experiments would also be useful in the area of planetary protection, as tests of the prevailing view that the present-day Venusian clouds represent a sterilizing environment for organisms likely to be inhabiting spacecraft exteriors.

As long as our ignorance of Venus remains large, notions of exotic extant organisms will inhabit the gaps in our knowledge. The best way to put these speculations to rest is to more fully explore our sister planet. For example, once we have fully characterized the "mode 3" cloud particles and determined the composition of their core materials, then (assuming they do not turn out to be the home of Venusian bacterial communities) astrobiological speculation about their nature will greatly diminish.

Future missions to Venus can help to address astrobiological issues with observations clarifying the longevity of potentially habitable surface conditions as well as observations clarifying the conditions and history of the clouds of Venus and the closely coupled physical and chemical history of the interior/surface/atmosphere/cloud/climate system. Of particular interest for the astrobiological questions discussed in this chapter will be:

- Observations of stable isotopes, escape fluxes and their responses to the solar wind to test assumptions about similar Venus-Earth early water endowments and to constrain the subsequent divergent histories of hydrogen escape and climate change.
- Geochemical and geochronological observations constraining the global resurfacing history and the sulfur cycle, to test whether a cloud niche has been continuous, and whether disequilibrium chemistry in the clouds is compatible with more mundane geochemical and aeronomical explanations.
- Characterization of the unknown UV absorber and its relationship with sulfur chemistry.
- A more full characterization of the global atmospheric circulation and its effect on cloud particle lifetimes.
- Reliable measurements of trace constituents in the clouds and the surrounding atmosphere.
- Characterization of the surface rocks in the plains and tesserae, including their mineralogy, and searches for metastable hydrated minerals, zircons, and possible isotopic biomarkers.
- Determination of the ages of the major surface units.

This list of needed astrobiological observations from future missions is nearly identical to that desired to achieve a more general understanding of the planet and its evolutionary divergence from Earth. Venus is the only other example of an Earth-sized planet that we can hope to study up-close within the foreseeable future. We still don't understand how and when Venus diverged from a more Earth-like past, or how long Venus or Mars retained habitable conditions. If life begins easily on warm, wet planets (which seems to be the case, judging from Earth) then Venus probably once harbored Venusians. Did they die out as the climate changed? Venus can help us to understand the evolutionary paths that habitable planets can take. Further exploration of Venus will greatly advance our understanding of terrestrial planet habitable zones and "continuously habitable zones", and provide needed context for extrasolar terrestrial planet discoveries. What is good for planetary exploration of such a promising target—so Earth-like in bulk qualities but so alien in environmental evolution—is also good for astrobiology.

Acknowledgements. We would like to thank Dirk Schulze-Makuch, Richard Hoover, Kirk Nordstum, Bob Carlson, Chris McKay and Victoria Meadows for helpful conversations and suggestions. The authors acknowledge support from the NASA Planetary Geology and Geophysics, Planetary Atmospheres, and Exobiology Research Programs.

REFERENCES:

Adelman, S.J. (1982) "Can Venus Be Transformed into an Earth-Like Planet?" JBIS, 35, 3–8

Battista, J.R. (1997) Against all odds: the survival strategies of *Deinococcus radiodurans*. *Am. Rev. Microbiol.* 51, 203–224.

Benner, S.A., A. Ricardo and M.A. Carrigan (2004) Is there a common chemical model for life in the universe? Current Opinion in Chemical Biology 8, 672–689.

Bercovici, D. (2003) The generation of plate tectonics from mantle convection. *Earth Plan. Sci. Lett.* 205. 107–121.

Berteau, J.L., Widemann, T., Hauchecorne, A., Moroz, V.I., and Ekonomov, A.P. (1996) Vega-1 and Vega-2 entry probes: an investigation of UV absorption (220–400 nm) in the atmosphere of Venus. *J. Geophys. Res.* 101, 12,709–12,745.

Bryantseva, I.A., Gorlenko, V.M., Tourova, T.P., Kuznetsov, B.B., Lysenko, A.M., Bykova, S.A., Gal'chemko, V.F., Mityushina, L.L., and Osipov, G.A. (2000) *Heliobacterium sulfidophilum sp. nov.* and *Heliobacterium undosum sp. nov.*: sulfide oxidizing heliobacteria from thermal sulfidic springs. *Microbiology/Mikrobiologiya* 69, 325–334.

Bullock, M.A., D.H. Grinspoon and J.W. Head (1993). Venus resurfacing rates: constraints provided by 3-D Monte Carlo simulations. *Geophys. Res. Lett.* 20, 2147–2150.

Bullock, M.A. and D.H. Grinspoon (2001) The Recent Evolution of Climate on Venus. *Icarus,* 150, 19–37.

Bullock, M.A., and J.M. Moore, Atmospheric conditions on early Mars and the lack of carbonate deposits, in *Vernadsky-Brown Microsymposium 41*, Lunar and Planetary Institute, Houston, 2005.

Cimino, J. (1982) The composition and vertical structure of the lower cloud deck on Venus. *Icarus* 51, 334–357 (1982).

Clark, B.C., R.V. Morris, S.M. McLennan, R. Gellert, B. Jolliff, A.H. Knoll, S.W. Squyres, T.K. Lowenstein, D.W. Ming, N.J. Tosca, A. Yen, P.R. Christensen, S. Gorevan, J. Bruckner, W. Calvin, G. Dreibus, W. Farrand, G. Klingelhoefer, H. Waenke, J. Zipfel, J.F. Bell III, J. Grotzinger, H.Y. McSween, and R. Rieder, Chemistry and mineralogy of outcrops at Meridiani Planum, *Earth and Planetary Science Letters*, 240, 73–94, 2005.

Cockell, C.S. (2000) The ultraviolet history of the terrestrial planets—implications for biological evolution. *Planetary and Space Science* 48, 203–221.

Cockell, C.S. (1999) Life on Venus. *Planet. Space Sci.* 47, 1487–1501.

Donahue, T.M., J.H. Hoffman, R.R. Hodges, and A.J. Watson (1982). Venus was wet: A measurement of the ratio of deuterium to hydrogen. Science 216: 630.

Donahue, T.M., D.H. Grinspoon, R.E. Hartle, and R.R. Hodges, Jr. (1997). Ion/neutral escape of hydrogen and deuterium: Evolution of water. In *Venus II Geology, Geophysics, Atmosphere and Solar Wind Environment* (S.W. Bougher, D.M. Hunten, and R. J. Phillips, Eds.), pp. 585–414. Univ. of Arizona Press, Tucson.

Donahue, T.M. (1999). New analysis of hydrogen and deuterium escape from Venus. Icarus 141, 226–235.

Dyson, F.J. (1989) "Terraforming Venus," correspondence in JBIS, 42, 593.

Eddington, A.S. *The Nature of the Physical World.* (1926–27 Gifford lectures) MacMillan. 1935.

Edwards, K.J., Bond, P.L., Gihring, T.M., and Banfield, J.F. (2000) An archael iron-oxidizing extreme acidophile important in acid mine drainage. *Science* 287, 1796–1799,

Ehling-Schulze, M. and Scherer, S. (1999) UV protection in cyanobacteria. European *Journal of Phycology* 34, 329–338.

Elwood Madden, M.E., R.J. Bodnar, and J.D. Rimstidt, (2004) Jarosite as an indicator of water-limited chemical weathering on Mars, *Nature, 431*, 821–823.

Esposito, L.W. Knollenberg, R.G., Marov, Y.A., Toon, O.B., and Turco, R.P. (1983) The clouds and hazes of Venus. In Venus, edited by D.M. Hunten et al., Univ. of. Arizona Press, pp.484–564.

Fegley, B., and R.G. Prinn, Estimation of the rate of volcanism on Venus from reaction rate measurements, *Nature, 337*, 55–58, 1989.

Fogg, M.J. (1987) "The Terraforming of Venus," JBIS, 40, 551–564.

Gendrin, A., N. Mangold, J.-P. Bibring, Y. Langevin, B. Gondet, F. Poulet, G. Bonello, C. Quantin, J. Mustard, R. Arvidson, and S. LeMouelic (2005). Sulfates in Martian Layered Terrains: The OMEGA/Mars Express View, *Science, 307*, 1587–1591.

Gladman, B.J., Burns, J.A., Duncan, M ., Lee P.C., Levison, H.F. (1996). The exchange of impact ejecta between terrestrial planets. *Science* 271. 1387–1392.

Grinspoon, D.H. (1987). Was Venus wet?: Deuterium reconsidered. *Science* 238: 1702–1704.

Grinspoon, D.H. (1993). Implications of the high deuterium-to-hydrogen ratio for the sources of water in Venus' atmosphere. Nature, 363: 428–431.

Grinspoon, D.H., J.B. Pollack, B.R. Sitton, R.W. Carlson, L.W. Kamp, K.H. Baines, T. Encrenaz and F.W. Taylor (1993). Probing Venus' cloud structure with Galileo NIMS. *Planet. Space Sci.* 41, 515–542.

Grinspoon, D.H. (1997). *Venus Revealed: A New Look Beneath the Clouds of our Mysterious Twin Planet.* Addison-Wesley. Reading, Mass.

Grinspoon, D.H. (2003). *Lonely Planets: the Natural Philosophy of Alien Life.* Ecco/HarperCollins.

Grinspoon, D.H. and Bullock, M. (2003) Did Venus experience one great transition or two?, B.A.A.S. 35.

Grotzinger, J.P., R.E. Arvidson, J.F. Bell III, W. Calvin, B.C. Clark, D.A. Fike, M. Golombek, R. Greeley, A. Haldemann, and K.E. Herkenhoff, Stratigraphy and sedimentology of a dry to wet eolian depositional system, Burns formation, Meridiani Planum, Mars, *Earth and Planetary Science Letters, 240*, 11–72, 2005.

Hoover, R. (2006). Personal communication.

Hynek, B.M., R.E. Arvidson, and R.J. Phillips, Geologic setting and origin of Terra Meridiani hematite deposit on Mars, *Journal of Geophysical Research, 107*, 5088, doi:10.1029/2002JE001891, 2002.

Imamura, T., and G.L. Hashimoto, Venus cloud formation in the meridional circulation, *Journal of Geophysical Research, 103*, 31349–31366, 1998.

Ip W. H. and Fernandez, J.A. (1998) Exchange of condensed matter among the outer and terrestrial protoplanets and the effect on surface impact and atmospheric accretion. Icarus 74, 47–61.

Irwin, L. N. and D, Schulze-Makuch (2001). Assessing the plausibility of life on other worlds. *Astrobiology* 1: 143–160.

James, E.P., Toon, O.B., and Schubert, G. (1997) A numerical microphysical model of the condensational Venus cloud. *Icarus* 129, 147–171.

Kasting, J.F., J.B. Pollack, and T.P. Ackerman, Response of Earth's atmosphere to increases in solar flux and implications for loss of water from Venus, *Icarus, 57*, 335–355, 1984.

Kasting, J.F., Runaway and moist greenhouse atmospheres and the evolution of Earth and Venus, *Icarus, 74*, 472–494, 1988.

Kasting, J.F., O.B. Toon, and J.B. Pollack, How Climate Evolved on the Terrestrial Planets, *Scientific American, 258*, 90–97, 1988.

Kasting, J.F. and D.H. Grinspoon (1991). The faint young sun problem. in *The Sun in Time* (C.P. Sonett, M.S. Giampapa, and M.S. Matthews, Eds.) University of Arizona Press, Tucson, pp. 447–462.

Kasting, J.F., Planetary atmosphere evolution: Do other habitable planets exist and can we detect them?, *Astrophysical and Space Science*, 241, 3–24, 1996.

Kumar, S., Hunten, D.M., and Pollack, J.B. (1983) Non-thermal escape of hydrogen and deuterium from Venus and implications for loss of water. *Icarus* 55, 369–375.Lederberg, J. (1960) Exobiology: Approaches to life beyond the Earth. *Science*, 132, 393

Lim, K.F. (2006) Negative pH does exist. *J. Chem. Education.* 83. 1465.

Lomonosov, M.V. (1955). *The Complete Works.* Moscow: Acad. Nauk USSR.

Marcano, V., Benitez, P., and Palacios-Pru, E. (2002) Growth of a lower eukaryote in non-aromatic hydrocarbon media greater than or equal to C_{12} and its exobiological significance. *Planet. Space Sci.* 50, 693–709.

Marchant, H.J., Da Vidson, A.T., and Kelly, G.J. (1991) UV-B protecting compounds in the marine alga Phaeocystis pouchetii from Antarctica. *Marine Biology* 109, 391–395.

Marov , M. Ya., V.S. Avduevsky, N.F. Borodin, A.P. Ekonomov, V.V. Kerzhanovich, V.P. Lysov, B.Ye. Moshkin, and M.K. Rozhdestvensky (1973). Preliminary results on the Venus atmosphere from the Venera 8 descent module. *Icarus* 20: 407–421.

Marov, M. Ya. and D.H. Grinspoon (1998) The Planet Venus. Yale University Press.

Mayer, C.H., R.M. Sloanaker, and T.P. McCullugh (1957). Radiation from Venus at 3.15 cm wavelength. *Astron. J.* 62: 26–27.

McKinnon, W.B., K.J. Zahnle, B.A. Ivanov and H.J. Melosh (1997) Cratering on Venus: Modeling and observations. In *Venus II Geology, Geophysics, Atmosphere and Solar Wind Environment* (S.W. Bougher, D.M. Hunten, and R. J. Phillips, Eds.), pp. 585–414. Univ. of Arizona Press, Tucson.

Melosh, H.J. (1988) The rocky road to panspermia. *Nature* 332, 687–688.

Meyer, B., Gouterman, M., Jensen, D., Oommen, T.V., Spitzer, K., and Stroyer-Hansen, T. (1972) The spectrum of sulfur and its allotropes. *Advan. Chem. Ser.* 110, 53–72.

Morbidelli, A., J. Chambers, J. I. Lunine, J.M. Petit, F. Robert, G. B. Valsecchi and K.E. Cyr. Source regions and timescales for the delivery of water to the Earth. *Meteoritics & Planetary Science* 35, 1309–1320.

Morrison, D. (2001) The NASA Astrobiology Program. *Astrobiology*, 1. 3–13.

Nordstrum, D.K, C.N. Alpers, C.J. Ptacek and D.W. Blowes. (2000) Negative pH and extremely acidic mine waters from Iron Mountain, California. *Environ. Sci. Technol.* 2000, 34, 254–258.

Nordstrum, D.K. (2005) personal communication.

Pace, N. (2001) The universal nature of biochemistry. *PNAS.* 98, 805–808.

Pavlov, A.A., Ono, S., and Kasting, J.F. (2002) MIF in Archean rocks: an evidence of S_8 aerosols in the Archean atmosphere. *EOS. Trans. AGU* 83, Fall Meet. Suppl., Abstract B71A-0734.

Phillips, R.J., M.T. Zuber, S.C. Solomon, M.P. Golombek, B.M. Jakosky, W.B. Banerdt, D.E. Smith, R.M.E. Williams, B.M. Hynek, O. Aharonson, and S.A. Hauck (2001). Ancient geodynamics and global-scale hydrology on Mars, *Science, 291*, 2587–259.

Pierson, B.K., Mitchell, H.K., and Ruffroberts, A.L. (1993) *Chloroflexus aurantiacus* and ultraviolet-radiation-implications for Archean shallow-water stromatolites. *Origin of Life and Evolution of the Biosphere* 23, 243–260.

Pick, U. (1999). In *Enigmatic Microorganisms and Life in Extreme Environments* (ed. Seckbach, J.), Kluwer, Dordrecht.

Pollack, J.B. and C. Sagan, "Planetary Engineering", in J. Lewis, and M. Matthews (Eds), Resources of Near-Earth Space, pp. 921–950, University of Arizona Press, Tucson, (1994).

Prinn, R.G., The sulfur cycle and clouds of Venus, in *Recent Advances in Planetary Meteorology*, edited by G.E. Hunt, Cambridge University Press, Cambridge, 1985.

Prinn, R.G., Climate change on Venus, *Nature, 412*, 36–37, 2001.

Raymond, S.N., A. V. Mandell and S.. Sigurdsson. (2006) Forming habitable worlds with giant planet migration. *Science* 313. 141–1416.

Reid, I.N. (2007) Extrasolar planets: A galactic perspective. To appear in *A decade of planets around normal stars*, M. Livio Ed. Cambridge University Press.

Sagan, C. (1961). The Planet Venus. *Science*, 133: 849–858.

Sagan, C. and H. Morowitz (1967). Life in the clouds of Venus. *Nature*, 215, 1259–1260.

Sagan, C. (1969). Microwave radiation from Venus: Thermal versus non-thermal models. *Astrophys. and Space Phys.* 1: 94–100.

Sattler, B., H. Puxbaum, and R. Psenner (2001)Bacterial Growth in Supercooled Cloud Droplets *Geophys. Res. Lett.* 28, 239–242

Schleper, C., Pühler, G., Kühlmorgen, B., and W. Zillig. (1995). Life at extremely low pH. *Nature* 375: 741–742.

Schleper, C., G. Pühler, H. P. Klenk, and W. Zillig. (1996). Picrophilus oshimae and Picrophilus torridus fam. nov., gen. nov., sp. nov., two species of hyperacidophilic, thermophilic, heterotrophic, aerobic archaea. *Int. J. Syst. Bacteriol.*, 46: 814–816.

Schulze-Makuch, D. and L.N Irwin (2002). Reassessing the possibility of life on Venus: Proposal for an astrobiology mission. *Astrobiology* 2, 197–202.

Schulze-Makuch, D. and Irwin, L.N. (2004) Life in the Universe: Expectations and Constraints, Springer-Verlag, Berlin, Germany.

Schulze-Makuch, D., D.H. Grinspoon, O. Abbas, L.N Irwin and M.A. Bullock. (2004). A sulfur-based survival strategy for putative phototrophic life in the Venusian atmosphere. *Astrobiology* 4. 11–18.

Segura, T.L., O.B. Toon, A. Colaprete and K. Zahnle. (2002) Environmental effects of large impacts on Mars. *Science.* 298. 1977–1980.

Space Science Board (1970), Venus: Strategy for Exploration, Report of a Study by the Space Science Board of the National Research Council, National Academy of Sciences, Washington, D.C. June 1970, pp. 12–13.

Space Science Board (1972), ad hoc Committee for Review of Planetary Quarantine Policy, Report, February 14, 1972, pp. 3–4.

Space Studies Board (2006), Committee on the Origin and Evolution of Life, Task Group on Planetary Protection Requirements for Venus Missions, "Assesment of Planetary Protection Requirements for Venus Missions". Letter report. Available online at http://www.nap.edu/catalog/11584.html.

Squyres, S.W., R.E. Arvidson, J.F. Bell, III, J. Bruckner, N.A. Cabrol, W. Calvin, M.H. Carr, P.R. Christensen, B.C. Clark, L. Crumpler, D.J.D. Marais, C. d'Uston, T. Economou, J. Farmer, W. Farrand, W. Folkner, M. Golombek, S. Gorevan, J.A. Grant, R. Greeley, J. Grotzinger, L. Haskin, K.E. Herkenhoff, S. Hviid, J. Johnson, G. Klingelhofer, A.H. Knoll, G. Landis, M. Lemmon, R. Li, M.B. Madsen, M.C. Malin, S.M. McLennan, H.Y. McSween, D.W. Ming, J. Moersch, R.V. Morris, T. Parker, J.W.

Rice, Jr., L. Richter, R. Rieder, M. Sims, M. Smith, P. Smith, L.A. Soderblom, R. Sullivan, H. Wanke, T. Wdowiak, M. Wolff, and A. Yen (2004a). The Opportunity Rover's Athena Science Investigation at Meridiani Planum, Mars, *Science, 306,* 1698–1703.

Squyres, S.W., J.P. Grotzinger, R.E. Arvidson, J.F. Bell, III, W. Calvin, P.R. Christensen, B.C. Clark, J.A. Crisp, W.H. Farrand, K.E. Herkenhoff, J.R. Johnson, G. Klingelhofer, A.H. Knoll, S.M. McLennan, H.Y. McSween, Jr., R.V. Morris, J.W. Rice, Jr., R. Rieder, and L.A. Soderblom (2004b). In Situ Evidence for an Ancient Aqueous Environment at Meridiani Planum, Mars, *Science, 306,* 1709–1714.

Squyres, S.W., and A.H. Knoll (2005). Sedimentary rocks at Meridiani Planum: Origin, diagenesis, and implications for life on Mars, *Earth and Planetary Science Letters, 240,* 1–10.

Toon, O.B., Turco, R.P., and Pollack, J.B. (1982) The ultraviolet absorber on Venus: amorphous sulfur, *Icarus* 51, 358–373.

Tortora, G., Funke, B., and Case, C. (2001) Microbiology: An Introduction. Addison Wesley Longman Publishers, San Francisco CA. 7th edition.

Vethanayagam, V.R. (1991) Purple photosynthetic bacteria from a tropical mangrove environment. *Marine Biology* 110, 161–163.

Walker, J.C.G., P.B. Hays, and J.F. Kasting. (1981) A negative feedback mechanism for the long-term stabilization of Earth's surface temperature. *J. Geophys. Res.* 86, 9776–9782.

Wynn-Williams, D.D. and Edwards, H.G.M (2000) Proximal analysis of regolith habitats and protective biomolecules in situ by laser Raman spectroscopy: overview of terrestrial Antarctic habitats and Mars analogs. *Icarus* 144: 486–503.

Wynn-Williams, D.D., Edwards, H.G.M., Newton, E.M., and Holder, J.M. (2002) Pigmentation as a survival strategy for ancient and modern photosynthetic microbes under high ultraviolet stress on planetary surfaces. *Int. J. of Astrobiology* 1, 39–49.

David H. Grinspoon. Department of Space Sciences, Denver Museum of Nature & Science, 2001 Colorado Blvd., Denver, CO 80205, Ph: (303) 370-6469, FAX: 303-370-6005, Email: dgrinspoon@dmns.org

Mark A. Bullock, Department of Space Studies, Southwest Research Institute, 1050 Walnut Street, Suite 400, Boulder, CO 80203, Ph: (303) 546-9027, FAX: 303-546-9687, Email: bullock@boulder.swri.edu

Technology Perspectives in the Future Exploration of Venus

James A. Cutts[1], Tibor S. Balint[1], Eric Chassefiere[2], Elizabeth A. Kolawa[1]

Science goals to understand the origin, history and environment of Venus have been driving international space exploration missions for over 40 years. Today, Venus is still identified as a high priority science target in NASA's Solar System Exploration Roadmap, and clearly fits scientific objectives of ESA's Cosmic Vision Program in addition to the ongoing Venus Express mission, while JAXA is planning to launch its own Venus Climate Orbiter. Technology readiness has often been the pivotal factor in mission prioritization. Missions in all classes—small, medium or large—could be designed as orbiters with remote sensing capabilities, however, the desire for scientific advancements beyond our current knowledge point to in-situ exploration of Venus at the surface and lower atmosphere, involving probes, landers, and aerial platforms. High altitude balloons could circumnavigate Venus repeatedly; deep probes could operate for extended periods utilizing thermal protection technologies, pressure vessel designs and advancements in high temperature electronics. In situ missions lasting for over an Earth day could employ a specially designed dynamic Stirling Radioisotope Generator (SRG) power system, that could provide both electric power and active thermal control to the spacecraft. An air mobility platform, possibly employing metallic bellows, could allow for all axis control, long traversing and surface access at multiple desired locations, thus providing an advantage over static lander or rover based architectures. Sample return missions are also featured in all planetary roadmaps. The Venus exploration plans over the next three decades are anticipated to greatly contribute to our understanding of this planet, which subsequently would advance our overall knowledge about Solar System history and habitability.

1. INTRODUCTION

Science objectives for Venus exploration are considered key to answer questions about Solar System formation and habitability. Venus, a member of the inner triad, is one of the first planets to be visited by spacecraft. Over the past 40 years, more than 20 missions succeeded in exploring it,

[1] Jet Propulsion Laboratory, California Institute of Technology, Pasadena, California, USA.

[2] Service d'Aéronomie/ Pôle Système Solaire de l'IPSL, Service d'Aéronomie, Université P & M Curie, Paris, France.

Exploring Venus as a Terrestrial Planet
Geophysical Monograph Series 176
10.1029/176GM13

through flybys, orbiters, probes or landers, not counting failed attempts. Past missions included the Magellan and Pioneer-Venus missions by the US; the Venera Program missions by the USSR; and the USSR Vega missions with extensive international cooperation.

From a US perspective, in 2003, the National Research Council (NRC) published an Integrated Exploration Strategy for the study of the Solar System, called the Decadal Survey [NRC, 2003]. It identified three criteria in establishing NASA mission priorities: scientific merit, opportunity (e.g. favorable budgetary or orbital configurations), and technology readiness. It also identified Venus as a high priority science target. In response to the NRC recommendation, the 2006 NASA Solar System Exploration Roadmap [NASA, 2006] included a number of potential Venus missions, ranging through all mission classes from small (Discovery), to medium (New Frontiers) and to large (Flagship) missions. From a European perspective, ESA's Venus Express orbiter is currently performing scientific investigations through remote sensing measurements. Future exploration of the Solar System, including Venus, will be addressed through the Cosmic Vision Program [ESA, 2006], for which the Announcement of Opportunity (AO) was released in March 2007. One of the proposed missions for the Cosmic Vision call is the European Venus Explorer (EVE) concept, with an expected earliest launch date of 2016. If selected, this mission would include significant international collaboration between ESA and Russia, and further potential contributions from NASA and JAXA. [Chassefière et al., 2007]. From JAXA, the Venus Climate Orbiter, also known as Planet-C, is planned for launch in 2010 [Imamura et al., 2006].

This chapter addresses technology perspectives for the future exploration of Venus, by first discussing the extreme environmental conditions, followed by mission architectures and technology needs.

2. EXTREME ENVIRONMENTS ON VENUS

Understanding the formation of our Solar System and answering questions about habitability require a focused exploration program. At the target destinations, however, we have to deal with extreme conditions, such as high temperatures combined with high pressures; extreme cold; high radiation; and thermal-cycling. Venus represents one of the most hostile environments in our Solar System. Its super rotating atmosphere consists mainly of carbon dioxide (CO_2 ~96.5%) and nitrogen (N_2 ~3.5%), with small amounts of noble gases (e.g., He, Ne, Ar, Kr, Xe) and small amounts of reactive trace gases (e.g., SO_2, H_2O, CO, OCS, H_2S, HCl, SO, HF). The cloud layer is composed of aqueous sulfuric acid droplets at ~45 to ~70 km attitude. The zonal winds near the surface are ~1 m/s, increasing up to 120 m/s at an altitude of ~65 km. The greenhouse effect results in very high surface temperatures, ~460°C to 480°C, while the surface pressure can reach as high as ~92 bars. At these conditions near the surface, the CO_2 becomes supercritical, which can further complicate missions planned to explore these regions. Therefore, mission architectures and related technologies must address ways to mitigate these environmental conditions.

3. MISSION ARCHITECTURES FOR VENUS EXPLORATION

Venus exploration is clearly an important element of the Solar System Exploration program for national and international space agencies. Parallel exploration strategies address both orbital and in situ explorations of Venus. In this section we provide a brief historical overview of previous Venus exploration missions, and then outline various candidate mission architectures planned for the next three decades by the space agencies, concluding with discussion of additional mission architectures currently not in the roadmaps.

3.1. Brief Overview of Historical and Ongoing Missions

The Soviet space agency executed a highly successful program for Venus exploration, reaching the surface with multiple landers. NASA's Pioneer-Venus multi-probe mission also performed in-situ exploration of the planet, while the Vega missions were conducted through international partnership under the leadership of the Soviet space agency. The major mission characteristics are summarized in Table 1.

Venus landers have historically been designed with spacecraft subsystems operating within conventional temperature ranges and protected passively inside pressure vessels. The success of this thermal design approach depends on its capability to both exclude the exterior heat and to absorb the heat generated internally by the operating electronics. The pressure vessel also requires a minimum number of ports, windows, and exterior connections in order to minimize the number of potential thermal leaks. Beside the protected components, the Venera 13 and 14 landers, as well as the Vega landers, included a sample acquisition system with sample processing that operated in the Venus environment throughout the collection of multiple samples. Designs used on these historic missions would not be compatible with some of the proposed missions requiring controlled, multiple sample acquisition and processing, since the acquisition system would need to be placed outside the vessel, operating in the Venus environment.

3.1.1. Pioneer-Venus Multiprobes.
To date, NASA's in situ Venus exploration program involved only the Pioneer-Venus Multiprobe mission, which sequentially deployed four atmospheric probes from the same Multi Probe bus. The probes were launched as a cluster on a single craft in 1978, while the orbiter was launched ten weeks earlier. Three of the 94 kg Small Probes were identical (including 61 kg pressure vessels and 33 kg deceleration modules), while the fourth probe, known as the Large Probe, was 302 kg (including a 193 kg pressure vessel and a 109 kg deceleration module). These probes were not designed to carry out surface observations, but were equipped to survive the descent to the surface with thermal controls, other than Phase Change Materials (PCM).

Table 1. Summary of historical Venus in situ missions

Missions	Launch	Module Mass (kg)	Descent (min)	Surface (min)	Design	Comments
Venera 3	1965	337			80°C, 5 bar	Entered Venus atmosphere
Venera 4	1967	380	93		300°C, 20 bar	Stopped transmitting at 25 km (battery)
Venera 5	1969	405	53		300°C, 20 bar	Stopped transmitting at ~20 km (320°C, 27 bar)
Venera 6	1969	405	51		300°C, 25 bar	Stopped transmitting at ~20 km (320°C, 27 bar)
Venera 7	1970	500	35	23	540°C, 150 bar	Parachute failure, rough landing, landed on the side
Venera 8	1972	495	55	50	490°C, 100 bar	Performed as designed
Venera 9	1975	1560 (660)	20+55	53	490°C, 100 bar	Orbiter out of the radio range
Venera 10	1975	1560 (660)	20+55	65	490°C, 100 bar	Orbiter out of the radio range
Venera 11	1978	1600 (760)	60	95	490°C, 100 bar	Transmission stopped
Venera 12	1978	1600 (760)	60	110	490°C, 100 bar	Transmission stopped
Venera 13	1981	1600 (760)	55	127	490°C, 100 bar	Transmission stopped
Venera 14	1981	1600 (760)	55	57	490°C, 100 bar	Transmission stopped

Missions	Launch	Module Mass (kg)	Descent (min)	Surface (min)	Design	Comments
Vega 1 lander	1984	750	60	20	490°C, 100 bar	Loss of contact after 20 min Transmission
Vega 2 lander	1984	750	60	56	490°C, 100 bar	Flyby out of range

Missions	Launch	Module Mass (kg)	Flight (hrs)	Design	Comments
Vega 1 balloon	1984	21 kg incl. 6.9 kg gondola	48	Designed to operate at 54-55 km altitude where temp. is about 40°C and press. About 0.5-1 bar	Transmission stopped after batteries were exhausted and balloon reached a day side
Vega 2 balloon	1984	Same	48	Same as Vega 1 balloon	Same as Vega 1 balloon

Pioneer Small Probes: Three of the probes were identical 0.47 m diameter spherical pressure vessels, housed inside a 0.76 m aeroshell that did not separate from the probe during descent. The pressure vessel was machined from Titanium with a heat shield of carbon phenolic, and filled with xenon to reduce the heat flow through the walls to the instruments. The instruments were mounted on heat-absorbing beryllium shelves. The interior of the vessel was protected from external heat by a thick 61-layer blanket of metallized insulation lining, while the exterior was covered by low emittance coating. The small probes did not carry a parachute, instead used aerodynamic braking during the descent to the surface.

The housing doors for the three instruments opened at 70 km above the surface, in a controlled manner, using an actuation mechanism with adequate torque to overcome aerodynamic forces resulting from the descent of the probes through the atmosphere. The small probes were each targeted at different parts of the planet and were named accordingly (North, Day, Night). Following the 53 to 56 minutes descent, only the Day Probe survived landing with its antenna pointed towards Earth, and continued sending radio signals for 67 minutes after impact. The electrical anomalies experienced by all four Pioneer-Venus Probes resulted in partial loss of science data below that altitude. This so called "12.5 km anomaly" [Seiff et al., 1993]—manifesting in electrical failures—is further discussed in the "Testing for Extreme Environments" section.

Pioneer Large Probe: The Pioneer-Venus large probe had a diameter of 1.42 m, with a 0.78 m diameter Titanium pressure vessel, filled with Nitrogen. After deceleration from initial atmospheric entry at about 11.5 km/s near the equator on the Venus night side, a parachute was deployed at 47 km altitude. The sealed spherical pressure vessel was encased in a nose cone and aft protective cover and contained seven science experiments. The vessel was lined with 41 layers of a dimpled and flat metallized kapton insulation blanket (with a ~2.5 cm total thickness), used 37 kg of beryllium for two equipment mounting shelves acting as a heat sink, and with the outside covered with low emittance coating. Electrical heaters helped remove the condensates from the two sapphire science instrument windows. For the infrared radiometer the prime and backup windows were fabricated from 200 and 31 carats natural diamond stones, and sealed using a preloaded system of a graphoil sealing surface with anviloy and inconel alloys. The Large Probe took 54 minutes to descend through the atmosphere and did not survive the landing.

3.1.2. The Venera Program.

The former Soviet Union carried out an extensive program over two decades as they successfully delivered landers, labeled sequentially as "Venera". (Venera is Russian for Venus.) These missions conducted a range of measurements through the upper atmosphere, atmosphere, and on the surface. Among the 10 successful landings the Venera 13 lander survived for 127 minutes after reaching the surface of Venus in 1982. These landers used a hermetically sealed pressure vessel with a phase change material (starting with Venera-8), as well as more traditional thermal insulation both inside and outside. The pressure vessel, containing all of the instrumentation and electronics, was mounted on a ring-shaped landing platform and was topped by an antenna. Another breakthrough was the development of a lightweight honeycomb composite outside insulator that was able to withstand high temperatures and pressures. These new materials resulted in a large mass savings. On Venera 9 and 10, some of the sensors designed to study clouds and atmosphere during descent, specifically nephelometers and spectrometer sensors, were mounted outside of the pressure vessel with their own insulation and PCM for thermal protection. The spectrometer sensors were connected to the spectrometer through fiber optic pipes to view upward and downward. The temperature, pressure, and density sensors were designed to operate and survive in the Venusian environment. The platinum temperature sensors were enameled to protect them from chemical corrosion and from the absorption of gas into metal.

3.1.3. The Vega Landers and Balloons.

Two Vega missions were executed under international partnership, led by the Russian Space Research Institute (IKI). The complex mission architecture called for an orbiter that released a lander and a balloon at Venus prior to continuing on to rendezvous with Comet Halley. The two spacecraft, Vega 1 and Vega 2, were launched on December 15 and December 21, 1984, respectively. Arriving at Venus in June 1985, each spacecraft deployed a 1500 kg descent module towards Venus and the main spacecraft were then retargeted for a Comet Halley encounter in March 1986. The descent modules split into two parts, a lander and a balloon package, and entered into the atmosphere of Venus on June 11 and June 15, respectively. Both landers reached the surface of Venus and returned data about the atmosphere of Venus and soil composition data was obtained by Vega 2 (the soil acquisition device failed on Vega 1). The Vega 1 lander survived for 20 minutes and the Vega 2 lander transmitted data for 56 minutes. The two balloons performed 4 science experiments, while the two landers executed 9 experiments. (While they were designed to perform 12 experiments, the temperature and pressure units and the soil device failed on Vega 1, and the mass spectrometer failed on Vega 2.) The Vega balloons were designed to deploy and float high in the atmospheric clouds where the temperature was near Earth ambient. However, the balloon materials had to be tolerant of the corrosive effects of concentrated sulfuric acid and other chemical agents present in the clouds and to ultraviolet radiation. In addition, the balloon materials as well as deployment system had to survive for months folded up and exposed to vacuum during cruise to Venus. The Vega balloons had a 6.9 kg payload suspended 12 meters below a fluoropolymer-coated Teflon fabric balloon, floating in the most active layer of the Venus three-tiered cloud system at ~54 km altitude. The tether and straps between gondola compartments were made of nylon-6 type of material called kapron. The deployment system included bottles of compressed helium and a 35 square meter parachute that was used during the filling of the balloons with helium. The filling process took about 230 sec and was controlled with barometric sensors. The batteries' lifetime of 48 hours limited the mission duration to 47 hours, during which time data was transmitted back to Earth. One-way Doppler and very long baseline interferometry (VLBI) were used to track the motion of the balloon to provide the wind velocity in the clouds. The tracking was done by a 6-station network on Soviet territory, and by a global network of 12 stations organized by France and the NASA Deep Space Network. Both Vega 1 and Vega 2 balloons survived for 47 hours, and traveled over 10,000 km until depletion of the batteries.

3.1.4. Flyby and Orbiter Missions. The US Mariner 2 mission was launched in 1962 and verified the planet's high temperatures. In 1974 Mariner 10 (US) bound for Mercury, flew by Venus (2/5/74), and tracked global atmospheric circulation with visible and violet imagery. The Pioneer-Venus Orbiter (U.S.) radar mapped Venus (12/78), and dropped four probes through the Venusian clouds, as discussed above. The most recent US Venus orbiter mission, Magellan was launched in 1989. It arrived at Venus in 1990 and mapped 98% of the planet. The mission ended in 1994. ESA's Venus Express was launched in 2005, as a follow on to Mars Express. Scientific investigations during its 500-day nominal missions in Venus orbit aim to enhance our knowledge of the composition, circulation and evolution of the atmosphere of the planet. The surface properties of Venus and the interaction between the atmosphere and the surface will be examined and evidence of volcanic activity will be also explored. NASA's MESSENGER spacecraft passed Venus twice, in October 2006 and June 2007. During the first Venus flyby very limited observations were made since the spacecraft was in solar conjunction. The science opportunity during the second flyby could include taking representative sets of VIS and IR image mosaics as a practice for the Mercury flybys. Other measurements covered atmospheric profiles, the exosphere of Venus on departure, observation of acceleration of energetic charged particles at the bow shock of Venus, and measurements of the magnetic field before, during and after the flyby.

3.1.5. Lessons Learned From Historic in Situ Missions. Looking at historic missions, several conclusions can be drawn from the experience of Soviet and U.S. spacecraft, conducting exploration of Venus. For example, it is feasible to reach the surface of Venus and conduct scientific measurement in the Venus environment for time periods of the order of an hour or so, limited by communication issues and the extreme environment. The descent systems, such as probes or landers, require careful and in some cases creative designs, as well as the selection of materials that can not only survive but operate in these environments. During descent through the 12.5 km level there are certain atmospheric phenomena that can induce transient effects on the vehicle, which can have a serious impact on the subsequent execution of the mission. The effects of long duration exposure in the lower atmosphere of Venus are unknown and more research is needed to understand the chemical and physical processes that occur. Finally, some instruments can be operated while exposed to the Venus environment, but most of them must be contained in a temperature and pressure controlled vehicle.

3.2. Mission Architecture Options for the Future Exploration of Venus

Technology readiness has often been the pivotal factor in mission prioritization. In particular, the technology challenges may account in significant part for the lack of new Venus in situ NASA missions and the slow pace of exploration over the past decade. While missions in all classes could be designed as orbiters with remote sensing capabilities, the desire for scientific advancements beyond our current knowledge point to in-situ exploration of Venus. The specific technology needs, such as high temperature electronics, requires the understanding of the planned mission architectures. Therefore, the various exploration strategies are bounded by relevant mission architectures.

Mission architectures for Venus exploration could include orbiters, high and low altitude balloons, microprobes, entry probes, long lived landers or mobile explorers, sample return and network missions. Some of these architectures are proposed by various space agencies, while others are included for completeness. Typically, these missions are optimized to reach the best balance between key trade space options, such as science objectives, programmatics, mission architectures and technologies. Driven by programmatic considerations, the missions can be also categorized by their size, namely small, medium or large. Moving from smaller to larger missions the science returns and mission complexities are expected to increase. At Venus, the harshness of the environment amplifies with the depth in the atmosphere. Thus, the technology challenges to enable these missions are predominantly influenced by two factors, the distance from the surface and the time spent in that environment. The various mission architectures described here reflect the increasing requirements for tolerating these extreme conditions, while in the next section we discuss the relevant system architecture approaches and technologies planned to enable these missions.

3.2.1. Proposed Roadmap Mission Architectures. Space agency roadmaps identified Venus exploration missions in all mission classes from small to large. While the descriptions below will not detail science objectives, they will provide generic mission architecture elements and technology drivers.

Orbiters and Flybys. Based on simple architectures, these types of missions can be designed with or without in situ elements. Due to the closeness to the Sun, they can be powered by solar panels. From orbit only remote sensing observations are available. However, they can act as telecommunication relays between in situ elements and Earth. Examples for flyby and orbiter missions include the MESSENGER spacecraft and the Venus Express and Venus Climate Orbiter orbiters.

High Altitude Balloon. Balloon platforms are of high interest for studying the atmosphere and in a limited way the surface. By staying above the clouds at 60 to 65 km from the surface, the pressures and temperature conditions are Earth-like (0.1 to 1 bar, and -50 to -70°C), providing simplicity to this architecture. At this altitude the temperature and pressure conditions would not require extreme environment technologies. The potentially solar powered payload could address optical, contact and analytical measurements, devoted to the characterization of clouds; chemical sensors to measure atmospheric composition at cloud level; and electromagnetic wave analyzers to monitor the electromagnetic activity of the planet below the ionosphere. A balloon could be deployed from an orbiter or flyby spacecraft, or by direct atmospheric entry. After entry the payload would release the aeroshell and inflate the balloon at an appropriate altitude, then drift in the super rotating atmosphere with limited mobility control. A balloon could also serve as a carrier to deploy microprobes. An ESA study described a platform, consisting of a gondola and balloon with a floating mass of 32 kg, including 8 kg of science instruments and microprobes. While the balloon circumnavigate Venus at an equilibrium floating altitude of ~55 km and analyze the middle cloud layer, it would also release 15 drop-sondes at scientifically interesting locations. For inflation gas, hydrogen, helium and ammonia could be considered, however, hydrogen is preferred for its overall system mass efficiency. Various high altitude balloon concepts are studied and proposed by both NASA and ESA, targeting small to medium class missions.

Microprobes. To understand the composition and dynamics of the atmosphere, it is important to obtain atmospheric measurements of the vertical structure down to the surface of Venus. Due to inhomogeneity, these observations should be performed at multiple locations, requiring multiple descent probes. An ESA proposed multi-probe mission would utilize 10 to 50 microprobes, deployed in short separation intervals of ~5 minutes, obtaining atmospheric profiles at both the day and night sides. The microprobes would communicate with the balloon, reducing the high power requirements for the telecom system on the probes. The collected data would be relayed to Earth from the balloon. Each of the 100 grams microprobes would carry a core payload of miniaturized sensors, encased inside an evacuated glass sphere, and based on commercial off the shelf (COTS) technology. The probe descent would be monitored from the balloon by a probe returned "ping" signal, and could be used for trajectory reconstruction. Initial estimates indicate that the probes could operate down to altitudes of 10–20 km, how-

ever, this requires further detailed thermal analysis, since at these altitudes the temperature reaches about 300 to 400°C. (In comparison, the temperature limit for COTS and military components is around +85°C and +125°C, respectively. Therefore, a suitable thermal management system on the microprobes would likely increase their mass above the current estimates.) An ESA funded study determined that an accompanying balloon subsystem, floating at an altitude of ~65 km, would weight less than 1.5 kg. The concept is proposed by ESA as part of a small to medium class mission. [Chassefière et al., 2006]

Low Altitude Balloons. At a 10 km altitude 2 cubic meter pressurized balloon could carry up to 100 kg of total payload. The Lavoisier project was submitted to ESA [Chassefière et al., 2002], as a follow up to and in the spirit of the balloon deployed at cloud level by the Soviet Vega mission in 1986. It is composed of a descent probe, for detailed atmosphere composition analysis, and of a network of 3 balloons. In the Lavoisier concept, each balloon probe is made of a gondola, protected by a thermal shield during the entry phase. Balloons have a complex and variable thermal environment during their operational phase. All the parts that can withstand this environment are not thermally protected in order to reduce the size of the temperature-controlled area. This is the case in particular for the hardware components of the telecom subsystem and for the batteries. The temperature controlled area is designed to keep all the thermally sensitive equipments below 40°C. Taking into account the limited power resources, a concept of a passive thermal cocoon made of concentric layers of PCM (Phase Change Materials) has been validated for the balloon probes. The concept of the gondola consists of a spherical cocoon with a series of concentric layers having the following characteristics (from the outer edge to the inner edge): insulating layer, liquid water layer, insulating layer, paraffin layer, inner compartment with the payload. The thermal shield would absorb the thermal flux during atmospheric entry and probe descent from the upper atmosphere down to the altitude measurement. The shield concept could be based on the technology used for the Huygens probe front shield sub-system. At an altitude of about 20 km, the thermal shield would be released, and the 2 m^3 balloon deployed from the gondola, then inflated with helium. This would stabilize the gondola at the measurement altitude of 10 km for about 20 hours. Following the balloon deflation, the gondola would descend to the surface for a 2 hours surface operation, completing the mission. Additional issues related to the balloon material that could tolerate high temperatures at these altitudes, is discussed in Section 5.7.2.

Entry Probes. Probes are simple and can utilize flight heritage from US and Soviet missions discussed above. They would have limited lifetimes of 1 to 2 hours, while descending from an altitude of ~80 km to the surface. Surface access and science return are expected to be limited. In the concept proposed for ESA's Lavoisier mission [Chassefière et al., 2002], a scientific payload of 15 kg would operate during the descent. Optical instruments would be expected to survive soft landing and continue operating for some time, similarly to the Huygens probe. Recent ESA discussions identified a potential multiprobe mission to Venus, with up to 4 probes descending to the surface. One probe would enter the night side and three the day side, at the equator, mid latitude and the pole. This architecture requires a trajectory design for the flyby spacecraft and the three entry probes, addressing telecommunication strategies. Entry probe concepts are proposed by ESA for small to medium class missions. [Chassefière et al., 2006]

Short Lived Lander. In the Decadal Survey the Venus In-Situ Explorer (VISE) mission concept was envisioned by the NRC as a balloon mission that would study Venus' atmospheric composition in detail and descend briefly to the surface to acquire samples that could be then analyzed at a higher altitude, where the temperature is less extreme. This mission would perform compositional and isotopic measurements of the surface and atmosphere of Venus. VISE was initially classified as a New Frontiers mission (although further assessments indicated that the mission architecture conceived in the Decadal Survey would be closer in size to a small Flagship class mission) [NASA, 2006]. The duration of surface operation could range from a few minutes up to several hours on the surface of Venus, acquiring and characterizing a core sample to study the mineralogy of the surface. Short lived landers offer simplicity and flight heritage from Soviet missions. However, the science return would be limited and the platform would not allow for mobility. Concepts are proposed by both ESA and NASA for a medium class missions.

Long Lived Mobile Explorer. A Venus Mobile Explorer (VME), proposed in NASA's SSE Roadmap [NASA, 2006], is proposed as a second decade Flagship class mission with a launch date as early as 2025 (see Plate 2). VME would explore and characterize the surface with a wheeled or aerial vehicle and would further acquire and characterize a core sample. It would need to operate in the Venus surface environment for up to 90 Earth days. After passing through the sulfuric acid clouds discussed previously, this mission is envisioned to require sample acquisition from about 10–20 cm depth and in situ sensing. Of even greater interest to the scientific community are long-term (1 year or longer) missions. In this case, robust active refrigeration, coupled with a long lived radioisotope power system, would be critical for the more fragile subsystems. Rovers could provide good

surface and subsurface access for sampling and seismic measurements, but limited traversing over the mission lifetime. An aerial platform could offer all axes mobility, go-to capability, good traversing and surface mapping at high resolution imaging from near the surface, but also involves significant technology challenges to mitigate the extreme environment. The entire project, from start to end-of-mission, could be accomplished in 6–7 years, including a surface stay time of days or weeks (although certain enabling technologies, such as the radioisotope power system with active cooling may require additional development time).

Atmospheric Sample Return. This architecture supports in-situ chemical and isotopic measurements of the upper atmosphere of Venus. Understanding the distribution of isotopes in the Solar System, such as those of oxygen, sulfur and noble gases, we require very high precision measurements up to 0.01%. Since current mass spectrometers available for in situ exploration have a sensitivity of about 1%, it is desirable to return the collected samples and conduct laboratory investigations on them using highly sensitive Earth based instruments. The preferred, and likely the simplest, concept for atmospheric sample return consists of a sample collection flyby pass through the upper atmosphere of Venus (~130 km). The bus would be equipped with a collector coupled with a cryogenic device. The collected samples would be returned to Earth on a free return ballistic trajectory. The concept is under consideration by ESA for a small to medium class mission.

Surface Sample Return. A Venus Sample Return is considered to be a very difficult mission that would certainly follow a successful Mars Surface Sample Return and an effective Venus Mobile Explorer mission. As for Mars, answers to detailed questions about the past suitability of Venus for the origin and sustenance of life can only be answered by bringing samples back to terrestrial laboratories. These include definitive interpretations of the petrology and mineralogy of crustal samples over scales from the regional to the microscopic and determinations of the C, O, S and H isotopic abundances in the crust and lower atmosphere. The implementation challenge lies not so much with Venus environmental issues (although they are not trivial) as it does with the mission energetics. There would need to be a buoyant ascent stage to collect the sample either from the surface or from another vehicle (deployed to the surface and back into the atmosphere) and then carried to an altitude from which atmospheric density is low enough for a launch to be feasible. At this point the propulsion needed is equivalent to an inner planet mission starting at Earth's surface. Needless to say, even with a very small sample return payload a multi-stage ascent system would be required. In line with current thinking, it would consist of a metallic bellows from the surface to ~15 km, a zero pressure balloon from ~15 km to ~65 km, and an ascent rocket carrying the sample to orbit. The Earth Return Vehicle would have to be waiting to rendezvous with the ascent stage, to transfer the sample for a return flight to Earth. Sample recovery at Earth would be similar to the proposed Mars sample return concept with a direct entry to a suitable recovery site. Advanced airborne systems and high-energy in-space propulsion are key capabilities needed for this mission. Although it is recommended in the Decadal Survey, the Venus Surface Sample Return (VSSR) mission is beyond the planning timeframe of the roadmaps, therefore, it may not require technology development within the next ten years. However, it is important to see the progression from short-term to longer-term missions and on to sample return in order to understand the importance of various technologies, particularly high temperature sample acquisition.

3.2.2. Additional Mission Architectures. The following mission concepts are currently not in agency roadmaps, but included here for completeness.

Mid Altitude Balloon. Due to the large pressure and the Archimedes buoyant force, Venus balloons could transport substantial scientific payloads, although harsh environmental conditions at mid altitudes (p>50 bars, T>400°C) the concepts might prove to be too difficult to implement. At about 35 to 45 km from the surface the balloons would be below the cloud layer, but still too far from the surface for high resolution remote sensing measurements or surface access. For example, due to the high atmospheric opacity, only low altitude balloons below 10 km, with a resulting visibility over 10%, would be able to perform efficient visual and infra-red sounding of the surface and near-surface atmosphere. The sulfuric acid droplets in the clouds would introduce material challenges, and the reduced solar availability could make power generation difficult. This would limit mission lifetime, and the concept is currently not in any of the roadmaps.

Long Lived Lander. This concept provides good surface access for both sample acquisition and seismic measurements, but it would face the same challenges as those encountered by static Soviet landers under the Venera program. Furthermore, the long mission duration would necessitate and internal power source, such as a radioisotope power system, with active cooling to the instruments and subsystems. Static landers, by definition, do not provide mobility. For long lived in situ explorations mobile platforms provide scientifically more rewarding architectures, therefore, static landers are currently not considered in the roadmaps.

Network Missions. These network mission architectures could be considered for balloons and seismic measurements. They would have similar characteristics to single missions of the same type with an added benefit of allowing access to multiple in situ sampling or measurement locations. The multiple elements would provide redundancy, but introduce additional mission architecture and technology challenges with deployments, communications, and issues related to extreme environments. The cost for these architectures would also be significantly higher than for single missions. Consequently, these concepts are currently not in any of the roadmaps.

4. SYSTEMS ARCHITECTURES FOR MISSIONS TO EXTREME ENVIRONMENTS

Systems architectures for extreme environments may be categorized by the isolation of sensitive materials from hazardous conditions; development of sensitive materials tolerant to hazardous conditions; and appropriate combinations of isolation and tolerance.

4.1. Environmental Isolation

One possible solution for extreme environment system architectures is to simply maintain all electronics and sensitive components in an environmentally controlled vessel. While this sounds feasible, the environmental protection is seldom complete or comes without cost. A good example is the active refrigeration needed for thermal isolation in hot environments. Such systems will require additional mass and power resources to maintain a heat sink. As a result, environmental isolation architectures typically require added resources and may not provide ideal solutions for all missions to extreme environments. Furthermore, in situ missions require certain components to be directly exposed to the environment (e.g., sample acquisition systems and sensors), making the implementation of this approach even more difficult [Kolawa et al., 2007].

4.2. Environmental Tolerance

The alternative extreme in missions to extreme environments is the development of hardware components that can reliably operate and survive in extreme temperature/pressure conditions, thus eliminating the need for environmental control. While this approach is ideal on a purely technological level, some of the key technologies would require a large investment to achieve the desired performance (e.g., avionics systems are capable of operating at ~500°C). Therefore, environmentally tolerant technologies may pose elegant solutions to some technology challenges (e.g., by removing the requirements for a pressure vessel and thermal management subsystem), but technology development may not be able to answer problems posed by fundamental physical limitations or impractical investment strategies.

4.3. Hybrid Systems

In a hybrid architecture, hardened components would be exposed directly to the environment and not hardened components would be protected. Depending on the mission duration, passive or active cooling would be applied only for components that cannot be hardened to tolerate the Venus environment (e.g., CPUs and visible imagers), while high temperature tolerant components would be used where practical (e.g., for certain mechanisms and RF systems). Consequently, some temperature-sensitive components would be maintained inside an insulated thermal enclosure, while other more tolerant components would remain outside. This approach requires simpler and lighter thermal control, enabling more functionality with less cabling. In addition, this approach is also cost-effective if technologies are selected for systems engineering capabilities, as well as for tolerance. The integration of isolation and tolerance to form a hybrid system is illustrated in Plate 1.

For Venus landers, a hybrid architecture is promising if technologies are appropriately sorted according to their suitability for environmental tolerance, versus the requirement for environmental protection. It is likely that thermal/pressure control zones will be required for sub-systems such as avionics, advanced instruments, and low temperature energy storage. On the other hand, sub-systems that dissipate most of the heat (e.g., telecom, power components) should be placed outside of the thermal control zone. In-situ sensors, drills, and sample acquisition mechanisms would also be fully exposed to the extreme environment.

5. TECHNOLOGIES FOR FUTURE EXPLORATION OF VENUS

Technology needs depend strongly on mission goals and mission architectures. The principal challenge, where new technology may play a role, is observation and access to the surface and lower atmosphere. Thus, new technologies will make the greatest impacts on in-situ missions, involving probes, landers and aerial platforms. While the Soviet VEGA balloons lasted for only two days in the upper atmosphere of Venus, future proposed small (Discovery) class missions could include entry probes as well as super pressure long-lived balloons, operating in the upper atmosphere above ~60km, potentially permitting balloon missions to circumnavigate Venus repeatedly, over many weeks. Deep probes and landers with current technologies are limited to a few hours of operation due to the environment. For life-

times on the surface and in the lower atmosphere of Venus of more than an Earth day, a specially designed dynamic Stirling Radioisotope Generator (SRG) power system could provide both electric power and active thermal control to the spacecraft. An air mobility platform, possibly employing metallic bellows, could allow for all axis control, long traversing and surface access at multiple desired locations, thus providing an advantage over static lander or rover based architectures. If this major technological hurdle can be overcome, long duration fixed and mobile floating platforms, such as the NASA's proposed Venus Mobile Explorer mission, will be enabled. This will be the key to intensive in-situ geological and geophysical investigation of the planet.

This section details technologies to enable future Venus exploration missions.

5.1. Orbiter Technologies

For orbital missions, much of the electromagnetic spectrum is blocked by the dense atmosphere, preventing the use of imagers in the visual range. Radar imaging deployed on Magellan has provided a global map. Advances in radar can provide improved topographic maps and potentially detection of surface changes using INSAR (Interferometric Synthetic Aperture Radar). Advances in passive infrared and millimeter spectroscopic techniques enable more effective probing of this part of the planet. However, much of the technologies associated with orbiters and flyby spacecraft can be handled with existing technologies, therefore, these are not discussed further in this chapter.

5.2. Protection Systems

In general, protection systems refer to those systems providing isolation from the extreme environment. Hypervelocity entry protection is a form of thermal control at extremely high peak heat fluxes. Pressure and thermal controls are also protection systems, although they are likely to be integrated into a joint pressure vessel for missions to Venus (see Plate 3).

5.2.1. Hypervelocity Entry. Hypervelocity atmospheric entry produces extreme heating (measured as heat flux in kW/cm^2) though both convective and radiative processes. Thermal Protection Systems (TPS) are composed of insulation layers that allow only a small fraction of the heat to penetrate the spacecraft surface conductively, and are designed to reject the majority of the heat through re-radiation and ablation. The success of hypervelocity entry is measured by two quantities: (1) the peak heat flux tolerated by the entry vehicle, and (2) the mass fraction dedicated to a thermal protection system. Current heritage carbon-phenolic family of materials can tolerate ~1 kW/cm^2, thus requiring mass fractions ranging from 12% for Venus missions to as high as 50% to 70% for missions to Jupiter. It is anticipated that technology development can improve the tolerated heat flux by an order of magnitude, thus allowing for reductions in the thermal protection mass fraction of anywhere from 25% to 50%. The rotational period of Venus is -243 days (retrograde), and at a 70 km altitude the super-rotating atmosphere is ~60 times faster than the solid surface. This results in a negligible ±390 km/h rotational difference at the equator between prograde and retrograde atmospheric entry experienced by an entry probe. Therefore, the planetary entry is not limited by the trajectory option.

5.2.2. Pressure Control. New advances in materials technology will enable advanced, lightweight pressure vessels that can be layered with insulating materials. The Soviet Venera and Vega missions, described above, used titanium pressure vessels surrounded by a rigid, porous, silicon-containing material that served as the outer thermal insulation. In order to compare emerging pressure vessel materials, three basic mechanical parameters may be used to estimate pressure vessel mass for a given shell diameter. The pressure vessel shell must satisfy these criteria at a temperature of 500°C:

(1) Buckling should not occur at the ultimate load of 150 atm pressure. Buckling is a catastrophic failure mode, caused by lateral instability in a structural element that is subjected to high compressive stresses.
(2) Yielding should not occur at the proof load of 125 atm pressure. Material yielding occurs if the stress on the material results in a permanent deformation.
(3) The total allowable creep in 10 hours and under 100 atm external load must be less than 0.5%. Material creep describes the tendency of a material to deform over an extended period of time in order to relieve stresses.

These criteria are evaluated based on compressive yield strength, compressive modulus, and creep strain rates. In addition, pressure vessel material must be impermeable to gases, while retaining compatibility with the Venus chemical environment. Low thermal conductivity would be also desirable; however, this requirement could be mitigated against through better insulation. Other factors to be considered in selecting shell materials include: fracture toughness, heat capacity, and thermal expansion coefficient. Candidate pressure vessel materials could include metallics (e.g., nickel-chromium alloys and beryllium), and composite materials (e.g, silicon carbide fiber reinforced titanium matrix or epoxy

polymer matrix composites). A very important consideration in the selection of materials is its manufacturability into a spherical pressure vessel shell that needs to include windows and feedthroughs. Monolithic shells can be fabricated from titanium or beryllium, which has been the traditional manufacturing process for spacecraft landing on Venus' surface. Composite wrapped shells are commonly seen in pressure cylinders and the technology is well developed. This manufacturing technique could be used, for example, for aluminum/silicon. Honeycomb sandwich shells are often formed into curved geometries (e.g., for aircraft engine cowlings). This could be an appropriate fabrication technique for Inconel pressure vessels. [Pauken et al., 2006]

Additional mission architectures may exist, where mission duration is extended by equilibrating to the ambient pressure, while protecting against the temperature increase. Like many of the systems analyses facing technologies for extreme environments, the understanding of such an approach is also incomplete.

5.2.3. Thermal Management and Control. Advanced thermal protection provides the highest benefit to the missions in terms of survivability, regardless of configuration or mission duration. Advanced thermal protection technologies may extend in situ mission lifetimes significantly and would considerably enhance the scientific yield of a proposed medium class (New Frontiers) Venus in situ mission.

Passive Thermal Control:

Aerogel. With a thermal conductivity approximately that of a gas at 0.1 W/mK, aerogel provides good insulation without convection. The current state of the art, has a density of approximately 20 kg/m^3.

Multi-layer insulation. Multi-layer insulation (MLI) reduces the radiated heat flux between a hot and a cold boundary surface, thus preventing large heat leaks. It typically consists of closely spaced shields of Mylar (polyester) or Kapton (polyimide), coated on one or both sides with thin films of aluminum, silver or gold. MLI blankets often contain spacers, such as coarse-netting material, to keep the layers properly separated. Possible next generation insulating materials include a cocoon of high-temperature multi-insulation, manufactured by stacking and sewing together crinkled reflective metal-alloy foils, separated by ceramic fabric and/or insulated with xenon gas, although MLI only provides significant performance improvements when used in a high vacuum. In the more external part of a pressure vessel, metallic, ceramic or PBO (Poly-p-phenylenebenzo-bisoxazole) materials could be used directly, although they would need to be fabricated in layers thicker than films because films are too thin to provide insulation. While Kapton (polyimide) is not suitable because it will not tolerate Venus surface temperatures, aramid or other polymers could in principle be combined with silica fabrics.

Phase change materials (PCMs). with high thermal inertia may be used to absorb the additional heat dissipated when the components are in operation, as was done with Pioneer-Venus. The current state of the art is the lithium salts with specific heat 296 kJ/kg used in the Soviet Venera and Vega landers. Materials are available with high transformation temperatures, high latent heat, and low density. However, the need for low volumetric change limits the transformations to solid-liquid and solid-solid transitions. A higher density PCM may be more appropriate if it could require a smaller volume and consequently requires less container or filler mass. The state-of-art PCM is a paraffin ($C_{16}H_{34}$) or paraffin-like polymeric material that dissipates about 250 kJ/kg during its solid-to-liquid phase transition.

Active Refrigeration:

For any high temperature surface mission lasting more than several days, a proactive approach to thermal control is required. Active thermal control or refrigeration systems would be described generally by the efficiency, defined as the ratio of the output heat to the removed energy. Limited theoretically by the Carnot limit of 72%, it is likely that such a system would operate at about 30% efficiency relative to a Carnot engine, or at about 20% total efficiency. No such system currently exists. One potential architecture for a Venus surface mission is a three-stage refrigeration system, isolating the internal electronics enclosure with a series of cylindrical vessels. To minimize conductive heat transfer, electronic components would be housed in an evacuated inner vessel and maintained at a cooler temperature with a refrigeration system.

5.3. Component Hardening

"Component hardening" refers to technologies expected to be exposed to the ambient environment, with general categories of electronics, electro-mechanical systems, and energy storage. Component hardening is the process of developing technologies capable of tolerating the external environment. This section describes the individual elements considered key to mission operation and survival.

5.3.1. High Temperature Electronics.
Active components. Terrestrial industries currently employ high temperature electronics in the fields or oil and gas

exploration, automotive and aviation industries, geothermal wells and defense related fields. The temperature limits of today's commercial and military application needs are below 300°C. These temperatures are well below the temperatures encountered at the surface of Venus, indicating that the need for electronics, which can tolerate the extreme high temperature and pressure environments are unique to space agencies, planning to explore these target locations [Kolawa et al., 2006]. The most promising high-temperature semiconductor technology, capable of operating in the Venus environment, is silicon carbide. Silicon carbide devices have been demonstrated to operate as high as 500°C for limited periods of time (tens of hours), but the availability of silicon carbide-based integrated circuits is extremely limited. In missions with passive thermal control, such as VISE, high power electronic and telecom systems act as internal heat sources. Placing these systems outside the thermally protected vessel may reduce internal heating and extend the life of the mission. Small scale integrated SiC high temperature technologies and heterogeneous high temperature packaging can support this need and produce components for power conversion, electronic drives for actuators and sensors amplifiers.

Another approach to high temperature electronics is related to further development of the original solid state vacuum devices. Vacuum electron devices are well suited for extreme temperatures, especially in the application to the high temperature telecom system where they can provide a noise-performance advantage over SiC. This is because it is possible to design high temperature micro vacuum based power amplifiers that can directly operate in the Venus surface environment without any penalty in noise and linear performance. It should be noted, however, that careful design and improvement of the device packaging (vacuum enclosure) and materials are required in order to allow operation within the 500 °C ambient environment. Furthermore, these vacuum devices are inherently unlikely to achieve the high levels of integration and functionality that is possible with semiconductor transistors. While both wide-bandgap semiconductor and vacuum tube approaches to high-temperature electronics have been demonstrated to operate at temperatures in excess of ~500°C, there are no strong commercial drivers for this technology.

In missions with active thermal control, medium-temperature large-scale integrated electronics can be used for fabricating all essential functions of the spacecraft. Electronics operating at medium temperatures (300°C) can reduce the temperature difference between the outside environment and inside of the thermally protected system. The smaller temperature difference will significantly reduce the corresponding power needed for cooling. Also, this type of electronics can enable Venus balloon missions to operate for prolonged duration down to 20 km altitudes where temperature does not exceeds 300°C. For intermediate (300°C) temperatures, it is more feasible to extend traditional silicon based electronics. Historical investments by the US Department of Energy, the oil industry, and the automotive industry have resulted in some silicon electronic components operational at 250–300°C. They are fabricated using silicon-on-insulator (SOI) technology to reduce device leakage current at higher temperatures. It is feasible that low-power medium-temperature Si-based electronics could be developed for hybrid systems for in situ Venus missions. [Del Castillo et al., 2006] [Spry et al., 2004]

Passive components. Passive elements for high temperature applications depend not only on the survivability of resistive elements, dielectrics or magnetic core materials, but also on the component packaging and interconnection technology. This is the most common source of failure for devices that have not been designed for high temperature operation. With respect to resistors, thin film and thick film resistors deposited on ceramic substrates provide the best performance and miniaturization of currently available resistor technologies. Potential problems include oxidation of thin film resistive elements with time at temperature and resistance value drift in thick film resistors. Degradation of potential candidate resistors must be characterized in detail, in order to make use of them in 500 °C circuits.

Capacitors are particularly challenging for high temperature operation, since they tend to vary in capacitance with greater temperature, particularly as the dielectric constant and dielectric dissipation are increased. At elevated temperatures, the leakage currents of these capacitors become very high, making it difficult for the capacitor to hold charge. The most promising candidates for high temperature (500 °C) capacitors are NP0 ceramic capacitors and piezoelectric based capacitors. NP0 capacitors have minimal variation in capacitance with temperature, but unfortunately they exhibit a significant increase in dissipation above 300 °C. Piezoelectric capacitors are designed for operation at a specific temperature, and therefore exhibit optimum properties at the desired temperature. Unfortunately, the temperature window for this peak performance is very narrow, and these capacitors change significantly with increasing temperature. Various capacitor technologies, such as diamond capacitors and other alternative dielectric materials, are currently under development and may eventually offer superior properties throughout the entire temperature range from 23 to 500 °C.

5.3.2. Electronic Packaging

. Critical to the implementation of any high temperature electronics system will be the packaging approach and materials selected. Development of materials for interconnects, metallizations, conductive (solders) and non-conductive (ceramics) bonding materials, wire bonding systems, mechanical protection, and thermal management) must be integrated with the development of proper packaging systems for each application.

5.4. Electro-Mechanical Systems for High Temperatures

High temperature motors and actuators are the key components of the sample acquisition and transfer system. The acquisition of un-weathered samples from at least 20 cm below the surface layer of Venus is required for VISE mission. In addition, motors and actuators are required for a variety of functions during space system operations like opening and closing valves, deployment of landing gear, operation of robotic arms, antenna gimbals and many others. For the VME mission motors and actuators may be crucial for mobility systems if a rover is selected as a mission baseline. In such a case the motors will be required to operate reliably for hundreds of hours.

The required technologies to be developed include motors and gearboxes, position sensors, high temperature electrical cabling, mechanical devices related to drilling and containing a sample, and mechanical sample transfer devices. A first step in the process is a selection of materials. For example, magnetic materials with high Curie temperature need to be identified and tested for the motor itself as well as suitable materials have to be identified for gearboxes and lubricants. Based on the results from materials testing at high temperatures, the motor assembly needs to be built and tested at high temperatures.

There are several approaches to reliably implement controlled motion and actuation in a high temperature environment, including electromagnetic systems, spring-based systems, thermally actuated switches, and piezoelectric devices.

5.5. Testing for Extreme Environments

The electrical anomalies experienced by all four Pioneer Venus Probes, starting around ~12.5 km above the surface, resulted in partial loss of science data below that altitude. While multiple options were considered, the most likely reason for the failures were contributed to condensation of conductive vapors on the external sensors in the deep atmosphere, leading to shorted electrical circuits. Other theories included chemical interactions of atmospheric constituents with the probe and sensors, such as reaction of residual sulfuric acid from the clouds with harness or sensor materials or carbon dioxide oxidation of titanium parts or polymers; and probe charging followed by electrical breakdown of the atmosphere, leading to sparks that could possibly ignite probe external fires. [Seiff et al., 1993]

The anomalies experienced by the Venus descent probes points to the need to simulate the Venus environment as accurately as possible. The Pioneer-Venus probes were tested in nitrogen at temperatures up to 500°C and pressures up to 100 bars. They were never tested under these conditions in a carbon dioxide environment, because it was assumed that both carbon dioxide and nitrogen are chemically inert, and consequently the substitution of carbon dioxide by nitrogen was acceptable. Recent work on the properties of carbon dioxide at high pressures and temperatures, when it enters a supercritical state, indicate that these assumptions are not correct. Therefore, testing in a relevant environment is imperative.

5.6. Power Storage and Power Generation

Power systems, either in the form or power storage or power generation are key enabling technologies in any space mission. For orbiter and high altitude Venus missions power can be generated with solar panels. Since in this section power technologies are considered for the extreme environments of Venus, solar power generation for orbiters and high altitude mission elements are not discussed further. Instead, we discuss relevant technologies for short and long lived in situ Venus missions, namely batteries and radioisotope power system.

5.6.1. Energy Storage for High Temperature Environments.
In the US, over the past five decades, several high-temperature energy storage technologies have been developed by and for NASA, the Department of Energy (DoE), and the Department of Defense (DoD). Development efforts, led by researchers at Argonne National Laboratory, resulted in thermally regenerative galvanic cells for the direct conversion of heat to electricity. Although not fully developed, several battery chemistries were created and qualified, which operated at or above 400°C. Without the need for high temperature batteries at the time, development was virtually stopped in 1995, due to the overwhelming interest in Li-ion batteries, offering high performance at 25°C. In ESA studies, the use of high temperature batteries operating at 425°C aboard a balloon is an important consideration to solve the thermal control requirement by reducing the mass and volume allocation of the cold compartment. Lithium and sodium batteries under development allow stable energy storage at ambient temperatures. It means that no significant energy loss is expected after integration of the power subsystem and during the cruise to Venus. These batteries could be operated in a temperature range of 325 to 480°C. Similarly, lithium-sulfur

batteries could operate in the range of 350–400°C. Practical energy densities in the range of 100–150 Wh/kg are reported on the optimized couple Li/FeS2 batteries. The sodium-sulfur (Na/S_2) batteries have a similar operating temperature range and practical energy density of ~100 Wh/kg. (In technology trade studies a mean energy density of 100 Wh/kg, compatible with the Venus environmental constraints, can be used for the power subsystem mass and performance estimation.) Rechargeable batteries could be beneficial for atmospheric cyclers, however, there are a few rechargeable battery systems capable of operating at high temperatures. These systems are typically based on molten salts based on alkali halides and/or solid electrolytes, especially sodium beta-alumina ceramic. The most promising, Na-NiCl$_2$ battery with molten salt electrolyte, is functional at ~460°C and can deliver up to 130 Wh/kg. This battery chemistry is still under development in Europe, and could likely be transitioned into an environment specific package in less than 3 years. High specific energy batteries could positively impact the proposed in situ missions to Venus and to the Giant Planets identified in roadmaps, such as Jupiter, Saturn and Neptune, by reducing mass and volume requirements for the power subsystem. Further details on energy storage can be found in [Mondt et al., 2004].

5.6.2. Radioisotope Power System for Long Lived In Situ Venus Missions.
As discussed earlier, the roadmaps identified in situ missions with various mission durations and technology difficulties. Driven by science requirements, longer missions provide greater science return. While short lived missions could be designed with power storage systems (batteries), long lived in situ mission require external or internal power sources, such as solar panels or radioisotope power systems (RPS). For high-altitude balloon missions solar power is readily available, but for long lived surface or low altitude aerial missions a specially designed RPS is required. RPS enabled missions could operate continuously near or at the surface for many months, as long as other issues related to the extreme environments, such as pressure and temperature, are addressed. The RPS, and the rest of the spacecraft, would also need to tolerate the highly corrosive supercritical carbon dioxide environment. For Venus conditions dynamic power conversion (e.g., Stirling converters) may provide an advantage over static conversion systems, because for the latter power conversion efficiency is strongly coupled with the temperature difference between the hot and cold sides of the thermocouples. This would result in a very low conversion efficiency, higher mass and volume for the static conversion based power system and higher plutonium requirement for the heat source. Dynamic conversion systems, have significantly higher conversion efficiencies, and due to the lower

plutonium requirement less excess heat to reject. In addition to power generation, an RPS for Venus would also require to power an active cooling system, in order to maintain a quasi steady state thermal environment for the payload. Current Stirling Radioisotope Generator (SRG) development work does not include a requirement to operate in the 480°C Venus environment. Therefore, future development should include work on special dynamic conversion systems including a power generator and an active cooler to address the need for sustained power at or near the surface of Venus. The power system should also utilize a suitable coating to minimize the impact of the corrosive environment while maintaining or if possible improving heat rejection performance. Furthermore, an aerial platform would impose mass and volume constraints to the RPS design, which should be balanced with the power requirement of the mission. [Balint, 2006]

RPSs must be designed for all mission phases, namely earth storage, launch, cruise, entry, descent and landing (EDL), and operations. Conditions between these mission phases may vary and so as the heat transfer mechanisms to reject the excess heat generated by the radioisotopic decay of the plutonium fuel. For example, operating in planetary environments with atmospheres heat is rejected through convection, conduction and radiation. For Venus in situ missions, during the approximately 6 months cruise phase the RPS would be encapsulated inside an aeroshell, during which time heat would be removed by a fluid loop and rejected to space through external radiators. Sizing of the fluid loop and the radiators for Venus missions would be different than similar in situ missions to Titan, because of the extreme environmental conditions at the destinations (480°C at Venus versus -178°C at Titan). RPS sizing should also account for the atmospheric entry phase, when the probe or lander would be still inside the aeroshell, but forced circulation would be no longer available. [Balint, 2006]

RPS technology is considered enabling for the proposed long lived in situ Roadmap missions. [NASA, 2006]

5.7. Mobility Technologies

Roadmap missions call for aerial mobility at low, medium and high latitudes. Each of these altitudes represent different technology challenges and approaches to mitigate the environmental effects, including mobility related technologies and material selection options.

5.7.1. Aerial and Surface Mobility Technologies. NASA's proposed Venus Mobile Explorer mission is currently baselined with aerial mobility, but surface mobility options were also considered. An aerial platform may not require deployment from a landed platform, and the mobility is not limited by planetary surface roughness or bearing strength. The surface traversing and access could be over a range of 1000 km, utilizing the super rotating atmosphere of Venus. Vertical control of a long lived low altitude metallic bellows configuration would allow for repeated aerial imaging of the surface, although the horizontal directional control of the platform would be limited. Sample collection would be limited to grab-sampling only, due to the short contact period with the surface. Anchoring or longer stay at the surface is not recommended, due to the high risk of snagging, which could result in mission failure. In comparison, a surface rover on Venus would have significant constraints regarding landing sites. Rough terrains at scientifically interesting regions, such as at Tessera, could significantly impact traversing. Mobility would also be impacted by the low bearing strength. Rovers require significant power for mobility, which is also a function of surface roughness. Since power availability for long lived in situ missions is expected to be limited, the actual traversing capability of such a mission would be below 10 km over the lifetime of the mission, which makes this type of in situ exploration architecture less desirable. However, a rover platform would allow for precise sampling, and even drilling, thus, as discussed before, subsurface access could answer important questions related to the age and composition of Venus. Based on the significant technology challenges, long lived in situ missions would falls under the large mission class. Furthermore, benefits provided by the various in situ platforms, aerial mobility could provide the largest operational flexibility and highest science return. To tolerate the extreme environments close to the surface, the platform is envisioned as a metallic bellows shown in Plate 2.

5.7.2. Materials for Balloons and Parachutes. High altitude balloons experience Earth-like conditions, and stay above the cloud layer. Therefore, the balloon material selection is not affected by extreme environments. A prototype high altitude balloon has been built at JPL in 2006 in support of a Discovery proposal. This type of balloon would inflate after atmospheric entry, and during its lifetime would be protected from the environment by a layer of Teflon film. At mid to low altitudes, the dense atmosphere of Venus may provide higher buoyancy for aerial platforms, and controlled descent for probes suspended under a parachute. Finding a single balloon material that could withstand the high temperature and pressure at these altitudes is challenging. Based on NASA studies, one of the polymer film materials known to work at Venus surface temperatures of 480°C is Poly-p-phenylenebenzobisoxazole (PBO). However, there is no experience with making gas-tight seams, needed for balloon construction involving initially flat sheets of material. An additional problem is that PBO requires a Teflon coating for

Protection Systems

Use conventional components; Develop protection systems
(Thermal vessel; pressure vessel, radiation shielding etc.)

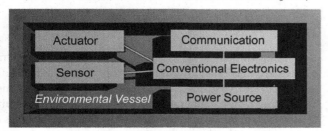

Impractical for planned missions

Component Hardening

Develop technologies tolerant
of extreme environments

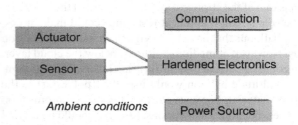

Ambient conditions

Prohibitively expensive for some technologies

Hybrid Architecture

Combination of hardened components and environmental protection

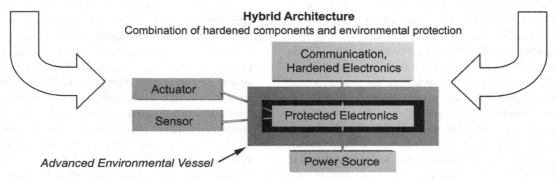

Requires development of innovative architectures

Plate 1. Hybrid Systems Architectures for Extreme Environments

Plate 2. Venus Mobile Explorer concept

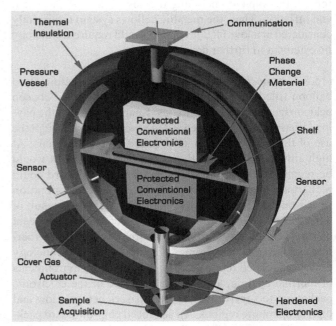

Plate 3. Illustration of the pressure vessel and thermal management for a Venus in situ mission

protection from the atmospheric sulfuric acid. While Teflon is acid resistant, it may not survive the near surface portion of the mission, since it becomes brittle at high temperatures. Zylon is one of the strongest synthetic fibers in the world. Its high tensile strength and heat resistance could make it perfect for balloons that have to survive the harsh conditions on Venus. However, at medium altitudes, below the cloud layer, Zylon could be corroded by sulfuric acid. For an ESA design, the high altitude balloon would use PET (polyester) as the baseline material, because it offers a good compromise on properties. For additional options, research should be carried out into the potential use of PPTA aramid. A potential solution for accessing both low and mid altitudes was proposed as a two-balloon system. The first balloon would operate near the planet's surface and would look like a cylindrical metallic bellows made of extremely thin sheets of stainless steel or another suitable alloy [Kerzhanovich et al., 2005]. The bellows would be flexible enough that it could be compressed like an accordion for storage on the way to Venus, and sturdy enough that it could survive the acidic clouds. The helium filled thin metal balloon would rise from the surface of Venus, carrying a suitable science payload. After reaching an altitude of about 10 to 15 kilometers, it would release a second balloon, which would climb to higher altitudes. Areal densities of 1 kg/m^2 are available with current technology; this will suffice for missions up to approximately 10 km altitude at Venus because of the very high atmospheric densities. The systems engineering of metal bellows balloons remain incomplete, including the key issue of deployment and inflation. Thus, the metallic bellows system is currently considered at a low TRL level, and would require technology investement to further develop it.

5.7.3. Atmospheric Cycler Concept.

A reversible-fluid balloon filled with ammonia or water could cycle between near surface and 55 km altitudes. Near the surface, the payload lifetime could be extended to several hours through the use of thermal insulation and Phase Change Materials. At high altitudes, following the ascent phase, the PCM would be regenerated and the secondary battery onboard could recharge using solar energy. The total surface operation time would be limited to less than 100 hours, as a result of a limited number of altitude cycling. The cycling of ascent and descent phases would predominantly affect the thermal cycling of the electronics. The most critical technology issues are related to electronics, operating in environments of thermal cycling, and the performance of the electronic packaging to survive the repeated passes between low and high temperature regions. An advanced chip-on-board packaging technology is currently under development in a Mars Focused Technology Program, with the goal to survive 1500

cycles in the –120 to +80°C temperature range. At Venus, this temperature range would be wider towards the hot side, which would require further technology development.

5.8. Telecommunication Technologies and Strategies

Telecommunication technologies and data transfer strategies are necessary to guarantee the scientific data return from in situ Venus mission. The issues involved with implementing a high temperature interplanetary telecommunication system are intimately related to both the electronics and actuation technologies.

5.8.1. Telecommunications for high temperatures.

In situ telecommunication systems on Venus would be likely limited to a fixed antenna, operating at S-band or X-band frequencies, since a precision gimbal system is not likely to be developed with high-gain pointing functionality. This is the result of the extreme temperature environment, where many of the sensors, electronic, and precision actuator elements would have to function reliably at 480°C. Each of the operating frequencies would pose its own challenge. Not only does a 10 W S-band communication link limit data transfer to 500 bps, the number of Earth-based S-band receivers is very limited (the Deep Space Network offers this functionality only at Goldstone), and they may be phased out by 2020. The X-band communications would provide even lower data transfer rates of 100 bps, and this design would result in a strong dependence on the X-band signal intensity on the elevation angle and proximity to Earth.

5.8.2. Telecommunication Strategies.

Telecommunication system sizing is dependent on power and antenna sizing on the sending and receiving ends, their separation distance, the chosen frequency, environmental effect, and mission architectures. Communications between an in situ asset and a flyby spacecraft is time limited to about 1 to 2 hours. An orbiter would have similar constraints, but would allow re-visits for subsequent communications. This, however, would require a long lived mission, which would increase technology requirements on the in situ element. Direct to Earth (DTE) communication would be also limited by terrestrial constraints, for example DSN antenna size, weather, and line of sight. A multi probe mission to Venus could have a mixed communication architecture, where a probe could use DTE, while the others would communicate through an orbiter of flyby spacecraft. Due to antenna and power limitations on the probe the DTE data rate would be low. Even with the largest planned array, the radio-astronomy Square Kilometer Array (SKA), the data rate would be lower than using proximity links and relay data through a flyby or orbiter spacecraft. Furthermore, the SKA in currently not designed for lower frequencies and for planetary program support. Therefore, it is suggested that if the planetary community is planning to utilize the SKA, this request should be communicated to the radio-astronomy community in order to include lower frequency capabilities into the SKA design. Frequencies also play an important role in communication strategies. For proximity communication, lower frequencies could be beneficial and less affected by atmospheric attenuation, while higher frequencies could provide higher data rates over larger distances, if not affected by environmental conditions. Since the telecommunication system is responsible to relay all of the collected science data to Earth, its performance is essential to achieve mission success and the communication strategies should be assessed carefully.

5.9. Cross Cutting Technologies for in Situ Exploration of the Giant Planets

Understanding Solar System formation requires in situ measurements of Giant Planets atmospheric composition. Deep entry probes to Jupiter are proposed to measure atmospheric constituents and dynamics down to about 100 bars. At this pressure elevation on Jupiter the pressure and temperature conditions are similar to those at the surface of Venus. Therefore, many of the technologies are directly relevant between Jupiter and Venus entry probe technologies. Influenced by trajectories and telecommunication constraints, a Jupiter probe would descent into the atmosphere for about 70 minutes before reaching 100 bars. This again is comparable to the 1 to 2 hours Venus probe descent. Such short lived missions would employ Phase Change Materials for thermal management inside the pressure vessel, and primary batteries for the power subsystem. From the probe's side the data rate would be limited by

the power system, antenna size and environmental condition, such as atmospheric attenuation. This would make direct to Earth communication from a Jupiter highly constrained or even not feasible, and from Venus it would severely limit the data rate.

6. SUMMARY

Surface missions to Venus encounter particular challenges, because they need to operate in extremely harsh environments (480°C and 90 bars) and prior to reaching the surface, a lander would face the additional hurdle of passing through the extremely corrosive sulfuric acid clouds at higher altitudes. For short-duration in situ missions passive thermal control approaches may be adequate, but very long-duration missions would require active cooling to "refrigerate" the thermally controlled avionics and instruments. Current states of practice technologies do not support long lived in situ Venus missions. Therefore, an aggressive early program of systems analysis is important to define the best approach and to determine realistic technology performance goals.

Table 2 summarizes technology development impacts on Venus surface missions. This table suggests that high-temperature sample acquisition is an enabling technology for all surface missions. Even short missions require hardened sample acquisition systems because of the necessary environmental exposure, although the exposure duration will determine the technology requirements. Accordingly, refrigeration technology would be needed for long duration missions, involving anything beyond the science that could be provided with current communications systems. Furthermore, certain functions will remain impractical for implementation at high temperatures and pressure; this group includes items such as most scientific sensors and microprocessors.

Table 2. Technology development impacts on Venus surface missions

Design Reference Mission	Venus In-Situ Explorer (VISE)	Venus Mobile Explorer (VME)	Venus Surface Sample Return (VSSR)
Earliest Launch Date	2015	2025	2035
Architecture	Short-lived (hours) surface	Extended (days 92)Surface	Sample return
HT electronics and communications	Medium	Very High	Low
HT sample acquisition	Very High	Very High	Very High
HT energy storage	Low	Medium	Medium
Passive thermal control	Very High	Low/Medium	Very High
Active thermal control	Low	Very High	Low
High pressure control	Very High	Very High	Very High
Aerial mobility	N/A	High	Very High

7. CONCLUSIONS AND RECOMMENDATIONS

Venus exploration plans over the next three decades—as identified in the NASA 2006 SSE Roadmap—are anticipated to greatly enhance our understanding of Venus. Subsequently, it will also advance our overall knowledge about Solar System history and habitability. Parallel exploration plans by ESA' Cosmic Vision for 2015–2025 and JAXA are expected to significantly contribute to this effort.

The proposed in situ exploration mission architectures range from smaller high altitude balloons and descent probes to large long-lived mobile near-surface missions. The success of the Mars Exploration Rovers has demonstrated that long-lived mobile in-situ vehicles can provide substantially improved science returns. A similar strategy, based on long-lived mobile platforms, could highly benefit the Venus Exploration Program as well, but the mission elements must be tailored for environmental and surface conditions. This necessitates new technologies and capabilities for tolerating and in some cases exploiting the severe environmental conditions of Venus. In general, in situ missions to Venus would benefit from a number of technologies for high temperatures, including active thermal cooling, pressure vessels, high-temperature electronics, energy storage, and high-temperature mechanisms.

Advances in thermal control, electronics, sensors, actuators, materials, power storage, power generation and other technologies are expected to enable these potential future missions. Systems architectures will be key in establishing which technologies will enable systems exposed to the environment, and which technologies will require consistent protection. Sample acquisition systems will clearly require environmental tolerance and appropriate systems engineering, since they openly interface with the environment. On the other hand, for certain sensors and microprocessors it will remain impractical to increase tolerance levels, thus these will need to remain in a protected environment. Materials research will continue to play an important role in developing technologies for aerial mobility. Balloons are envisioned as possible modes of exploration for Venus, but the environmental constraints must be addressed. Development of these new capabilities may require substantial technology investments. Thus, a credible long range technology investment strategy could animate a set of prior missions, some of which would permit validation of technologies needed for future missions. Since planetary extreme environments and related technologies are unique to space agency driven missions, agencies are expected to take the lead in the development of these critical technologies, with support from industry and academia.

Acknowledgments. This work has been performed at the Jet Propulsion Laboratory, California Institute of Technology, Pasadena, California, USA, under contract to NASA, and at the Université P & M Curie, Paris, France. The authors wish to thanks Adriana Ocampo from NASA HQ and Thomas Thompson from JPL, and Peter Falkner from ESA for their support. Further thanks to the Extreme Environments Team members at JPL, including Linda del Castillo, J. Hall, Mohammad Mojarradi, Michael Pauken, and Jay Whitacre. The opinions expressed here those of the authors and do not necessarily reflect the positions of the various space agencies.

REFERENCES

Balint, T., Radioisotope Power Systems for In-Situ Exploration of Titan and Venus. *4th International Planetary Probe Workshop.* Pasadena, California, June 27–30 (2006).

Chassefière, E., et al., European Venus Explorer (EVE), An in-situ mission to Venus, ESA Cosmic Vision proposal, June 29 (2007).

Chassefière, E., et al., Proceedings of the Venus Entry Probe workshop, ESA/ESTEC, Noordwijk 19–20 January 2006, Note du Ptle de Planitologie de l'IPSL n015, February (2006).

Chassefière, E., Berthelier, J.J., Bertaux, J.L., Quemerais, E., Pommereau, J.P., Rannou, P., Raulin, F., Coll, P., Coscia, D., Jambon, A., Sarda, P., Sabroux, J.C., Vitter, G., Le Pichon, A., Landau, B., Lognonne, P., Cohen, Y., Vergniole, S., Hulot, G., Mandea, M., Pineau, J.F., Bezard, B., Keller, H.U., Titov, D., Breuer, D., Szego, K., Ferencz, CS., Roos-Serote, M., Korablev, O., Linkin, V., Rodrigo, R., Taylor, F.W. and Harri, A.M., The Lavoisier mission: a system of descent probe and balloon flotilla for geochemical investigation of the deep atmosphere and surface of Venus, *Advances in Space Research*, Volume 29, Issue 2, pp. 255–264 (2002).

Del Castillo, L.Y., Johnson, T.W., Hatake, T., Mojarradi, M.M., Kolawa, E.A., Sensor Amplifier for the Venus Ground Ambient, *IMAPS International Conference and Exhibition on High Temperature Electronics (HiTEC 2006).* Hilton of Santa Fe, NM, USA, May 16 (2006)

ESA. Cosmic Vision, Space Science for Europe 2015–2025, *ESA publication BR-247*, October (2005)

Imamura, T., Nakamura, M., Ueno, M., Iwagami, N., Satoh, T., Watanabe, S., Taguchi, M., Takahashi, Y., Suzuki, M., Abe, T., Hashimoto, G.L., Sakanoi, T., Okano, S., Kasaba, Y., Yoshida, J., Yamada, M., Ishii, N., Yamada, T., Oyama, K., PLANET-C: Venus Climate Orbiter Mission of Japan. Submitted to Planetary and Space Science (2006).

Kerzhanovich, V., Hall, J., Yavrouian, A., Cutts, J., Dual Balloon Concept For Lifting Payloads From The Surface Of Venus, AIAA 5th Aviation, Technology, Integration, and Operations Conference (ATIO), 16th Lighter-Than-Air Systems Technology Conference and Balloon Systems Conference, AIAA-2005-7322, September 26–28 (2005)

Kolawa, E., Mojaradi, M., Balint, T. Applications of High Temperature Electronics in Space Exploration. *IMAPS International Conference and Exhibition on High Temperature Electronics (HiTEC 2006).* Hilton of Santa Fe, NM, USA, May 16 (2006).

Kolawa, E., et al., Extreme Environments Technologies for Future Space Science Missions, Technical Report # JPL D-32832,

National Aeronautics and Space Administration, Jet Propulsion Laboratory, California Institute of Technology, Pasadena, CA, September 19 (2007).

Mondt, J., Burke, K., Bragg, B., Rao, G., Vukson, S., Energy Storage Technology for Future Space Science Missions, National Aeronautics and Space Administration, Technical Report, JPL D-30268, Rev.A., November (2004).

NASA. Solar System Exploration—This is the Solar System Exploration Roadmap for NASA's Science Mission Directorate. *Technical Report JPL D-35618*. National Aeronautics and Space Administration, Jet Propulsion Laboratory, California Institute of Technology, Pasadena, CA. May (2006).

NRC. New Frontiers in the Solar System, an Integrated Exploration Strategy. *Technical Report*, Space Studies Board, National Research Council, Washington, D.C. (2003).

Pauken, M., Kolawa, E., Manvi, R., Sokolowski, W., Lewis, J., Pressure vessel technology development, *4th International Planetary Probe Workshop*. Pasadena, California, June 27–30 (2006).

Seiff, A., Sromovsky, L., Borucki, W., Craig, R., Juergens, D., Young, R.E., Ragent, B., Pioneer Venus 12.5 km Anomaly Workshop Report (Volume I), *Proceedings of a workshop held at Moffett Field, California*. NASA-CP-3303, September 28–29 (1993).

Spry, D., Neudeck, P., Okojie, R., Chen, L.Y., Beheim, G., Meredith, R., Mueller, W., Ferrier, T., Electrical Operation of 6H-SiC MESFET at 500 C for 500 Hours in Air Ambient, *2004 IMAPS International High Temperature Electronics Conference*, Santa Fe, NM, May 18–20 (2004).